ACOUSTICS

ACOUSTICS

Leo L. Beranek

1954 Edition: Acoustics Laboratory
Massachusetts Institute of Technology
Bolt Beranek and Newman, Inc.
1986 Edition: 975 Memorial Drive
Cambridge, MA 02138

**Published for the Acoustical Society of America
by the American Institute of Physics**

Library of Congress Catalog Card Number: 86-70671
International Standard Book Number: 0-88318-494-X

Published by the American Institute of Physics, Inc.
335 East 45 Street, New York, New York 10017

Printed in the United States of America

third printing 1990

PREFACE

Acoustics is a most fascinating subject. Music, architecture, engineering, science, drama, medicine, psychology, and linguistics all seek from it answers to basic questions in their fields. In the Acoustics Laboratory at M.I.T. students may be found working on such diversified problems as auditorium and studio design, loudspeaker design, subjective perception of complex sounds, production of synthetic speech, propagation of sound in the atmosphere, dispersion of sound in liquids, reduction of noise from jet-aircraft engines, and ultrasonic detection of brain tumors. The annual meetings of the Acoustical Society of America are veritable five-ring shows, with papers and symposia on subjects in all the above-named fields. Opportunities for employment are abundant today because management in industry has recognized the important contributions that acoustics makes both to the improvement of their products and to the betterment of employee working conditions.

There is no easy road to an understanding of present-day acoustics. First the student must acquire the vocabulary that is peculiar to the subject. Then he must assimilate the laws governing sound propagation and sound radiation, resonance, and the behavior of transducers in an acoustic medium. Last, but certainly not of least importance, he must learn to understand the hearing characteristics of people and the reactions of listeners to sounds and noises.

This book is the outgrowth of a course in acoustics that the author has taught to seniors and to first-year graduate students in electrical engineering and communication physics. The basic wave equation and some of its more interesting solutions are discussed in detail in the first part of the text. The radiation of sound, components of acoustical systems, microphones, loudspeakers, and horns are treated in sufficient detail to allow the serious student to enter into electroacoustical design.

There is an extensive treatment of such important problems as sound in enclosures, methods for noise reduction, hearing, speech intelligibility, and psychoacoustic criteria for comfort, for satisfactory speech intelligibility, and for pleasant listening conditions.

The book differs in one important respect from conventional texts on acoustics in that it emphasizes the practical application of electrical-circuit theory in the solution of a wide variety of problems. Wherever possible, the background of the electrical engineer and the communication physicist is utilized in explaining acoustical concepts.

v

The high-fidelity expert will find the chapters on loudspeaker enclosures, horns, and rooms particularly interesting because they show how the performance of loudspeakers either in baffles or attached to horns may be accurately and simply calculated. These chapters also illustrate the necessity of considering in design the over-all system, including the amplifier, the loudspeaker, the baffle or horn and considering also the room in which they are to be operated. Numerical examples and summary charts are given to facilitate application of this material to music-reproduction systems.

In view of the increased interest in noise control, the author has kept this subject in mind in writing Chapters 1, 2, 4, and 10 to 13. These chapters served as the basis of a special summer program on noise reduction at M.I.T. in 1953. The material of Chapters 11 and 13 is new, and it is hoped that it will be of value to those interested in noise and its effect on human beings.

In short, the engineer or scientist who wishes to practice in the field of acoustics and who does not intend to confine his efforts to theoretical matters must know the material of this text.

Problems for each chapter are included at the end of the text for use by the student. References to collateral reading in English are given in the text, although no attempt has been made to give a bibliography of the primary sources of material. Suggestions to instructors for best use of the text are given immediately after this preface.

The author wishes to express his deep appreciation to Francis M. Wiener and Rudolph H. Nichols, Jr., for their assistance in the detailed review and editing of the text and the preparation of some original material. Many members of the Acoustics Laboratory at M.I.T. have read one or more chapters and have given valuable assistance to the author. Of these, particular mention is made of Mary Anne Summerfield, Walter A. Rosenblith, Kenneth N. Stevens, Jerome R. Cox, Jordan J. Baruch, Joanne J. English, and Norman Doelling.

The illustrations are due to the highly capable and untiring efforts of Clare Twardzik. The author is deeply indebted to his typist, Elizabeth H. Jones, to his secretary, Lydia Bonazzoli, and to his wife, Phyllis, who made it possible for him to complete the text within a reasonably short span of time.

LEO L. BERANEK

SUGGESTIONS FOR INSTRUCTORS

This text is divided into thirteen chapters, comprising thirty-two parts. Each part is intended to be approximately 1 week's work, although this will vary among students owing to differences in their previous training.

If the entire class expects to take a full year of acoustics, the parts should be taught in sequence, with the exception of Part XXVIII, Measurement of Acoustic Levels, which may be referred to in associated laboratory experiments and demonstrations throughout the course. If only a part of the class plans to continue through both terms, the fundamental material should be taught in the first term and the more applied material in the second. One suggested division, in this case, is as follows:

First Term		Second Term	
Part I.	Introduction	Part V.	Energy Density and Intensity
Part II.	Terminology	Part IX.	Circuit Theorems, Energy and Power
Part III.	The Wave Equation		
Part IV.	Solutions of the Wave Equation	Part XI.	Directivity Index and Directivity Factor
Part VI.	Mechanical Circuits	Part XIV.	General Characteristics of Microphones
Part VII.	Acoustical Circuits	Part XV.	Pressure Microphones
Part VIII.	Transducers	Part XVI.	Gradient and Combination Microphones
Part X.	Directivity Patterns		
Part XII.	Radiation Impedances	Part XVIII.	Design Factors Affecting Direct-radiator Loudspeakers
Part XIII.	Acoustic Elements	Part XX.	Bass Reflex Enclosures
Part XVII.	Basic Theory of Direct-radiator Loudspeakers	Part XXI.	Horn Driving Units
		Part XXII.	Horns

A course in acoustics should be accompanied by a set of well-planned laboratory experiments. For example, the material of the first few chapters will be more significant if accompanied by a laboratory experiment on noise measurement. This will familiarize the student with the measurement of sound pressure and with the use of a frequency analyzer. He will appreciate more fully the meaning of sound pressure, sound intensity, decibels, sound energy density, and power level; and he will understand the accuracy with which noise can be measured.

A suggested minimum of 10 experiments, listed both numerically for a year's course and by term, is as follows:

First Term		Second Term	
No. 1.	Noise measurement	No. 3.	Free-field calibration of microphones
No. 2.	Measurement of the constants of an electromechanical transducer	No. 5.	Design and testing of a loudspeaker baffle
No. 4.	Measurement of free-field response of a loudspeaker	No. 8.	Prediction and control of noise in a ventilating system
No. 6.	Study of sound fields in a small rectangular enclosure	No. 9.	Audiometric testing of hearing
No. 7.	Study of sound fields in a large irregular enclosure	No. 10.	Application of psychoacoustic criteria in the design of an auditorium

An assignment of two problems per week should provide sufficient application of the material of the text. The short list of problems for each chapter should be supplemented by timely problems derived from the instructor's experience.

CONTENTS

PREFACE TO THE PAPERBACK EDITION

With the advent of the compact disc, with miniature high-fidelity systems ambulating everywhere, and with emphasis on combination voice and data in worldwide digital and telephone networks, electroacoustics is a subject more vital today than it was three decades ago.

The heart of *Acoustics*, the first three-fourths of the text, Parts I–XXIV, is still valid. These parts encompass fundamental acoustics, principles of electro-mechano-acoustical circuits, radiation of sound, acoustic elements, microphones, loudspeakers, and their enclosures, and sound in rooms. Even so, the literature on microphones, loudspeakers and rooms has increased several times over since publication of the original volume. In Appendix III I have assembled for all chapters a representative selection of textbooks and articles that have appeared since about 1950, which contains, at various technical levels, the accomplishments leading to and the trends of today. I have chosen not to list every article published, nor to include literature in foreign languages. The intent is to supply supplementary reading in English.

Originally, a primary desire of the author was to help the student, engineer and acoustical consultant visualize better how to design an audio system to achieve the elusive goal of "high fidelity" sound reproduction. The medium I chose for achieving that result is the schematic circuit, analogous to that used in electronics, but differing from prior literature by combining into one diagram the necessary electrical, mechanical, and acoustical components, including the transduction process. An examination of the literature seems to indicate that one result of that effort was to stimulate the development of small loudspeaker enclosures, which in most locations have replaced the once ever-expanding "woofer" boxes.

A large proportion of the leading writers on loudspeaker system design in the last fifteen years, including E. M. Villchur, A. N. Thiele, R. H. Small, J. R. Ashley, A. D. Broadhurst, S. Morita, N. Kyouno, A. L. Karminsky, J. Merhaut, R. F. Allison, R. Berkovitz, and others, have used the middle chapters of this book as their starting point. In other words, a knowledge of the principles taught here has been a preface to their progress.

I hold no particular brief for Chapter 11, part XXV, through Chapter 13. My later text, *Noise and Vibration Control* (McGraw-Hill, New York, 1971), treats this material in more detail benefiting from nearly twenty years of intervening progress in the field. The engineer interested in noise control should, perhaps, consider *Acoustics* and *Noise and Vibration Control* as Volumes I and II on the subject. The supplementary literature of the last 15 years on noise control, much of which is listed in Appendix III of this reprint, is not too formidable to peruse. Finally, I have made corrections to all the known errata in the book.

I wish to thank the Acoustical Society of America for their interest in reprinting *Acoustics*. I hope their faith in this portion of the acoustical literature is substantiated by the assistance it may give students and engineers in learning and practicing in the field of electroacoustics.

Leo L. Beranek
June 1986

CHAPTER 1

INTRODUCTION AND TERMINOLOGY

PART I *Introduction*

1.1. A Little History. Acoustics is entering a new era—the precision-engineering era. One hundred years ago acoustics was an art. For measuring instruments, engineers in the field used their ears primarily. The only controlled noise sources available were whistles, gongs, and sirens. Microphones consisted of either a diaphragm connected to a mechanical scratcher that recorded the shape of the wave on the smoked surface of a rotating drum or a flame whose height varied with the sound pressure. About that time the great names of Rayleigh, Stokes, Thomson, Lamb, Helmholtz, König, Tyndall, Kundt, and others appeared on important published papers. Their contributions to the physics of sound were followed by the publication of Lord Rayleigh's two-volume treatise, "Theory of Sound" (1877 and 1878). Acoustics rested there until W. C. Sabine, in a series of papers (1900–1915), advanced architectural acoustics to the status of a science.

Even though the contributions of these earlier workers were great, the greatest acceleration of interest in the field of acoustics followed the invention of triode-vacuum-tube circuits (1907) and the advent of radio-broadcasting (1920). With vacuum-tube amplifiers available, loud sounds of any desired frequency could be produced, and the intensity of very faint sounds could be measured. Above all it became feasible to build measuring instruments that were compact, rugged, and insensitive to air drafts.

The progress of communication acoustics was hastened, through the efforts of the Bell Telephone Laboratories (1920*ff*), by the development of the modern telephone system in the United States.

Architectural acoustics received a boost principally from the theory and experiments coming out of Harvard, the Massachusetts Institute of Technology, and the University of California at Los Angeles (1930–1940), and several research centers in England and Europe, especially Germany.

1

In this period, sound decay in rectangular rooms was explained in detail, the impedance method of specifying acoustical materials was shown to be useful, and the computation of sound attenuation in ducts was put on a precise basis. The advantages of skewed walls and of using acoustical materials in patches rather than on entire walls were demonstrated. Functional absorbers were introduced to the field, and a wider variety of acoustical materials came on the market.

The science of psychoacoustics was also developing. At the Bell Telephone Laboratories, under the splendid leadership of Harvey Fletcher, the concepts of loudness and masking were quantified, and many of the factors governing successful speech communication were determined (1920–1940). Acoustics, through the medium of ultrasonics, entered the fields of medicine and chemistry. Ultrasonic diathermy was being tried, and acoustically accelerated chemical reactions were reported.

Finally, World War II came, with its demand for the successful detection of submerged submarines and for highly reliable speech communication in noisy environments such as aircraft and armored vehicles. Great laboratories were formed in England, Germany, France, and in the United States at Columbia University, Harvard, and the University of California to deal with these problems. Research in acoustics reached proportions undreamed of a few years before and has continued unabated.

Today, acoustics is passing from being a tool of the telephone industry, a few enlightened architects, and the military into being a concern in the daily life of nearly every person. International movements are afoot to legislate and to provide quiet housing. Labor and office workers are demanding safe and comfortable acoustic environments in which to work. Architects in rapidly increasing numbers are hiring the services of acoustical engineers as a routine part of the design of buildings. In addition there is the more general need to abate the great noise threat from aviation—particularly that from the jet engine, which promises to ruin the comfort of our homes. Manufacturers are using acoustic instrumentation on their production lines. Acoustics is coming into its own in the living room, where high-fidelity reproduction of music has found a wide audience.

This book covers first the basic aspects of acoustics: wave propagation in the air, the theory of mechanical and acoustical circuits, the radiation of sound into free space, and the properties of acoustic components. Then follow chapters dealing with microphones, loudspeakers, enclosures for loudspeakers, and horns. The basic concepts of sound in enclosures are treated next, and practical information on noise control is given. The text deals finally with measurements and psychoacoustics. Throughout the text we shall speak to *you*—the student of this modern and interesting field.

1.2. What Is Sound? In reading the material that follows, your goal should be to form and to keep in mind a picture of what transpires when the diaphragm on a loudspeaker, or any surface for that matter, is vibrating in contact with the air.

A sound is said to exist if a disturbance propagated through an elastic material causes an alteration in pressure or a displacement of the particles of the material which can be detected by a person or by an instrument. Because this text deals primarily with devices for handling speech and music, gases (more particularly, air) are the only types of elastic material with which we shall concern ourselves. Fortunately, the physical properties of gases are relatively easy to express, and we can describe readily the nature of sound propagation in such media.

Imagine that we could cut a tiny cubic "box" out of air and hold it in our hands as we would a block of wood. What physical properties would it exhibit? First, it would have weight and, hence, mass. In fact, a cubic meter of air has a mass of a little over one kilogram. If a force is applied to it, the box will then accelerate according to Newton's second law, which says that force equals mass times acceleration.

If we exert forces compressing two opposing sides of the little cube, the four other sides will bulge outward. The incremental pressure produced in the gas by this force will be the same throughout this small volume. This obtains because pressure in a gas is a scalar, *i.e.*, a nondirectional quantity.

Imagine the little box of air to be held tightly between your hands. Still holding the box, move one hand forward so as to distort the cube into a parallelepiped. You find that no opposition to the distortion of the box is made by the air outside the two distorted sides. This indicates that air does not support a shearing force.†

Further, if we constrain five sides of the cube and attempt to displace the sixth one, we find that the gas is elastic; *i.e.*, a force is required to compress the gas. The magnitude of the force is in direct proportion to the displacement of the unconstrained side of the container. A simple experiment will convince you of this. Close off the hose of an automobile tire pump so that the air is confined in the cylinder. Push down on the plunger. You will find that the confined air behaves as a simple spring.

The spring constant of the gas varies, however, with the method of compression. A force acting to compress a gas necessarily causes a displacement of the gas particles. The incremental pressure produced in the gas will be directly proportional to the incremental change in volume. If the displacement takes place slowly one can write

$$\Delta P = -K \, \Delta V \qquad \text{slow process} \qquad \text{isothermal}$$

where K is a constant. If, on the other hand, the displacement, and

---

† This is only approximately true, as the air does have viscosity, but the shearing forces are very small compared with those in solids.

hence the change in volume, takes place rapidly, and further if the gas is
air, oxygen, hydrogen, or nitrogen, the incremental pressure produced is
equal to $1.4K$ times the incremental change in volume.

$$\Delta P = -1.4K \, \Delta V \qquad \text{fast process, diatomic gas} \quad \textit{adiabatic}$$

Note that a positive increment (increase) in pressure produces a negative
increment (decrease) in volume.   Processes which take place at intermediate rates are more difficult to describe, even approximately, and
fortunately need not be considered here.

What is the reason for the difference between the pressure arising from
changes in volume that occur rapidly and the pressure arising from
changes in volume that occur slowly?   For slow variations in volume the
compressions are *isothermal*.   By an isothermal variation we mean one
that takes place at constant temperature.   There is time for the heat
generated in the gas during the compression to flow to other parts of the
gas or, if the gas is confined, to flow to the walls of the container.   Hence,
the temperature of the gas remains constant.   For rapid variations in
volume, however, the temperature rises when the gas is compressed and
falls when the gas is expanded.   There is not enough time during a cycle
of compression and expansion for the heat to flow away.   Such rapid
alternations are said to be *adiabatic.*

In either isothermal or adiabatic processes, the pressure in a gas is due
to collisions of the gas molecules with container walls.   You will recall
that pressure is force per unit area, or, from Newton, time rate of change
of momentum per unit area.   Let us investigate the mechanism of this
momentum change in a confined gas.   The container wall changes the
direction of motion of the molecules which strike it and so changes their
momentum; this change appears as a pressure on the gas.   The *rate* at
which the change of momentum occurs, and so the magnitude of the pressure, depends on two quantities.   It increases obviously if the number of
collisions per second between the gas particles and the walls increases, or
if the amount of momentum transferred per collision becomes greater, or
both.   We now see that the isothermal compression of a gas results in an
increase of pressure because a given number of molecules are forced into a
smaller volume and will necessarily collide with the container more
frequently.

On the other hand, although the adiabatic compression of a gas results
in an increase in the number of collisions as described above, it causes also
a further increase in the number of collisions and a greater momentum
transfer per collision.   Both these additional increases are due to the
temperature change which accompanies the adiabatic compression.
Kinetic theory tells us that the velocity of gas molecules varies as the
square root of the absolute temperature of the gas.   In the adiabatic
process then, as contrasted with the isothermal, the molecules get hotter;

they move faster, collide with the container walls more frequently, and, having greater momentum themselves, transfer more momentum to the walls during each individual collision.

For a given volume change ΔV, the rate of momentum change, and therefore the pressure increase, is seen to be greater in the adiabatic process. It follows that a gas is stiffer—it takes more force to expand or compress it—if the alternation is adiabatic. We shall see later in the text that sound waves are essentially adiabatic alternations.

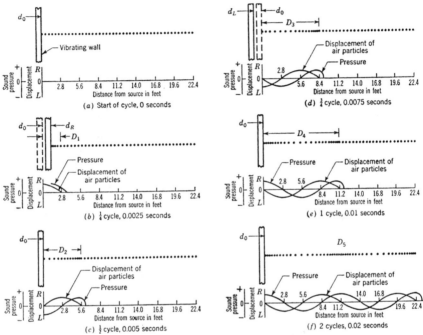

Fig. 1.1 Pressure and displacement in a plane sound wave produced by a sinusoidally vibrating wall. D_1 = one-fourth wavelength; D_2 = one-half wavelength; D_3 = three-fourths wavelength; D_4 = one wavelength; D_5 = two wavelengths. R means displacement of the air particles to the right, L means displacement to the left, and O means no displacement. Crowded dots mean positive excess pressure and spread dots mean negative excess pressure. The frequency of vibration of the piston is 100 cycles per second.

1.3. Propagation of Sound through Gas. The propagation of sound through a gas can be fully predicted and described if we take into account the factors just discussed, *viz.*, the mass and stiffness of the gas, and its conformance with basic physical laws. Such a mathematical description will be given in detail in later chapters. We are now concerned with a qualitative picture of sound propagation.

If we put a sinusoidally vibrating wall in a gas (see Fig. 1.1a), it will accelerate adjacent air particles and compress that part of the gas nearest to it as it moves forward from rest. This initial compression is shown in Fig. 1.1*b* as a crowding of dots in front of the wall. The dots represent air particles. These closely crowded air particles have, in addition to their random velocities, a forward momentum gained from the wall. They collide with their neighbors to the right and during the collision transfer forward momentum to these particles, which were at rest. These particles in turn move closer to their neighbors, with which they collide, and so on. Progressively more and more remote parts of the medium will be set into motion. In this way, through successive collisions, the force built up by the original compression may be transferred to distant parts of the gas.

When the wall reverses its motion, a rarefaction occurs immediately in front of it (see Figs. 1.1*c* and 1.1*d*). This rarefaction causes particles to be accelerated backward. and the above process is now repeated in the reverse direction, and so on, through successive cycles of the source.

It is important to an understanding of sound propagation that you keep in mind the relative variations in pressure, particle displacement, and particle velocity. Note that, at any one instant, the maximum particle displacement and the maximum pressure do not occur at the same point in the wave. To see this, consider Fig. 1.1*c*. The maximum pressure occurs where the particles are most tightly packed, *i.e.*, at $D_2 = 5.6$ ft. But at D_2 the particles have not yet moved from their original rest position, as we can see by comparison with Fig. 1.1*a*. At D_2, then, the pressure is a maximum, and the particle displacement is zero. At this instant, the particles next to the wall are also at their zero-displacement position, for the wall has just returned to its zero position. Although the particles at both D_2 and d_0 have zero displacement, their environments are quite different. We found the pressure at D_2 to be a maximum, but the air particles around d_0 are far apart, and so the pressure there is a minimum. Halfway between d_0 and D_2 the pressure is found to be at the ambient value (zero incremental pressure), and the displacement of the particles at a maximum. At a point in the wave where pressure is a maximum, the particle displacement is zero. Where particle displacement is a maximum, the incremental pressure is zero. Pressure and particle displacement are then 90° out of phase with each other.

At any given point on the wave the pressure and particle displacement are varying sinusoidally in time with the same frequency as the source. If the pressure is varying as $\cos 2\pi ft$, the particle displacement, 90° out of phase, must be varying as $\sin 2\pi ft$. Particle velocity, however, is the time derivative of displacement and must be varying as $\cos 2\pi ft$. At any one point on the wave, then, pressure and particle velocity are in phase.

We have determined the relative phases of the particle displacement, velocity, and pressure at a point in the wave. Now we ask, What phase relationship exists between values of, say, particle displacement measured at two different points on the wave? If the action originating from the wall were transmitted instantaneously throughout the medium, all particles would be moving in phase with the source and with each other. This is not the case, for the speed of propagation of sound is finite, and at points increasingly distant from the source there is an increasing delay in the arrival of the signal. Each particle in the medium is moved backward and forward with the same frequency as the wall, but not at the same time. This means that two points separated a finite distance from each other along the wave in general will not be moving in phase with each other. Any two points that are vibrating in exact phase will, in this example of a plane wave, be separated by an integral number of wavelengths. For example, in Fig. 1.1*f* the 11.2- and 22.4-ft points are separated by exactly one wavelength. A disturbance at the 22.4-ft point occurs at about 0.01 sec after it occurs at the 11.2-ft point. This corresponds to a speed of propagation of about 1120 ft/sec. Mathematically stated, a *wavelength* is equal to the speed of propagation divided by the frequency of vibration.

$$\lambda = \frac{c}{f} \qquad (1.1)$$

where λ is the wavelength in meters (or feet), c is the speed of propagation of the sound wave in meters (or feet) per second, and f is the frequency in cycles per second.

It is an interesting fact that sound waves in air are longitudinal; *i.e.*, the direction of the vibratory motion of air particles is the same as the direction in which the wave is traveling. This can be seen from Fig. 1.1. Light, heat, or radio waves in free space are transverse; *i.e.*, the vibrations of the electric and magnetic fields are perpendicular to the direction in which the wave advances. By contrast, waves on the surface of water are circular. The vibratory motion of the water molecules is in a small circle or ellipse, but the wave travels horizontally.

1.4. Measurable Aspects of Sound. Consider first what measurements might be made on the medium before a sound wave or a disturbance is initiated in it. The gas particles (molecules) are, on the average, at rest. They do have random motion, but there is no net movement of the gas in any direction. Hence, we say that the *particle displacement* is zero. It follows that the *particle velocity* is zero. Also, when there is no disturbance in the medium, the *pressure* throughout is constant and is equal to the *ambient pressure*, so that the *incremental pressure* is zero. A value for the ambient pressure may be determined from the readings of a barometer. The *density*, another measurable quantity in the medium, is defined as usual as the mass per unit volume. It equals the *ambient density* when there is no disturbance in the medium.

When a sound wave is propagated in the medium, several measurable changes occur. The particles are accelerated and as a result are displaced from their rest positions. The particle velocity at any point is not zero except at certain instants during an alternation. The temperature at a point fluctuates above and below its ambient value. Also, the pressure at any point will vary above and below the ambient pressure. This incremental variation of pressure is called the *sound pressure* or the *excess pressure*. A pressure variation, in turn, causes a change in the density called the *incremental density*. An increase in sound pressure at a point causes an increase in the density of the medium at that point.

The speed with which an acoustical disturbance propagates outward through the medium is different for different gases. For any given gas, the speed of propagation is proportional to the square root of the absolute temperature of the gas [see Eq. (1.8)]. As is the case for all types of wave motion, the speed of propagation is given by Eq. (1.1).

In later chapters of this book, we shall describe instruments and techniques for measuring most of the quantities named above. Sound will then seem real, and you will have a "feel" for intensity, sound-pressure level, power level, and the other terms of acoustics that we are about to define.

PART **II** *Terminology*

You now have a general picture of the nature of a sound wave. To proceed further in acoustics, you must learn the particular "lingo," or accepted terminology. Many common words such as pressure, intensity, and level are used in a special manner. Become well acquainted with the special meanings of these words at the beginning. They will be in constant use throughout the text. The list of definitions below is by no means exhaustive, and some additional terminology will be presented as needed in later chapters.[1] If possible, your instructor should have you make measurements of sounds with a sound-level meter and a sound analyzer so that the terminology becomes intimately associated with physical phenomena.

The mks system of units is used throughout this book. Although the practicing acoustical engineer may believe that this choice is unwarranted in view of the widespread use of cgs units, it will be apparent in Chap. 3, and again in Chap. 10, that great simplicity results from the use of the mks

---

[1] A good manual of terminology is "American Standard Acoustical Terminology," Z24.1—1951, published by the American Standards Association, Inc., New York, N.Y.

system. In the definitions that follow, the units in the cgs system are indicated in parentheses following the mks units. Conversion tables are given in Appendix II.

1.5. General. *Acoustic.* The word "acoustic," an adjective, means intimately associated with sound waves or with the individual media, phenomena, apparatus, quantities, or units discussed in the science of sound waves. Examples: "Through the acoustic medium came an acoustic radiation so intense as to produce acoustic trauma. The acoustic filter has an output acoustic impedance of 10-acoustic ohms." Other examples are acoustic horn, transducer, energy, wave, mobility, refraction, mass, component, propagation constant.

Acoustical. The word "acoustical," an adjective, means associated in a general way with the science of sound or with the broader classes of media, phenomena, apparatus, quantities, or units discussed in the science of sound. Example: "Acoustical media exhibit acoustical phenomena whose well-defined acoustical quantities can be measured, with the aid of acoustical apparatus, in terms of an acceptable system of acoustical units." Other examples are acoustical engineer, school, glossary, theorem, circuit diagram.

1.6. Pressure and Density. *Static Pressure* (P_0). The static pressure at a point in the medium is the pressure that would exist at that point with no sound waves present. At normal barometric pressure, P_0 equals approximately 10^5 newtons/m^2 (10^6 dynes /cm^2). This corresponds to a barometer reading of 0.751 m (29.6 in.) Hg mercury when the temperature of the mercury is 0°C. Standard atmospheric pressure is usually taken to be 0.760 m Hg at 0°C. This is a pressure of 1.013×10^5 newtons/m^2. In this text when solving problems we shall assume $P_0 = 10^5$ newtons/m^2.

Microbar (μb). A microbar is a unit of pressure commonly used in acoustics. One microbar is equal to 0.1 newton per square meter or 1 dyne per square centimeter. In this text its use is not restricted to the cgs system.

Instantaneous Sound Pressure $[p(t)]$. The instantaneous sound pressure at a point is the incremental change from the static pressure at a given instant caused by the presence of a sound wave. The unit is the microbar (0.1 newton per square meter or 1 dyne per square centimeter).

Effective Sound Pressure (p). The effective sound pressure at a point is the root-mean-square (rms) value of the instantaneous sound pressure, over a time interval at that point. The unit is the microbar (0.1 newton per square meter or 1 dyne per square centimeter). In the case of periodic sound pressures, the interval should be an integral number of periods. In the case of nonperiodic sound pressures, the interval should be long enough to make the value obtained essentially independent of small changes in the length of the interval.

$\mu bar = 0.1 N/m^2$

Density of Air (ρ_0).   The ambient density of air is given by the formulas

$\rho_0 = 1.18 \ kg/m^3$
@ Room

$$\rho_0 = 1.29 \frac{273}{T} \frac{P_0}{0.76} \qquad kg/m^3 \ (mks) \qquad (1.2)$$

Temperature

$$\rho_0 = 0.00129 \frac{273}{T} \frac{P_0}{0.76} \qquad gm/cm^3 \ (cgs) \qquad (1.3)$$

where T is the absolute temperature in degrees Kelvin and P_0 is the barometric pressure in meters of mercury.   At normal room temperature of $T = 295°K$ ($22°C$ or $71.6°F$), and for a static pressure $P_0 = 0.751$ m Hg, the ambient density is $\rho_0 = 1.18$ kg/m³.   This value of ρ_0 will be used in solving problems unless otherwise stated.

1.7. Speed and Velocity.   *Speed of Sound* (c).   The speed of sound in air is given approximately by the formulas

$$c = 331.4 + 0.607\theta \qquad m/sec \ (mks) \qquad (1.4)$$
$$c = 33,140 + 60.7\theta \qquad cm/sec \ (cgs) \qquad (1.5)$$
$$c = 1087 + 1.99\theta \qquad ft/sec \ (English\text{-}centigrade) \qquad (1.6)$$
$$c = 1052 + 1.106F \qquad ft/sec \ (English\text{-}Fahrenheit) \qquad (1.7)$$

where θ is the ambient temperature in degrees centigrade and F is the same in degrees Fahrenheit.   For temperatures above 30°C or below $-30°C$, the velocity of sound must be determined from the exact formula

$$c = 331.4 \sqrt{\frac{T}{273}} = 331.4 \sqrt{1 + \frac{\theta}{273}} \qquad m/sec \qquad (1.8)$$

where T is the ambient temperature in degrees Kelvin.   At a normal room temperature of $\theta = 22°$($F = 71.6°$), $c \doteq 344.8$ m/sec, or 1131.2 ft/ sec.   These values of c will be used in solving problems unless otherwise stated.

Instantaneous Particle Velocity (*Particle Velocity*) [$u(t)$].   The instantaneous particle velocity at a point is the velocity, due to the sound wave only, of a given infinitesimal part of the medium at a given instant.   It is measured over and above any motion of the medium as a whole.   The unit is the meter per second (in the cgs system the unit is the centimeter per second).

Effective Particle Velocity (u).   The effective particle velocity at a point is the root mean square of the instantaneous particle velocity (see *Effective Sound Pressure* for details).   The unit is the meter per second (in the cgs system the unit is the centimeter per second).

Instantaneous Volume Velocity [$U(t)$].   The instantaneous volume velocity, due to the sound wave only, is the rate of flow of the medium perpendicularly through a specified area S.   That is, $U(t) = Su(t)$, where $u(t)$ is the instantaneous particle velocity.   The unit is the cubic meter per second (in the cgs system the unit is the cubic centimeter per second).

1.8. Impedance. *Acoustic Impedance* (*American Standard Acoustic Impedance*). The acoustic impedance at a given surface is defined as the complex ratio† of effective sound pressure averaged over the surface to effective volume velocity through it. The surface may be either a hypothetical surface in an acoustic medium or the moving surface of a mechanical device. The unit is newton-sec/m⁵, or the mks acoustic ohm.‡ (In the cgs system the unit is dyne-sec/cm⁵, or acoustic ohm.)

pressure / *volume velocity*
$$Z_A = \frac{p}{U} \qquad \text{newton-sec/m}^5 \text{ (mks acoustic ohms)} \qquad (1.9)$$

Specific Acoustic Impedance (Z_s). The specific acoustic impedance is the complex ratio of the effective sound pressure at a point of an acoustic medium or mechanical device to the effective particle velocity at that point. The unit is newton-sec/m³, or the mks rayl.§ (In the cgs system the unit is dyne-sec/cm³, or the rayl.) That is,

pressure / *velocity*
$$Z_s = \frac{p}{u} \qquad \text{newton-sec/m}^3 \text{ (mks rayls)} \qquad (1.10)$$

Mechanical Impedance (Z_M). The mechanical impedance is the complex ratio of the effective force acting on a specified area of an acoustic medium or mechanical device to the resulting effective linear velocity through or of that area, respectively. The unit is the newton-sec/m, or the mks mechanical ohm. (In the cgs system the unit is the dyne-sec/cm, or the mechanical ohm.) That is,

force / *velocity*
$$Z_M = \frac{f}{u} \qquad \underline{\text{newton-sec/m}} \text{ (mks mechanical ohms)} \qquad (1.11)$$

Characteristic Impedance ($\rho_0 c$). The characteristic impedance is the ratio of the effective sound pressure at a given point to the effective particle velocity at that point in a free, plane, progressive sound wave. It is equal to the product of the density of the medium times the speed of sound in the medium ($\rho_0 c$). It is analogous to the characteristic impedance of an infinitely long, dissipationless transmission line. The unit is the mks rayl, or newton-sec/m³. (In the cgs system, the unit is the rayl, or dyne-sec/cm³.)

In the solution of problems in this book we shall assume for air that $\rho_0 c = 407$ mks rayls (or $\rho_0 c = 40.7$ rayls) which is valid for a temperature of 22°C (71.6°F) and a barometric pressure of 0.751 m (29.6 in.) Hg.

1.9. Intensity, Energy Density, and Levels. *Sound Intensity* (I). The sound intensity measured in a specified direction at a point is the average

---

† "Complex ratio" has the same meaning as the complex ratio of voltage and current in electric-circuit theory.

‡ This notation is taken from Table 12.1 of American Standard Z24.1—1951.

§ Named in honor of Lord Rayleigh.

(margin note: mks rayl = newton-sec / m³)

rate at which sound energy is transmitted through a unit area perpendicular to the specified direction at the point considered. The unit is the watt per square meter. (In the cgs system the unit is the erg per second per square centimeter.) In a plane or spherical free-progressive sound wave the intensity in the direction of propagation is

$$I = \frac{p^2}{\rho_0 c} \qquad \text{watts/m}^2 \tag{1.12}$$

NOTE: In the acoustical literature the intensity has often been expressed in the units of watts per square centimeter, which is equal to $10^{-7} \times$ the number of ergs per second per square centimeter.

Sound Energy Density (D). The sound energy density is the sound energy in a given infinitesimal part of the gas divided by the volume of that part of the gas. The unit is the watt-second per cubic meter. (In the cgs system the unit is the erg per cubic centimeter.) In many acoustic environments such as in a plane wave the sound energy density at a point is

$$D = \frac{p^2}{\rho_0 c^2} = \frac{p^2}{\gamma P_0} \tag{1.13}$$

where γ is the ratio of specific heats for a gas and is equal to 1.4 for air and other diatomic gases. The quantity γ is dimensionless.

Electric Power Level, or Acoustic Intensity Level. The electric power level, or the acoustic intensity level, is a quantity expressing the ratio of two electrical powers or of two sound intensities in logarithmic form. The unit is the decibel. Definitions are

$$\text{Electric power level} = 10 \log_{10} \frac{W_1}{W_2} \qquad \text{db} \tag{1.14}$$

$$\text{Acoustic intensity level} = 10 \log_{10} \frac{I_1}{I_2} \qquad \text{db} \tag{1.15}$$

where W_1 and W_2 are two electrical powers and I_1 and I_2 are two sound intensities.

Extending this thought further, we see from Eq. (1.14) that

$$\text{Electric power level} = 10 \log_{10} \frac{E_1^2}{R_1} \frac{R_2}{E_2^2}$$

$$= 20 \log_{10} \frac{E_1}{E_2} + 10 \log_{10} \frac{R_2}{R_1} \qquad \text{db} \tag{1.16}$$

where E_1 is the voltage across the resistance R_1 in which a power W_1 is being dissipated and E_2 is the voltage across the resistance R_2 in which a power W_2 is being dissipated. Similarly,

$$\text{Acoustic intensity level} = 20 \log_{10} \frac{p_1}{p_2} + 10 \log_{10} \frac{R_{s2}}{R_{s1}} \qquad \text{db} \tag{1.17}$$

where p_1 is the pressure at a point where the specific acoustic resistance (*i.e.*, the real part of the specific acoustic impedance) is R_{S1} and p_2 is the pressure at a point where the specific acoustic resistance is R_{S2}. We note that $10 \log_{10} (W_1/W_2) = 20 \log_{10} (E_1/E_2)$ only if $R_1 = R_2$ and that $10 \log_{10} (I_1/I_2) = 20 \log_{10} (p_1/p_2)$ only if $R_{S2} = R_{S1}$.

Levels involving voltage and pressure alone are sometimes spoken of with no regard to the equalities of the electric resistances or specific acoustic resistances. This practice leads to serious confusion. It is emphasized that the manner in which the terms are used should be clearly stated always by the user in order to avoid confusion.

Sound Pressure Level (SPL). The sound pressure level of a sound, in decibels, is 20 times the logarithm to the base 10 of the ratio of the measured effective sound pressure of this sound to a reference effective sound pressure. That is,

$$\text{SPL} = 20 \log_{10} \frac{p}{p_{\text{ref}}} \quad \text{db} \qquad (1.18)$$

In the United States p_{ref} is either

(a) $\qquad p_{\text{ref}} = 0.0002$ microbar $(2 \times 10^{-5} \text{ newton/m}^2)$

or

(b) $\qquad p_{\text{ref}} = 1$ microbar (0.1 newton/m^2)

Reference pressure (a) has been in general use for measurements dealing with hearing and for sound-level and noise measurements in air and liquids. Reference pressure (b) has gained widespread use for calibrations of transducers and some types of sound-level measurements in liquids. The two reference levels are almost exactly 74 db apart. The reference pressure must always be stated explicitly.

Intensity Level (IL). The intensity level of a sound, in decibels, is 10 times the logarithm to the base 10 of the ratio of the intensity of this sound to a reference intensity. That is,

$$\text{IL} = 10 \log_{10} \frac{I}{I_{\text{ref}}} \qquad (1.19)$$

In the United States the reference intensity is often taken to be $10^{-16} \text{ watt/cm}^2$ $(10^{-12} \text{ watt/m}^2)$. This reference at standard atmospheric conditions in a plane or spherical progressive wave was originally selected as corresponding approximately to the reference pressure $(0.0002$ microbar).

The exact relation between intensity level and sound pressure level in a plane or spherical progressive wave may be found by substituting Eq. (1.12) for intensity in Eq. (1.19).

$$\text{IL} = \text{SPL} + 10 \log_{10} \frac{p_{\text{ref}}^2}{\rho_0 c I_{\text{ref}}} \quad \text{db} \tag{1.20}$$

Substituting $p_{\text{ref}} = 2 \times 10^{-5}$ newton/m^2 and $I_{\text{ref}} = 10^{-12}$ watt/m^2 yields

$$\text{IL} = \text{SPL} + 10 \log_{10} \frac{400}{\rho_0 c} \quad \text{db} \tag{1.21}$$

It is apparent that the intensity level IL will equal the sound pressure level SPL only if $\rho_0 c = 400$ mks rayls. For certain combinations of temperature and static pressure this will be true, although for $T = 22°$C and $P_0 = 0.751$ m Hg, $\rho_0 c = 407$ mks rayls. For this common case then, the intensity level is smaller than the sound pressure level by about 0.1 db. Regardless of which reference quantity is used, it must always be stated explicitly.

Acoustic Power Level (PWL). The acoustic power level of a sound source, in decibels, is 10 times the logarithm to the base 10 of the ratio of the acoustic power radiated by the source to a reference acoustic power. That is,

$$\text{PWL} = 10 \log_{10} \frac{W}{W_{\text{ref}}} \quad \text{db} \tag{1.22}$$

In this text, W_{ref} is 10^{-13} watt. This means that a source radiating 1 acoustic watt has a power level of 130 db.

If the temperature is 20°C (67°F) and the pressure is 1.013×10^5 newtons/m^2 (0.76 m Hg), the sound pressure level in a duct with an area of 1 ft^2 cross section, or at a distance of 0.282 ft from the center of a "point" source (at this distance, the spherical surface has an area of 1 ft^2), is, from Eqs. (1.12) and (1.18)

$$\begin{aligned}
\text{SPL}_{1 \text{ ft}^2} &= 10 \log_{10} \frac{I \rho_0 c}{p_{\text{ref}}^2} = 10 \log_{10} \frac{W \rho_0 c}{S p_{\text{ref}}^2} \\
&= 10 \log_{10} \left[\frac{W}{0.093} \times 412.5 \times \frac{1}{(2 \times 10^{-5})^2} \right] \\
&= 10 \log_{10} \frac{W}{10^{-13}} + 0.5
\end{aligned}$$

where W = acoustic power in watts
$\rho_0 c$ = characteristic impedance = 412.5 mks rayls
S = 1 ft^2 of area = 0.093 m^2
p_{ref} = reference sound pressure = 2×10^{-5} newton/m^2

In words, the sound pressure level equals the acoustic power level plus 0.5 db under the special conditions that the power passes uniformly through an area of 1 ft^2, the temperature is 20°C (67°F), and the barometric pressure is 0.76 m (30 in.) Hg.

Sound Level. The sound level at a point in a sound field is the reading in decibels of a sound-level meter constructed and operated in accordance with the latest edition of "American Standard Sound Level Meters for the Measurement of Noise and Other Sounds."[2]

The meter reading (in decibels) corresponds to a value of the sound pressure integrated over the audible frequency range with a specified frequency weighting and integration time.

Band Power Level (PWL$_n$). The band power level for a specified frequency band is the acoustic power level for the acoustic power contained within the band. The width of the band and the reference power must be specified. The unit is the decibel. The letter n is the designation number for the band being considered.

Band Pressure Level (BPL$_n$). The band pressure level of a sound for a specified frequency band is the effective sound pressure level for the sound energy contained within the band. The width of the band and the reference pressure must be specified. The unit is the decibel. The letter n is the designation number for the band being considered.

Power Spectrum Level. The power spectrum level of a sound at a specified frequency is the power level for the acoustic power contained in a band one cycle per second wide, centered at this specified frequency. The reference power must be specified. The unit is the decibel (see also the discussion under Pressure Spectrum Level).

Pressure Spectrum Level. The pressure spectrum level of a sound at a specified frequency is the effective sound pressure level for the sound energy contained within a band one cycle per second wide, centered at this specified frequency. The reference pressure must be explicitly stated. The unit is the decibel.

DISCUSSION. The concept of pressure spectrum level ordinarily has significance only for sound having a continuous distribution of energy within the frequency range under consideration.

The level of a uniform band of noise with a continuous spectrum exceeds the spectrum level by

$$C_n = 10 \log_{10} (f_b - f_a) \qquad \text{db} \qquad (1.23)$$

where f_b and f_a are the upper and lower frequencies of the band, respectively. The level of a uniform noise with a continuous spectrum in a band of width $f_b - f_a$ cps is therefore related to the spectrum level by the formula

$$L_n = C_n + S_n \qquad (1.24)$$

where L_n = sound pressure level in decibels of the noise in the band of width $f_b - f_a$, for C_n see Eq. (1.23), S_n = spectrum level of the noise, and n = designation number for the band being considered.

[2] "American Standard Sound Level Meters for the Measurement of Noise and Other Sounds," Z24.3—1944, American Standards Association, Inc., New York, N.Y. This standard is in process of revision.

CHAPTER 2

THE WAVE EQUATION AND SOLUTIONS

PART III *The Wave Equation*

2.1. Introduction. We have already outlined in a qualitative way the nature of sound propagation in a gas. In this chapter we shall put the physical principles described earlier into the language of mathematics. The approach is in two steps. First, we shall establish equations expressing Newton's second law of motion, the gas law, and the laws of conservation of mass. Second, we shall combine these equations to produce a wave equation.

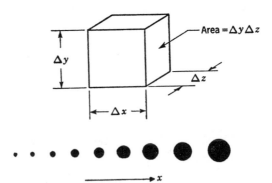

Fig. 2.1. The very small "box" of air shown here is part of a gaseous medium in which the sound pressure increases from left to right at a space rate of $\partial p/\partial x$ (or, in vector notation, grad p). The sizes of the dots indicate the magnitude of the sound pressure at each point.

The mathematical derivations are given in two ways: with and without use of vector algebra. Those who are familiar with vector notation will appreciate the generality of the three-dimensional vector approach. The two derivations are carried on in parallel; on the left sides of the pages, the one-dimensional wave equation is derived with the use of simple

16

differential notation; on the right sides, the three-dimensional wave equation is derived with the use of vector notation. The simplicity of the vector operations is revealed in the side-by-side presentation of the two derivations.

2.2. Derivation of the Wave Equation. *The Equation of Motion.* If

we write Newton's second law for a small volume of gas located in a homogeneous medium, we obtain the equation of motion, or the force equation as it is sometimes called. Imagine the small volume of gas to be enclosed in a box with weightless flexible sides.

One-dimensional Derivation[1]

Let us suppose that the box is situated in a medium where the sound pressure p increases from left to right at a space rate of $\partial p/\partial x$ (see Fig. 2.1).

Three-dimensional Derivation[2]

Let us suppose that the box is situated in a medium (see Fig. 2.1) where the sound pressure p changes in space at a space rate of $\mathbf{grad}\, p = \mathbf{i}\,\dfrac{\partial p}{\partial x} + \mathbf{j}\,\dfrac{\partial p}{\partial y} + \mathbf{k}\,\dfrac{\partial p}{\partial z}$ where $\mathbf{i}$, $\mathbf{j}$, and $\mathbf{k}$ are unit vectors in the x, y, and z directions, respectively, and p is the pressure at a point.

Assume that the sides of the box are completely frictionless; *i.e.*, any viscous drag between gas particles inside the box and those outside is negligible. Thus the only forces acting on the enclosed gas are due to the pressures at the faces of the box.

The difference between the forces acting on the two sides of our tiny box of gas is equal to the rate at which the force changes with distance times the incremental length of the box:

Force acting to accelerate the box in the

$$\text{positive } x \text{ direction } = -\left(\frac{\partial p}{\partial x}\Delta x\right)\Delta y\,\Delta z \tag{2.1a}$$

Force acting to accelerate the box in the

$$\text{positive direction } = -\left[\mathbf{i}\left(\frac{\partial p}{\partial x}\Delta x\right)\Delta y\,\Delta z \right.$$
$$\left. + \mathbf{j}\left(\frac{\partial p}{\partial y}\Delta y\right)\Delta x\,\Delta z + \mathbf{k}\left(\frac{\partial p}{\partial z}\Delta z\right)\Delta x\,\Delta y\right] \tag{2.1b}$$

Note that the positive gradient causes an acceleration of the box in the negative direction of x.

[1] Nonvector derivations of the wave equation are given in Rayleigh, "Theory of Sound," Vol. 2, pp. 1–15, Macmillan & Co., Ltd., London, 1896; P. M. Morse, "Vibration and Sound," 2d ed., pp. 217–225, McGraw-Hill Book Company, Inc., New York, 1948; L. E. Kinsler and A. R. Frey, "Fundamentals of Acoustics," pp. 118–137, John Wiley & Sons, Inc., New York, 1950; R. W. B. Stephens and A. E. Bate, "Wave Motion and Sound," pp. 32–43, 400–406, Edward Arnold & Co., London, 1950; and other places.

[2] A vector derivation of the wave equation is given in two papers that must be read together: W. J. Cunningham, Application of Vector Analysis to the Wave Equation, *J. Acoust. Soc. Amer.*, **22**: 61 (1950); and R. V. L. Hartley, Note on "Application of Vector Analysis to the Wave Equation," *J. Acous. Soc. Amer.*, **22**: 511 (1950).

Division of both sides of the above equation by $\Delta x\,\Delta y\,\Delta z = V$ gives the force per unit volume acting to accelerate the box,

$$\frac{f}{V} = -\frac{\partial p}{\partial x} \qquad (2.2a)$$

Division of both sides of the equation by $\Delta x\,\Delta y\,\Delta z = V$ gives the force per unit volume acting to accelerate the box,

$$\frac{f}{V} = -\operatorname{grad} p \qquad (2.2b)$$

By Newton's law, the force per unit volume (f/V) of Eq. (2.2) must be equal to the time rate of change of the momentum per unit volume of the box. We have already assumed that our box is a deformable packet so that the mass of the gas within it is always constant. That is,

$$\frac{f}{V} = -\frac{\partial p}{\partial x} = \frac{M}{V}\frac{\partial u}{\partial t} = \rho'\frac{\partial u}{\partial t} \quad (2.3a)$$

where u is the average velocity of the gas in the "box" in the x direction, ρ' is the space average of the instantaneous density of the gas in the box, and $M = \rho'V$ is the total mass of the gas in the box.

$$\frac{f}{V} = -\operatorname{grad} p = \frac{M}{V}\frac{D\mathbf{q}}{Dt} = \rho'\frac{D\mathbf{q}}{Dt} \quad (2.3b)$$

where $\mathbf{q}$ is the average vector velocity of the gas in the "box," ρ' is the average density of the gas in the box, and $M = \rho'V$ is the total mass of the gas in the box. D/Dt is not a simple partial derivative but represents the total rate of the change of the velocity of the particular bit of gas in the box regardless of its position, i.e.,

$$\frac{D\mathbf{q}}{Dt} = \frac{\partial \mathbf{q}}{\partial t} + q_x\frac{\partial \mathbf{q}}{\partial x} + q_y\frac{\partial \mathbf{q}}{\partial y} + q_z\frac{\partial \mathbf{q}}{\partial z}$$

where q_x, q_y, and q_z are the components of the vector particle velocity $\mathbf{q}$.

If the change in density of the gas due to the sound wave is small enough, then the instantaneous density ρ' is approximately equal to the average density ρ_0. Then,

EQUATION OF MOTION

$$-\frac{\partial p}{\partial x} = \rho_0\frac{\partial u}{\partial t} \qquad (2.4a)$$

If the vector particle velocity $\mathbf{q}$ is small enough, the rate of change of momentum of the particles in the box can be approximated by the rate of change of momentum at a fixed point, $D\mathbf{q}/Dt \doteq \partial\mathbf{q}/\partial t$, and the instantaneous density ρ' can be approximated by the average density ρ_0. Then,

$$-\operatorname{grad} p = \rho_0\frac{\partial \mathbf{q}}{\partial t} \qquad (2.4b)$$

The approximations just given are generally acceptable provided the sound pressure levels being considered are below about 110 db re 0.0002 microbar. Levels above 110 db are so large as to create hearing discomfort in many individuals, as we shall see in Chap. 13 at the end of this book.

The Gas Law. If we assume an ideal gas, the Charles-Boyle gas law applies to the box. It is

$$PV = RT \qquad (2.5)$$

where P is the total pressure in the box, V is the volume equal to $\Delta x\,\Delta y\,\Delta z$, T is the absolute temperature in degrees Kelvin, and R is a constant for the gas whose magnitude is dependent upon the mass of gas chosen.† Using this equation, we can find a relation between the sound pressure (excess pressure) and an incremental change in V for our box. Before we can establish this relation, however, we must know how the temperature T varies with changes in P and V and, in particular, whether the phenomenon is adiabatic or isothermal.

At audible frequencies the wavelength of a sound is long compared with the spacing between air molecules. For example, at 1000 cps, the wavelength λ equals 0.34 m, as compared with an intermolecular spacing of 10^{-9} m. Now, whenever a portion of any gas is compressed rapidly, its temperature rises, and, conversely, when it is expanded rapidly, its temperature drops. At any one point in an alternating sound field, therefore, the temperature rises and falls relative to the ambient temperature. This variation occurs at the same frequency as that of the sound wave and is in phase with the sound pressure.

Let us assume, for the moment, that the sound wave has only one frequency. At points separated by one-half wavelength, the pressure and the temperature fluctuations will be 180° out of phase with each other. Now the question arises, Is there sufficient time during one-half an alternation in the temperature for an exchange of heat to take place between these two points of maximally different temperatures?

It has been established[3] that under normal atmospheric conditions the speed of travel of a thermal diffusion wave at 1000 cps is about 0.5 m/sec, and at 10,000 cps it is about 1.5 m/sec. The time for one-half an alternation of 1000 cps is 0.0005 sec. In this time, the thermal wave travels a distance of only 0.00025 m. This number is very small compared with one-half wavelength (0.17 m) at 1000 cps. At 10,000 cps the heat travels 7.5×10^{-5} m, which is a small distance compared with a half wavelength $(1.7 \times 10^{-2}$ m). It appears safe for us to conclude, therefore, that there is negligible heat exchange in the wave in the audible frequency range. Gaseous compressions and expansions of this type are said to be adiabatic.

For adiabatic expansions, the relation between the total pressure and the volume is known to be[4]

$$PV^\gamma = \text{constant} \tag{2.6}$$

where γ is the ratio of the specific heat of the gas at constant pressure to the specific heat at constant volume for the gas. This equation is

---

† If a mass of gas is chosen so that its weight in grams is equal to its molecular weight (known to chemists as the gram-molecular weight, or the mole), then the volume of this mass at 0°C and 0.76 m Hg is the same for all gases and equals 0.02242 m³. Then $R = 8.314$ watt-sec per degree centigrade per gram-molecular weight. If the mass of gas chosen is n times its molecular weight, then $R = 8.314n$.

[3] See L. L. Beranek, "Acoustic Measurements," p. 49, John Wiley & Sons, Inc., New York, 1949.

[4] M. W. Zemansky, "Heat and Thermodynamics," 2d ed., pp. 104–114, McGraw-Hill Book Company, Inc., New York, 1943.

obtained from the gas law in the form of Eq. (2.5), assuming adiabatic conditions. For air, hydrogen, nitrogen, and oxygen, *i.e.*, gases with diatomic molecules,

$$\gamma = 1.4$$

Expressing Eq. (2.6) in differential form, we have

$$\frac{dP}{P} = \frac{-\gamma \, dV}{V} \tag{2.7}$$

Let

$$P = P_0 + p \qquad V = V_0 + \tau \tag{2.8}$$

where P_0 and V_0 are the undisturbed pressure and volume, respectively, and p and τ are the incremental pressure and volume, respectively, owing

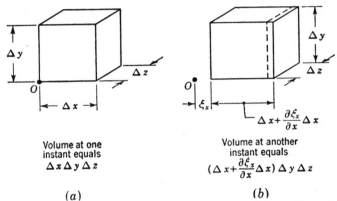

Volume at one
instant equals
$\Delta x \Delta y \Delta z$

(a)

Volume at another
instant equals
$(\Delta x + \frac{\partial \xi_x}{\partial x} \Delta x) \Delta y \Delta z$

(b)

Fig. 2.2. Change in volume of the box with change in position. From (a) and (b) it is seen that the incremental change in volume of the box equals $\tau = (\partial \xi_x / \partial x) \, \Delta x \, \Delta y \, \Delta z$.

to the presence of the sound wave. Then, to the same approximation as that made preceding Eq. (2.4) and because $p \ll P_0$ and $\tau \ll V_0$,

$$\frac{p}{P_0} = -\frac{\gamma \tau}{V_0} \tag{2.9}$$

The time derivative of this equation gives

CONTINUITY EQUATION = TOTAL MASS
OF GAS MUST REMAIN CONSTANT.

$$\frac{1}{P_0} \frac{\partial p}{\partial t} = \frac{-\gamma}{V_0} \frac{\partial \tau}{\partial t} \qquad \text{GAS LAW} \tag{2.10}$$

The Continuity Equation. The continuity equation is a mathematical expression stating that the total mass of gas in a deformable "box" must remain constant. Because of this law of conservation of mass, we are able to write a unique relation between the time rate of change of the incremental velocities at the surfaces of the box.

One-dimensional Derivation

Refer to Fig. 2.2. If the mass of gas within the box remains constant, the change in volume τ depends only on the difference of displacement of the air particles on the opposite sides of box. Another way of saying this is that unless the air particles adjacent to any given side of the box move at the same velocity as the box itself, some will cross in or out of the box and the mass inside will change.

In a given interval of time the air particles on the left-hand side of the box will have been displaced ξ_x. In this same time, the air particles on the right-hand side will have been displaced

$$\xi_x + \frac{\partial \xi_x}{\partial x}\, \Delta x$$

The difference of the two quantities above multiplied by the area $\Delta y\, \Delta z$ gives the increment in volume τ

$$\tau = \frac{\partial \xi_x}{\partial x}\, \Delta x\, \Delta y\, \Delta z \qquad (2.11a)$$

or

$$\tau = V_0 \frac{\partial \xi_x}{\partial x} \qquad (2.12)$$

Differentiating with respect to time yields,

CONTINUITY EQUATION

$$\frac{\partial \tau}{\partial t} = V_0 \frac{\partial u}{\partial x} \qquad (2.13a)$$

where u is the instantaneous particle velocity.

Three-dimensional Derivation

If the mass of gas within the box remains constant, the change in incremental volume τ depends only on the divergence of the vector displacement. Another way of saying this is that unless the air particles adjacent to any given side of the box move at the same velocity as the side of the box itself, some will cross into or out of the box and the mass inside will change; so

$$\tau = V_0 \operatorname{div} \boldsymbol{\xi} \qquad (2.11b)$$

Differentiating with respect to time yields,

$$\frac{\partial \tau}{\partial t} = V_0 \operatorname{div} \mathbf{q} \qquad (2.13b)$$

where $\mathbf{q}$ is the instantaneous particle velocity.

The Wave Equation in Rectangular Coordinates

COMBINATION OF EQUATION OF MOTION GAS LAW CONTINUITY EQUATION

One-dimensional Derivation

The one-dimensional wave equation is obtained by combining the equation of motion (2.4a), the gas law (2.10), and the continuity equation (2.13a). Combination of (2.10) and (2.13a) gives

$$\frac{\partial p}{\partial t} = -\gamma P_0 \frac{\partial u}{\partial x} \qquad (2.14a)$$

Differentiate (2.14a) with respect to t.

$$\frac{\partial^2 p}{\partial t^2} = -\gamma P_0 \frac{\partial^2 u}{\partial t\, \partial x} \qquad (2.15a)$$

Differentiate (2.4a) with respect to x.

$$-\frac{\partial^2 p}{\partial x^2} = \rho_0 \frac{\partial^2 u}{\partial x\, \partial t} \qquad (2.16a)$$

Three-dimensional Derivation

The three-dimensional wave equation is obtained by combining the equation of motion (2.4b), the gas law (2.10), and the continuity equation (2.13b). Combination of (2.10) and (2.13b) gives

$$\frac{\partial p}{\partial t} = -\gamma P_0 \operatorname{div} \mathbf{q} \qquad (2.14b)$$

Differentiate (2.14b) with respect to t.

$$\frac{\partial^2 p}{\partial t^2} = -\gamma P_0 \operatorname{div} \frac{\partial \mathbf{q}}{\partial t} \qquad (2.15b)$$

Take the divergence of each side of Eq. (2.4b).

$$-\operatorname{div}(\operatorname{grad} p) = \rho_0 \operatorname{div} \frac{\partial \mathbf{q}}{\partial t} \qquad (2.16b)$$

One-dimensional Derivation

Three-dimensional Derivation

Replacing the div (grad p) by $\nabla^2 p$, we get

$$-\nabla^2 p = \rho_0 \, \text{div} \, \frac{\partial \mathbf{q}}{\partial t} \qquad (2.17)$$

Assuming interchangeability of the x and t derivatives, and combining (2.15a) and (2.16a), we get

where ∇^2 is the operator called the Laplacian. Combining (2.15b) and (2.17), we get

$$\frac{\partial^2 p}{\partial x^2} = \frac{\rho_0}{\gamma P_0} \frac{\partial^2 p}{\partial t^2} \qquad (2.18a)$$

$$\nabla^2 p = \frac{\rho_0}{\gamma P_0} \frac{\partial^2 p}{\partial t^2} \qquad (2.18b)$$

Let us, by definition set,

$$c^2 \equiv \frac{\gamma P_0}{\rho_0} \qquad (2.19)$$

We shall see later that c is the speed of propagation of the sound wave in the medium.   *THREE DIMENSIONAL WAVE EQUATION*

We obtain the one-dimensional wave equation

We obtain the three-dimensional wave equation

$$\frac{\partial^2 p}{\partial x^2} = \frac{1}{c^2} \frac{\partial^2 p}{\partial t^2} \qquad (2.20a)$$

$$\nabla^2 p = \frac{1}{c^2} \frac{\partial^2 p}{\partial t^2} \qquad (2.20b)$$

In rectangular coordinates

$$\nabla^2 p \equiv \frac{\partial^2 p}{\partial x^2} + \frac{\partial^2 p}{\partial y^2} + \frac{\partial^2 p}{\partial z^2} \qquad (2.21)$$

We could also have eliminated p and retained u, in which case we would have

We could also have eliminated p and retained $\mathbf{q}$, in which case we would have

$$\nabla^2 \mathbf{q} = \frac{1}{c^2} \frac{\partial^2 \mathbf{q}}{\partial t^2} \qquad (2.22b)$$

$$\frac{\partial^2 u}{\partial x^2} = \frac{1}{c^2} \frac{\partial^2 u}{\partial t^2} \qquad (2.22a)$$

where $\nabla^2 \mathbf{q} = \text{grad} \, (\text{div} \, q)$ when there is no rotation in the medium.

Equations (2.20) and (2.22) apply to sound waves of "small" magnitude propagating in a source-free, homogeneous, isotropic, frictionless gas at rest.

The Wave Equation in Spherical Coordinates. The one-dimensional wave equations derived above are for plane-wave propagation along one dimension of a rectangular coordinate system. In an anechoic (echo-free) chamber or in free space, we frequently wish to express mathematically the radiation of sound from a spherical (nondirectional) source of sound. In this case, the sound wave will expand as it travels away from the source, and the wave front always will be a spherical surface. To apply the wave equation to spherical waves, we must replace the operators on the left side of Eqs. (2.20) and (2.22) by operators appropriate to spherical coordinates.

Assuming equal radiation in all directions, the wave equation in one-dimensional spherical coordinates is

$$\frac{\partial^2 p}{\partial r^2} + \frac{2}{r} \frac{\partial p}{\partial r} = \frac{1}{c^2} \frac{\partial^2 p}{\partial t^2} \qquad (2.23)$$

Simple differentiation will show that (2.23) can also be written

$$\frac{\partial^2 (pr)}{\partial r^2} = \frac{1}{c^2} \frac{\partial^2 (pr)}{\partial t^2} \tag{2.24}$$

It is interesting to note that this equation has exactly the same form as Eq (2.20a). Hence, the same formal solution will apply to either equation except that the dependent variable is $p(x,t)$ in one case and $pr(r,t)$ in the other case.

Example 2.1. In the steady state, that is, $\partial u / \partial t = j\omega u$, determine mathematically how the sound pressure in a plane progressive sound wave (one-dimensional case) could be determined from measurement of particle velocity alone.

Solution. From Eq. (2.4a) we find in the steady state that

$$\frac{-\partial p}{\partial x} = j\omega \rho_0 u$$

where p and u are now rms values of the sound pressure and particle velocity, respectively. Written in differential form,

$$-\Delta p = j\omega \rho_0 u \, \Delta x$$

If the particle velocity is 1 cm/sec, ω is 1000 radians/sec, and Δx is 0.5 cm, then

$$\Delta p = -j0.005 \times 1000 \times 1.18 \times 0.01$$
$$= -j0.059 \text{ newton/m}^2$$

We shall have an opportunity in Chap. 6 of this text to see a practical application of these equations to the measurement of particle velocity by a velocity microphone.

PART **IV** *Solutions of the Wave Equation*

2.3. General Solutions of the One-dimensional Wave Equation. The one-dimensional wave equation was derived with either sound pressure or particle velocity as the dependent variable. Particle displacement, or the variational density, may also be used as the dependent variable. This can be seen from Eqs. (2.4a) and (2.13a) and the conservation of mass, which requires that the product of the density and the volume of a small box of gas remain constant. That is,

$$\rho' V = \rho_0 V_0 = \text{constant} \tag{2.25}$$

and so

$$\rho' \, dV = -V \, d\rho' \tag{2.26}$$

Let

$$\rho' \equiv \rho_0 + \rho \tag{2.27}$$

where ρ is the incremental change in density. Then, approximately, from Eqs. (2.8) and (2.26),

$$\tau = \text{increment in volume} \qquad \rho_0 \tau = -V_0 \rho \tag{2.28}$$

Differentiating,

$$\frac{\partial \tau}{\partial t} = -\frac{V_0}{\rho_0}\frac{\partial \rho}{\partial t}$$

so that, from Eq. (2.13a),

$$\frac{\partial \rho}{\partial t} = -\rho_0 \frac{\partial u}{\partial x} \qquad (2.29)$$

2.13a

$\frac{\partial \tau}{\partial t} = V_0 \frac{\partial u}{\partial x}$

Also, we know that the particle velocity is the time rate of change of the particle displacement.

$$u = \frac{\partial \xi_x}{\partial t} \qquad (2.30)$$

Inspection of Eqs. (2.4a), (2.13a), (2.29), and (2.30) shows that the pressure, particle velocity, particle displacement, and variational density are related to each other by derivatives and integrals in space and time. These operations performed on the wave equation do not change the form of the solution, as we shall see shortly. Since the form of the solution is not changed, the same wave equation may be used for determining density, displacement, or particle velocity as well as sound pressure by substituting ρ, or ξ_x, or u for p in Eq. (2.20a) or ρ, ξ, or q for p in Eq. (2.20b), assuming, of course, that there is no rotation in the medium.

(2.20b)

$\frac{\partial^2 p}{\partial x^2} = \frac{1}{c^2}\frac{\partial^2 p}{\partial t^2}$

General Solution. With pressure as the dependent variable, the wave equation is

WAVE EQUATION

$$\frac{\partial^2 p}{\partial x^2} = \frac{1}{c^2}\frac{\partial^2 p}{\partial t^2} \qquad (2.31)$$

The general solution to this equation is a sum of two terms,

GENERAL SOLUTION

$$p = f_1\left(t - \frac{x}{c}\right) + f_2\left(t + \frac{x}{c}\right) \qquad (2.32)$$

where f_1 and f_2 are arbitrary functions. We assume only that they have continuous derivatives of the first and second order. Note that because t and x occur together, the first derivatives with respect to x and t are exactly the same except for a factor of $\pm c$.

The ratio x/c must have the dimensions of time, so that c is a speed. From $c^2 = \gamma P_0/\rho_0$ [Eq. (2.19)] we find that

$$c = \left(1.4 \times \frac{10^5}{1.18}\right)^{\frac{1}{2}} = 344.2 \text{ m/sec}$$

in air at an ambient pressure of 10^5 newtons/m^2 and at 22°C. This quantity is nearly the same as the experimentally determined value of the speed of sound 344.8 [see Eq. (1.8)], so that we recognize c as the speed at which a sound wave is propagated through the air.

From the general solution to the wave equation given in Eq. (2.32) we observe two very important facts:

1. The sound pressure at any point x in space can be separated into two components: an outgoing wave, $f_1(t - x/c)$, and a backward-traveling wave, $f_2(t + x/c)$.

2. Regardless of the shape of the outward-going wave (or of the backward-traveling wave), it is propagated without change of shape. To show this, let us assume that, at $t = t_1$, the sound pressure at $x = 0$ is $f_1(t_1)$. At a time $t = t_1 + t_2$ the sound wave will have traveled a distance x equal to t_2c m. At this new time the sound pressure is equal to $p = f_1(t_1 + t_2 - t_2c/c) = f_1(t_1)$. In other words the sound pressure has propagated without change. The same argument can be made for the backward-traveling wave which goes in the $-x$ direction.

It must be understood that inherent in Eqs. (2.31) and (2.32) are two assumptions. First, the wave is a *plane wave*, i.e., it does not expand laterally. Thus the sound pressure is not a function of the y and z coordinates but is a function of distance only along the x coordinate. Second, it is assumed that there are no losses of dispersion (scattering of the wave by turbulence or temperature gradients, etc.) in the air, so that the wave does not lose energy as it is propagated. Dissipative and dispersive cases are not treated analytically in this book, but are discussed briefly in Chaps. 10 and 11.

Steady-state Solution. In nearly all the studies that we make in this text we are concerned with the steady state. As is well known from the theory of Fourier series, a steady-state wave can be represented by a linear summation of sine-wave functions, each of which is of the form

$$\psi_\nu(t) = \sqrt{2}\, |\phi^\nu| \cos (\omega_\nu t + \theta_\nu) \tag{2.33a}$$

For example, if ψ_ν is sound pressure, we write

$$p(t) = \sum_\nu p_\nu(t) = \sum_\nu \sqrt{2}\, |p^\nu| \cos (\omega_\nu t + \theta_\nu) \tag{2.33b}$$

where $\omega_\nu = 2\pi f_\nu$; $f_\nu =$ frequency of vibration of the νth component of the wave; θ_ν is the phase angle of it; and $\sqrt{2}\, |\phi^\nu|$ (or $\sqrt{2}\, |p^\nu|$) is the peak magnitude of the component. Because the wave is propagated without change of shape, we need consider, in the steady state, only those solutions to the wave equation for which the time dependence at each point in space is sinusoidal and which have the same angular frequencies ω_ν as the source.

Borrowing from electrical-circuit theory, we represent a sinusoidal function with a frequency ω by the real part of a complex exponential function. Thus, at a fixed point in space x, we have the sound pressure,

$$\psi(x,t) = \sqrt{2}\, \text{Re}\, [\phi(x)e^{j\omega t}] \tag{2.34a}$$

or

$$p(x,t) = \sqrt{2}\, \text{Re}\, [p(x)e^{j\omega t}] \tag{2.34b}$$

where $p(x)$ is a complex function (*i.e.*, it has a real and an imaginary part) that gives the dependence of p on x. The product of $\sqrt{2}$ times the magnitude of $p(x)$ is the peak value of the sinusoidal sound pressure function at x. The phase angle of $p(x)$ is the phase shift measured from some reference position. Generally we omit writing Re although it always must be remembered that the real part must be taken when using the final expression for the sound pressure. In the steady state, therefore, we may replace f_1 and f_2 of Eq. (2.32) by a sum of functions each having a particular angular driving frequency ω_ν, so that

$$p(x,t) = \sum_\nu p_\nu(x,t) = \sum_\nu \sqrt{2}\ \text{Re}\ [(p_+{}^\nu e^{-j\omega_\nu x/c} + p_-{}^\nu e^{j\omega_\nu x/c})e^{j\omega_\nu t}] \quad (2.35)$$

The part of Eq. (2.35) within the brackets is the same as that within the brackets of Eq. (2.34). The factor $\sqrt{2}$ is introduced so that later $p_+{}^\nu$ and $p_-{}^\nu$ may represent complex rms functions averaged in the time dimension. The $+$ and $-$ subscripts indicate the forward and backward traveling waves respectively.

It is apparent that the first term of Eq. (2.35) represents an outward-traveling wave whose rms magnitude $|p_+{}^\nu|$ does not change with time t or position x. A similar statement may be made for the second term, which is the backward-traveling wave.

It is customary in texts on acoustics to define a *wave number* k,

$$k \equiv \frac{\omega}{c} = \frac{2\pi f}{c} = \frac{2\pi}{\lambda} \quad (2.36)$$

Also, let us drop Re and the subscript ν for convenience. Any one term of Eq. (2.35), with these changes, becomes

$$p(x,t) = \sqrt{2}\ [p(x)e^{j\omega t}] = \sqrt{2}\ (p_+ e^{jk(ct-x)} + p_- e^{jk(ct+x)}) \quad (2.37)$$

Similarly, the solution to Eq. (2.22a), assuming steady-state conditions, is

$$u(x,t) = \sqrt{2}\ (u_+ e^{jk(ct-x)} + u_- e^{jk(ct+x)}) \quad (2.38)$$

It is understood that the real part of Eqs. (2.37) and (2.38) will be used in the final answer. The complex magnitudes of p_+ and p_- or u_+ and u_- are determined from the boundary conditions.

The complex rms pressure and particle velocity are found directly from Eqs. (2.37) and (2.38) by canceling $\sqrt{2}\ e^{j\omega t}$ from the right-hand sides. When the remaining function is converted into magnitude and phase angle, the *magnitude is the quantity that would be indicated by a rms sound-pressure meter*. Note, however, that when we take the real part of $p(x,t)$ or $u(x,t)$, the quantity $\sqrt{2}\ e^{j\omega t}$ must be in the equation if the proper values for the instantaneous pressure and particle velocity are to be obtained.

Example 2.2.   Assume that for the steady state, at a point $x = 0$, the sound pressure in a one-dimensional *outward-traveling* wave has the recurrent form shown by the dotted curve in the sketch below.   This wave form is given by the real part of the equation

$$p(0,t) = \sqrt{2}\,(4e^{j628t} + 2e^{j1884t})$$

(*a*) What are the particle velocity and the particle displacement as a function of time at $x = 5$ m?   (*b*) What are the rms values of these two quantities?   (*c*) Are the rms values dependent upon x?

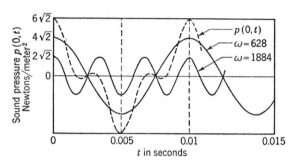

Solution.   *a.* We have for the solution of the wave equation giving both x and t [see Eq. (2.37)]

$$p(x,t) = \sqrt{2}\,(4e^{j628(t-x/c)} + 2e^{j1884(t-x/c)})$$

From Eq. (2.4*a*) we see that

$$u(x,t) = \frac{-1}{j\omega\rho_0}\,\frac{\partial p(x,t)}{\partial x}$$

or

$$u(x,t) = \frac{1}{\rho_0 c}\,p(x,t)$$

And from Eq. (2.30) we have

$$\xi(x,t) = \frac{\sqrt{2}}{j\rho_0 c}\left(\frac{4}{628}\,e^{j628(t-x/c)} + \frac{2}{1884}\,e^{j1884(t-x/c)}\right)$$

At $x = 5$ m, $x/c = 5/344.8 = 0.0145$ sec.

$$u(x,t) = \frac{\sqrt{2}}{407}\,(4e^{j628(t-0.0145)} + 2e^{j1884(t-0.0145)})$$

and

$$\xi_x(x,t) = \frac{\sqrt{2}}{407}\left(\frac{4}{628}\,e^{j[628(t-0.0145)-(\pi/2)]} + \frac{2}{1884}\,e^{j[1884(t-0.0145)-(\pi/2)]}\right)$$

Taking the real parts of the two preceding equations,

$$u(x,t) = \frac{\sqrt{2}}{407}\,[4\cos(628t - 9.1) + 2\cos(1884t - 27.3)]$$

$$\xi(x,t) = \frac{\sqrt{2}}{407}\left[\frac{4}{628}\sin(628t - 9.1) + \frac{2}{1884}\sin(1884t - 27.3)\right]$$

Note that each term in the particle displacement is 90° out of time phase with the velocity and that the wave shape is different.   As might be expected, differentiation emphasizes the higher frequencies.

These equations are plotted below:

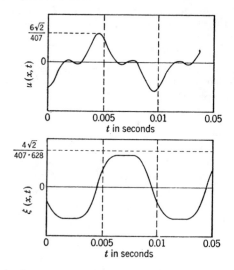

b. The rms magnitude of a sine wave is equal to its peak value divided by $\sqrt{2}$. This may be verified by squaring the sine wave and finding the average value over one cycle and then taking the square root of the result.   If two sine waves of different frequencies are present at one time, the rms value of the combination equals the square root of the sums of the squares of the individual rms magnitudes, so that

$$u = \frac{1}{407} \sqrt{4^2 + 2^2} = 0.011 \text{ m/sec}$$

$$\xi_x = \frac{1}{407} \sqrt{\left(\frac{4}{628}\right)^2 + \left(\frac{2}{1884}\right)^2} = 1.58 \times 10^{-5} \text{ m}$$

c. The rms values u and ξ_x are independent of x for a plane progressive sound wave.

2.4. Solution of Wave Equation for Air in a Rigidly Closed Tube.   For this example of wave propagation, we shall consider a hollow cylindrical

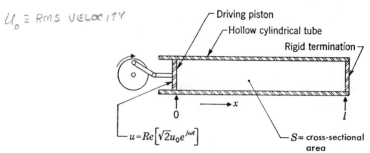

$u_0 \equiv$ RMS VELOCITY

FIG. 2.3. Rigidly terminated tube with rigid side walls.   The velocity at $x = 0$ has a value of $\sqrt{2}\, u_0 \cos \omega t$ m/sec.

tube, closed at one end by a rigid wall and at the other end by a flat vibrating piston (see Fig. 2.3).   The angular frequency of vibration of the piston is ω, and its rms velocity is u_0. We shall assume that the diameter of the tube is sufficiently small so that the waves travel down the tube with plane wave fronts. In order that this be true, the ratio of the wavelength of the sound wave to the diameter of the tube must be greater than about 6.

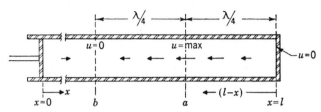

FIG. 2.4. Portion of the tube showing the direction and magnitude of movement of the air particles as a function of $l - x$. At position a, the particle velocity and displacement are a maximum. At position b, they are zero.

Particle Velocity. The form of solution we shall select is Eq. (2.38). Set u_0 equal to the rms velocity of the vibrating piston at $x = 0$, and set l equal to the length of the tube. The boundary conditions are

At $x = 0$, $u(0,t) = \sqrt{2}\, u_0 e^{j\omega t}$, so that

$$u_+ + u_- = u_0$$

At $x = l$, $u = 0$, so that

$$u_+ e^{-jkl} + u_- e^{jkl} = 0 \tag{2.39}$$

Remember that

$$\sin y = \frac{e^{jy} - e^{-jy}}{2j}$$

Hence

$$u_- = \frac{u_0 e^{-jkl}}{-2j \sin kl} \tag{2.40}$$

and

$$u_+ = \frac{u_0 e^{jkl}}{2j \sin kl} \tag{2.41}$$

which gives us

$$u(x,t) = \sqrt{2}\, u_0 e^{j\omega t}\, \frac{\sin k(l - x)}{\sin kl} \tag{2.42}$$

or

$$u = u_0 \frac{\sin k(l - x)}{\sin kl} \tag{2.43}$$

Note that the $\sqrt{2}$ and the time exponential have been left out of Eq. (2.43) so that both u_0 and u are complex rms quantities averaged over time.

Refer to Fig. 2.4. If the length l and the frequency are held constant, the particle velocity will vary from a value of zero at $x = l$ to a maximum at $l - x = \lambda/4$, that is, at $l - x$ equal to one-fourth wavelength. In the entire length of the tube the particle velocity varies according to a sine function. Between the end of the tube and the $\lambda/4$ point, the oscillatory motions are *in phase*. In other words, there is no progressive phase shift with x. This type of wave is called a standing wave because, in the equation, x and ct do not occur as a difference or a sum in the argument of the exponential function. Hence the wave is not propagated.

In the region between $l - x = \lambda/4$ and $l - x = \lambda/2$, the particle velocity still has the same phase except that its amplitude decreases sinusoidally. At $l - x = \lambda/2$, the particle velocity is zero. In the region between $l - x = \lambda/2$ and $l - x = \lambda$ the particle velocity varies with x according to a sine function, but the particles move 180° out of phase with those between 0 and $\lambda/2$. This is seen from Eq. (2.43), wherein the sines of arguments greater than π are negative.

If we fix our position at some particular value of x and if l is held constant, then, as we vary frequency, both the numerator and denominator of Eq. (2.43) will vary. When kl is some multiple of π, the particle velocity will become very large, except at $x = 0$ or at points where $k(l - x)$ is a multiple of π, that is, at points where $l - x$ equals multiples of $\lambda/2$. Then for $kl = n\pi$

$$l\Big|_{u=\infty} = \frac{n\lambda}{2} \qquad n = 1, 2, 3, \ldots \tag{2.44}$$

Equation (2.43) would indicate an infinite rms velocity under this condition. In reality, the presence of some dissipation in the tube, which was neglected in the derivation of the wave equation, will keep the particle velocity finite, though large.

The rms particle velocity u will be zero at those parts of the tube where $k(l - x) = n\pi$† and n is an integer or zero. That is,

$$x\Big|_{u=0} = l - \frac{\lambda}{2} n \qquad n = 0, 1, 2, \ldots \tag{2.45}$$

In other words, there will be planes of zero particle velocity at points along the length of the tube whenever l is greater than $\lambda/2$.

Some examples of the particle velocity for l *slightly greater than* various multiples of $\lambda/2$ are shown in Fig. 2.5. Two things in particular are apparent from inspection of these graphs. First, the quantity n determines the approximate number of half wavelengths that exist between the two ends of the tube. Secondly, for a fixed u_0, the maximum velocity

---

† For the type of source we have assumed and no dissipation, this case breaks down for $kl = n\pi$.

of the wave in the tube will depend on which part of the sine wave falls at $x = 0$. For example, if $l - n\lambda/2 = \lambda/4$, the maximum amplitude in the tube will be the same as that at the piston. If $l - n\lambda/2$ is very near zero, the maximum velocity in the tube will become very large.

Let us choose a frequency such that $n = 2$ as shown. Two factors determine the amplitude of the sine function in the tube. First, at $x = 0$ the sine curve must pass through the point u_0. Second, at $x = l$ the sine curve must pass through zero. It is obvious that one and only one sine wave meeting these conditions can be drawn so that the amplitude is determined. Similarly, we could have chosen a frequency such that

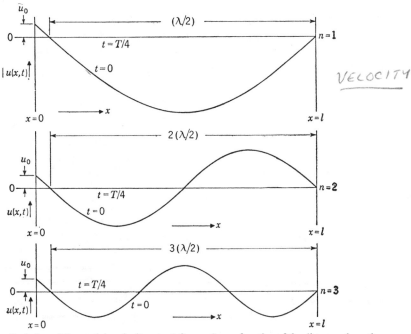

FIG. 2.5. Variation of the particle velocity $u(x,t)$ for $t = 0$, as a function of the distance along the tube of Fig. 2.3 for three frequencies, *i.e.*, for three wavelengths. At $x = 0$, the rms particle velocity is u_0, and at $x = l$, the particle velocity is zero. The period $T = 1/f$.

$n = 2$, but where the length of the tube is slightly less than two half wavelengths. If this case had been asked for, the sine wave would have started off with a positive instead of a negative slope at $x = 0$.

Sound Pressure. The sound pressure in the tube may be found from the velocity with the aid of the equation of motion [Eq. (2.4a)], which, in the steady state, becomes

$$p = -j\omega\rho_0 \int u \, dx \qquad (2.46)$$

The constant of integration in Eq. (2.46), resulting from the integration of Eq. (2.4a), must be independent of x, because we integrated with respect to x. The constant then represents an increment to the ambient pressure of the entire medium through which the wave is passing. Such an increment does not exist in our tube, so that in Eq. (2.46) we have set the constant of integration equal to zero. Integration of Eq. (2.46), after we have replaced u by its value from Eq. (2.42), yields

$$p(x,t) = -j\rho_0 c \sqrt{2}\, u_0 e^{j\omega t} \frac{\cos k(l - x)}{\sin kl} \tag{2.47}$$

or

$$p = -j\rho_0 c u_0 \frac{\cos k(l - x)}{\sin kl} \tag{2.48}$$

Note that the $\sqrt{2}$ and the time exponential have been left out of Eq. (2.48) so that both p and u_0 are complex rms quantities averaged in time.

The rms pressure p will be zero at those points of the tube where $k(l - x) = n\pi + \pi/2$, where n is an integer or zero.

$$x\Big|_{p=0} = l - \frac{\lambda}{2}\left(n + \frac{1}{2}\right) \tag{2.49}$$

The pressure will equal zero at one or more planes in the tube whenever l is greater than $\lambda/4$. Some examples are shown in Fig. 2.6. Here again, quantity n is equal to approximate number of half wavelengths in tube.

Refer once more to Fig. 2.5 which is drawn for $t = 0$. The instantaneous particle velocity is at its maximum (as a function of time). By comparison, in Fig. 2.6 at $t = 0$, the instantaneous sound pressure is zero. At a later time $t = T/4 = \frac{1}{4}f$, the instantaneous particle velocity has become zero and the instantaneous sound pressure has reached its maximum. Equations (2.42) and (2.47) say that whenever $k(l - x)$ is a small number the sound pressure *lags* by one-fourth period behind the particle velocity. At some other places in the tube, for example when $(l - x)$ lies between $\lambda/4$ and $\lambda/2$, the sound pressure *leads* the particle velocity by one-fourth period.

To see the relation between p and u more clearly, refer to Figs. 2.5 and 2.6, for the case of $n = 2$. In Fig. 2.5, the particle motion is to the right whenever u is positive and to the left when it is negative. Hence, at the $2\lambda/2$ point the particles on either side are moving toward each other, so that one-fourth period later the sound pressure will have built up to a maximum, as can be seen from Fig. 2.6. At $(l - x) = \lambda/2$, the particles are moving apart, so that the pressure is dropping to below barometric as can be seen from Fig. 2.6.

Figures 2.5 and 2.6 also reveal that, wherever along the tube the magnitude of the velocity is zero, the magnitude of the pressure is a maximum, and vice versa. Hence, for maximum pressure, Eq. (2.45) applies.

Specific Acoustic Impedance. It still remains for us to solve for the specific acoustic impedance Z_s at any plane x in the tube. Taking the ratio of Eq. (2.48) to Eq. (2.43) yields

SPECIFIC ACOUSTIC IMPEDANCE

$$Z_s = \frac{p}{u} = -j\rho_0 c \cot kl' = jX_s \qquad \text{mks rayls} \qquad (2.50)$$

where X_s is the reactance.

$X_s = REACTANCE$

Where we have set

$$l - x \equiv l' \qquad (2.51)$$

That is, l' is the distance between any plane x in Fig. 2.3 and the end

$-j\rho_0 c u_0 \dfrac{\cos k(l-x)}{\sin kl}$

$= \dfrac{u_0 \sin k(l-x)}{\sin kl}$

$= -j\rho_0 c \cot k(l-x)$

PRESSURE

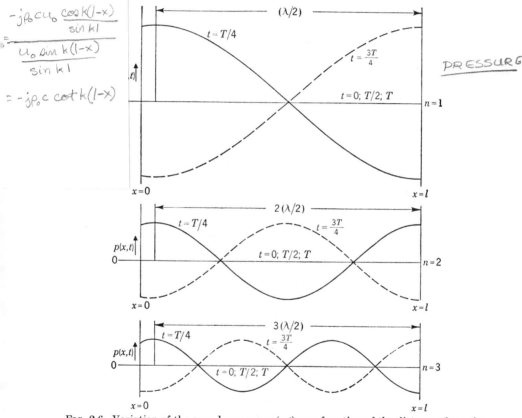

FIG. 2.6. Variation of the sound pressure $p(x,t)$ as a function of the distance along the tube for three frequencies, *i.e.*, for three wavelengths. At $x = 0$, the rms particle velocity is u_0, and at $x = l$, it is zero. The period T equals $1/f$.

of the tube at l. The $-j$ indicates that at low frequencies where $\cot kl' \doteq 1/kl'$ the particle velocity leads the pressure in time by 90° and the reactance X_s is negative. At all frequencies the impedance is reactive and either leads or lags the pressure by exactly 90° depending, respectively, on whether X_s is negative or positive. The reactance X_s varies as shown in Fig. 2.7.   If the value of kl' is small, we may approximate the cotangent by the first two terms of a series.

$$\cot kl' \doteq \frac{1}{kl'} - \frac{kl'}{3} \tag{2.52}$$

This approximation is valid whenever the product of frequency times the distance from the rigid end of the tube to the point of measurement is very small.   If the second term is very small, then it may be neglected with respect to the first.

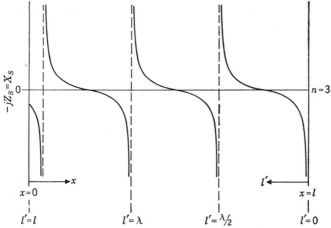

FIG. 2.7. The specific acoustic reactance (p_{rms}/u_{rms}) along the tube of Fig. 2.3 for a particular frequency, *i.e.*, a particular wavelength where $3(\lambda/2)$ is a little less than the tube length l.   For this case, the number of zeros is 3, and the number of poles is 4.

Let us see how small the ratio of the distance l' to the wavelength λ must be if the second term of Eq. (2.52) is to be 3 per cent or less of the first term.   That is, let us solve for l'/λ from

$$\frac{2\pi l'}{3\lambda} \leq 0.03 \, \frac{\lambda}{2\pi l'} \tag{2.53}$$

which gives us

$$\frac{l'}{\lambda} \lesssim 0.05 \tag{2.54}$$

In other words, if $\cot kl'$ is to be replaced within an accuracy of 3 per cent by the first term of its series expansion, l' must be less than one-twentieth wavelength in magnitude.

Assuming $l' < \lambda/20$, Eq. (2.50) becomes

$$Z_s = jX_s = -j\,\frac{\rho_0 c}{kl'} = \frac{1}{j\omega(l'/\rho_0 c^2)} \equiv \frac{1}{j\omega C_s} \qquad \text{mks rayls} \tag{2.55}$$

Hence, the *specific acoustic impedance* of a short length of tube can be

Handwritten margin notes:

COMPLIANCE
UNITS

$[m]$

$\dfrac{[m]}{\left[\dfrac{kg}{m^3}\right]\left[\dfrac{m}{s}\right]^2} = \left[\dfrac{m^2 \cdot s^2}{kg}\right]$

SPECIFIC ACOUSTIC COMPLIANCE

pacitance" called *specific acoustic compliance,* of mag-
Note also that $C_s = l'/\gamma P_0$, because of Eq. (2.19).
lance is of the same type, except that an area factor

ACOUSTIC IMPEDANCE

$$Z_A = \frac{p}{Su} = \frac{1}{j\omega(V/\rho_0 c^2)} \equiv \frac{1}{j\omega C_A} \qquad \text{mks acoustic ohms} \qquad (2.56)$$

where $V = l'S$ is the volume and S is the area of cross section of the tube.
C_A is called the *acoustic compliance* and equals $V/\rho_0 c^2$. Note also that
$C_A = V/\gamma P_0$, from Eq. (2.19). ACOUSTIC COMPLIANCE

Example 2.3. A cylindrical tube is to be used in an acoustic device as an impedance
element. (a) The impedance desired is that of a compliance. What length should
it have to yield a reactance of 1.4×10^3 mks rayls at an angular frequency of 1000
radians/sec? (b) What is the relative magnitude of the first and second terms of
Eq. (2.52) for this case?
Solution. The reactance of such a tube is

(a) $$X_s = 1.4 \times 10^3 = \frac{\gamma P_0}{\omega l'} = \frac{1.4 \times 10^5}{10^3 l'}$$

Hence, $l' = 0.1$ m.

(b) $$\frac{kl'}{3} \div \frac{1}{kl'} = \frac{k^2 l'^2}{3} = \frac{\omega^2 l'^2}{3c^2} = \frac{10^6 \times 10^{-2}}{(3)(344.8)^2} = 0.028$$

Hence, the second term is about 3 per cent of the first term.

2.5. Freely Traveling Plane Wave. *Sound Pressure.* If the rigid
termination of Fig. 2.3 is replaced by a perfectly absorbing termination, a
backward-traveling wave will not occur. Hence, Eq. (2.37) becomes

$$p(x,t) = \sqrt{2}\, p_+ e^{jk(ct-x)} \qquad (2.57)$$

where p_+ is the complex rms magnitude of the wave. This equation also
applies to a plane wave traveling in free space.
Particle Velocity. From Eq. (2.4a) in the steady state, we have

Handwritten margin note:
2.4a $-\dfrac{\partial p}{\partial x} = \rho_0 \dfrac{\partial u}{\partial t}$; $\dfrac{\partial u}{\partial t} = j\omega u$

$$u = -\frac{1}{j\omega \rho_0}\frac{\partial p}{\partial x} \qquad (2.58)$$

Hence,

$$u(x,t) = \frac{\sqrt{2}\, p_+}{\rho_0 c} e^{jk(ct-x)} = \frac{p(x,t)}{\rho_0 c} \qquad (2.59)$$

The particle velocity and the sound pressure are in phase. This is mathematical proof of the statement made in connection with the qualitative
discussion of the wave propagated from a vibrating wall in Chap. 1 and
Fig. 1.1.
Specific Acoustic Impedance. The specific acoustic impedance is

$$Z_s = \frac{p}{u} = \rho_0 c \qquad \text{mks rayls} \qquad (2.60)$$

This equation says that in a plane freely traveling wave the specific acoustic impedance is purely resistive and is equal to the product of the average density of the gas and the speed of sound. This particular quantity is generally called the *characteristic impedance of the gas* because its magnitude depends on the properties of the gas alone. It is a quantity that is analogous to the surge impedance of an infinite electrical line. For air at 22 °C and a barometric pressure of 10^5 newtons/m², its magnitude is 407 mks rayls.

2.6. Freely Traveling Spherical Wave. *Sound Pressure.* A solution to the spherical wave equation (2.24) is

$$\frac{\partial^2 (pr)}{\partial r^2} = \frac{1}{c^2} \frac{\partial^2 (pr)}{\partial t^2}$$

$$p(r,t) = \sqrt{2} \left(\frac{A_+ e^{-jkr}}{r} + \frac{A_- e^{+jkr}}{r} \right) e^{j\omega t} \tag{2.61}$$

where A_+ is the magnitude of the rms sound pressure in the outgoing wave at unit distance from the center of the sphere and A_- is the same for the reflected wave.

If there are no reflecting surfaces in the medium, only the first term of this equation is needed, *i.e.*,

$$p(r,t) = \frac{\sqrt{2}\, A_+ e^{-jkr}}{r} e^{j\omega t} \tag{2.62}$$

Particle Velocity. With the aid of Eq. (2.4b), solve for the particle velocity in the r direction.

$$u(r,t) = \frac{\sqrt{2}\, A_+}{\rho_0 c r} e^{j\omega t} \left(1 + \frac{1}{jkr} \right) e^{-jkr} \tag{2.63}$$

Specific Acoustic Impedance. The specific acoustic impedance is found from Eq. (2.62) divided by Eq. (2.63),

$$Z_s = \frac{p}{u} = \rho_0 c \frac{jkr}{1 + jkr} = \frac{\rho_0 c k r}{\sqrt{1 + k^2 r^2}} \underline{/90° - \tan^{-1} kr} \qquad \text{mks rayls} \tag{2.64}$$

Plots of the magnitude and phase angle of the impedance as a function of kr are given in Figs. 2.8 and 2.9. The real and imaginary parts, R_s and X_s, are plotted in Fig. 2.10.

For large values of kr, that is, for large distances or for high frequencies, this equation becomes, approximately,

$$Z_s \doteq \rho_0 c \qquad \text{mks rayls} \tag{2.65}$$

The impedance here is nearly purely resistive and approximately equal to the characteristic impedance for a plane freely traveling wave. In other words, the specific acoustic impedance a large distance from a spherical source in free space is nearly equal to that in a tube in which no reflections occur from the end opposite the source.

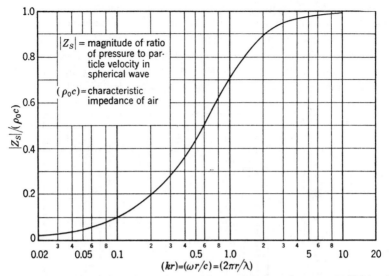

FIG. 2.8. Plot of the magnitude of the specific acoustic-impedance ratio $|Z_s|/\rho_0 c$ in a spherical freely traveling wave as a function of kr, where k is the wave number equal to ω/c or $2\pi/\lambda$ and r is the distance from the center of the spherical source. $|Z_s|$ is the magnitude of ratio of pressure to particle velocity in a spherical free-traveling wave, and $\rho_0 c$ is the characteristic impedance of air.

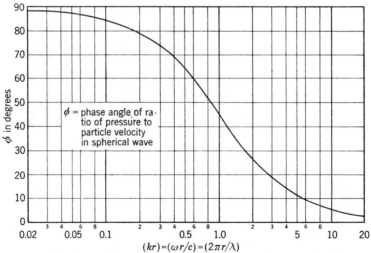

FIG. 2.9. Plot of the phase angle, in degrees, of the specific acoustic-impedance ratio $Z_s/\rho_0 c$ in a spherical wave as a function of kr, where k is the wave number equal to ω/c or $2\pi/\lambda$, and r is the distance from the center of the spherical source.

Equations (2.62) and (2.63) are significant because they reveal the difference between the responses of a microphone sensitive to pressure and a microphone sensitive to particle velocity as the microphones are brought close to a small spherical source of sound at low frequencies. As r is made smaller, the output of the pressure-responsive microphone will double for each halving of the distance between the microphone and the

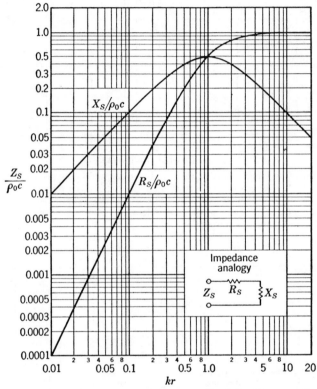

Fig. 2.10. Real and imaginary parts of the normalized specific acoustic impedance $Z_s/\rho_0 c$ of the air load on a pulsating sphere of radius r located in free space. Frequency is plotted on a normalized scale where $kr = 2\pi fr/c = 2\pi r/\lambda$. Note also that the ordinate is equal to $Z_M/\rho_0 c S$, where Z_M is the mechanical impedance; and to $Z_A S/\rho_0 c$, where Z_A is the acoustic impedance. The quantity S is the area for which the impedance is being determined, and $\rho_0 c$ is the characteristic impedance of the medium.

center of the spherical source. Expressed in decibels, the output increases 6 db for each halving of distance. For the velocity-responsive microphone, the output variation is not so simple. Only at sufficiently large distances ($k^2 r^2 \gg 1$) does the output increase 6 db for each halving of distance. For shorter distances the second term inside the parentheses on the right-hand side of Eq. (2.63) becomes large, and the magnitude of u increases at a rate exceeding $+6$ db for each halving of distance. For very short distances ($k^2r^2 \ll 1$), the rate of increase of u approaches a limit of $+12$ db for each halving of distance. It is for this reason that the voice of a radio crooner sounds "bassy" when he sings very near to a velocity-sensitive microphone which was designed to have its best response when located a large distance from the source of sound.

Another significant thing is to be learned from Eq. (2.64). At low frequencies it is very difficult to radiate sound energy from a small loudspeaker. A small loudspeaker may be likened to a pulsating balloon of some small radius r. The specific acoustic impedance Z_s of the air presented to each square centimeter of the balloon is given by Eq. (2.64) and Fig. 2.10. At low frequencies, the impedance becomes nearly purely reactive, and the resistance becomes very, very small. Hence, the power radiated by a small loudspeaker becomes very small. At high frequencies, $kr > 2$, the impedance Z_s becomes nearly purely resistive and has its maximum value of $\rho_0 c$, so that the power radiated for a given value of p_+ reaches its maximum.

The important steady-state relations derived in this chapter are summarized in Table 2.1.

TABLE 2.1.   General and Steady-state Relations for Small-signal Sound Propagation in Gases

Name	General equation	Steady-state equation
Wave equations	$\dfrac{\partial^2(\ )}{\partial x^2} = \dfrac{1}{c^2}\dfrac{\partial^2(\ )}{\partial t^2}$ $\nabla^2(\ ) = \dfrac{1}{c^2}\dfrac{\partial^2(\ )}{\partial t^2}$ $\dfrac{\partial^2(pr)}{\partial r^2} = \dfrac{1}{c^2}\dfrac{\partial^2(pr)}{\partial t^2}$	$\dfrac{\partial^2(\ )}{\partial x^2} = -\dfrac{\omega^2}{c^2}(\ )$ $\nabla^2(\ ) = -\dfrac{\omega^2}{c^2}(\ )$ $\nabla^2(pr) = -\dfrac{\omega^2}{c^2}(pr)$
Equation of motion	$\dfrac{\partial p}{\partial x} = -\rho_0\dfrac{\partial u}{\partial t}$ $\operatorname{grad} p = -\rho_0\dfrac{\partial \mathbf{q}}{\partial t}$	$u = \dfrac{-1}{j\omega\rho_0}\dfrac{\partial p}{\partial x}$ $p = -j\omega\rho_0\int u\,dx$ $\operatorname{grad} p = -j\omega\rho_0\mathbf{q}$
Displacement	$\xi = \int u\,dt$ $\boldsymbol{\xi} = \int \mathbf{q}\,dt$	$\xi = \dfrac{u}{j\omega}$ $\boldsymbol{\xi} = \dfrac{\mathbf{q}}{j\omega}$
Incremental density	$\rho = \dfrac{\rho_0}{\gamma P_0}p = \dfrac{p}{c^2}$ $\dfrac{\partial\rho}{\partial t} = -\rho_0\dfrac{\partial u}{\partial x}$	$\rho = \dfrac{\rho_0}{\gamma P_0}p = \dfrac{p}{c^2}$ $\rho = -\dfrac{\rho_0}{j\omega}\dfrac{\partial u}{\partial x}$
Incremental temperature	$\Delta T = \dfrac{T_0}{P_0}\dfrac{\gamma-1}{\gamma}p$	$\Delta T = \dfrac{T_0}{P_0}\dfrac{\gamma-1}{\gamma}p$

PART V *Energy Density and Intensity*

2.7. Energy Density. Energy density is an important concept in acoustics because, in dealing with sound in enclosures, it is necessary to study the flow of energy from a source to all parts of the room. The energy density, *i.e.*, watt-seconds per unit volume, is greater near the source than farther away and is the variable that appears in the equations describing the acoustical conditions. On the other hand, the ear and most sound-level meters respond to rms sound pressure. We need to ascertain, therefore, the relation between energy density and sound pressure in sound fields.

The energy density associated with the small "box" of gas at any particular instant is the sum of the kinetic and potential energies per unit volume of the air particles in the box. The kinetic energy density due to the excess pressure of the sound wave D_{KE} is

KINETIC ENERGY DENSITY

$$D_{KE} = \frac{1}{2}\frac{Mu^2}{V_0} = \frac{1}{2}\rho_0 u^2 \tag{2.66}$$

where u is the average instantaneous velocity of the air particles in the box, ρ_0 is the average density, and M/V_0 is the mass per unit volume.

The potential energy density due to the sound wave D_{PE} may be found from the gas law. For very small changes in the volume of the box, we may write [see Eqs. (2.8)]

POTENTIAL ENERGY DENSITY

$$D_{PE} = -\frac{\int p\, d\tau}{V_0} \tag{2.67}$$

If we differentiate Eq. (2.9) and substitute the resulting expression for $d\tau$, the potential energy density becomes

(2.9)

$\dfrac{p}{P_0} = \dfrac{-\gamma\tau}{V_0}$

$$D_{PE} = \frac{\int p\, dp}{\gamma P_0} = \frac{1}{2}\frac{p^2}{\gamma P_0} \tag{2.68}$$

When the sound pressure p is equal to zero, the potential energy due to the sound wave must be zero. The arbitrary constant of integration is therefore also equal to zero.

The total energy density due to the sound wave $D = D_{KE} + D_{PE}$, or

$$D(t) = \frac{1}{2}\left(\rho_0 u^2 + \frac{p^2}{\gamma P_0}\right) \tag{2.69}$$

This equation is true at any instant at a given point in space.

2.8. Energy Density in Plane Waves. *Energy Density in a Plane Free-progressive Wave.* From Eqs. (2.57) and (2.59) we have seen that the pressure and particle velocity in a plane free-progressive (outgoing) wave are equal to

$$p(x,t) = \text{Re } \sqrt{2}\, p_+ e^{jk(ct-x)} = \sqrt{2}\, |p_+| \cos [k(ct - x) + \theta] \qquad (2.70)$$

$$u(x,t) = \text{Re } \frac{\sqrt{2}\, p_+}{\rho_0 c} e^{jk(ct-x)} = \frac{\sqrt{2}\, |p_+|}{\rho_0 c} \cos [k(ct - x) + \theta] \qquad (2.71)$$

where $p_+ = |p_+| e^{j\theta}$.

The instantaneous energy density for such a wave in the steady state is, from Eqs. (2.68) and (2.19), equal to

(2.19) $c^2 \equiv \dfrac{\delta P_0}{P_0}$

ENERGY IS
BEING
TRANSPORTED,

NOT CONSERVED AS
WITH A PENDULUM

$$D(x,t) = |p_+|^2 \left(\frac{\rho_0}{\rho_0{}^2 c^2} + \frac{1}{\rho_0 c^2} \right) \cos^2 [k(ct - x) + \theta]$$

$$= \frac{2|p_+|^2}{\rho_0 c^2} \frac{1 + \cos 2\omega(t - x/c + \theta/\omega)}{2}$$

$$= \frac{|p_+|^2}{\rho_0 c^2} [1 + \cos 2\omega(t - x/c + \theta/\omega)] \qquad (2.72)$$

This equation says that for a plane free-progressive wave, at all times, the kinetic and potential energy densities are equal at a given point in space but that they vary with position or with time sinusoidally from zero to twice their average value. The situation here is different from that for a pendulum where the kinetic energy and the potential energy vary in opposite phase, *i.e.*, one is a maximum when the other is a minimum. Here, energy is being transported away from the source. Conversely, the pendulum is a conservative system.

When averaged over either a length of time equal to $t = T/2 = 1/2f$ or a distance in space $x = \lambda/2 - c/2f$, we find the *average energy density* to be equal to

AVERAGE ENERGY DENSITY

$$D_{\text{avg}} = \frac{|p_+|^2}{\rho_0 c^2} \qquad \text{watt-sec/m}^3 \qquad (2.73)$$

where $|p_+|$ is the magnitude of the rms value (in time) of the sound pressure measured at any point in the sound wave. Note also that $\rho_0 c^2 \equiv \gamma P_0$ as stated before. Inspection of Eq. (2.60) shows that we may let

$$\frac{p_+}{\rho_0 c} \equiv u_+ \qquad (2.74)$$

where u_+ is the rms value (in time) of the velocity at any point in the wave. Then,

$$D_{\text{avg}} = |u_+|^2 \rho_0 \qquad (2.75)$$

Equations (2.73) and (2.75) give the relations among rms sound pressure, particle velocity, and energy density.

Energy Density in a Plane Standing Wave. From Eqs. (2.42) and (2.47) we have that

$$p(x,t) = \sqrt{2}\, \rho_0 c\, |u_0| \frac{\sin (\omega t + \theta) \cos k(l - x)}{\sin kl} \qquad (2.76)$$

where θ is the phase angle of u_0.

$$u(x,t) = \sqrt{2}\,|u_0|\,\frac{\cos{(\omega t + \theta)}\sin{k(l-x)}}{\sin{kl}} \qquad (2.77)$$

In this case, the kinetic and potential energy are 90° out of time phase. The situation is analogous to that for a pendulum because in both cases the systems are conservative.

The instantaneous energy density for such a wave in the steady state is, from Eqs. (2.69) and (2.19), equal to

$$D(x,t) = |u_0|^2\rho_0\,\frac{1 - \cos{2(\omega t + \theta)}\cos{2k(l-x)}}{1 - \cos{2kl}} \qquad (2.78)$$

When averaged over either a length of time t equal to $T/2$ or a distance x in space equal to $\lambda/2$, we find the *average energy density* to be equal to

$$D_{avg} = \frac{|u_0|^2\rho_0}{1 - \cos{2kl}} \qquad \text{watt-sec/m}^3 \qquad (2.79)$$

where $|u_0|$ is the magnitude of the rms velocity of the piston at $x = 0$. This equation shows that, for a constant value of $|u_0|$, the average energy density varies from $|u_0|^2\rho_0/2$ to infinity depending on the value of $kl = 2\pi l/\lambda$.

A better way of representing the average energy density is in terms of the rms pressure. If, by definition, we let the rms value of the pressure be related to the rms velocity u_0 at $x = 0$ by the formula

$$p_1 \equiv \frac{\rho_0 c u_0}{\sin{kl}} \qquad (2.80)$$

we have

$$p(x,t) = \sqrt{2}\,|p_1|\sin{(\omega t + \theta)}\cos{k(l-x)} \qquad (2.81)$$

Then Eq. (2.79) becomes

$$D_{avg} = \frac{|p_1|^2}{2\rho_0 c^2} \qquad (2.82)$$

Here, $|p_1|$ is the magnitude of the rms value (in time) of the *maximum* value (in space) of the sound pressure. If we measure the rms value of the sound pressure *in space* by moving a microphone backward and forward over a wavelength and averaging the varying output in a rms rectifier, then $|p_{avg}| = |p_1|/\sqrt{2}$ and

$$D_{avg} = \frac{|p_{avg}|^2}{\rho_0 c^2} \qquad \text{watt-sec/m}^3 \qquad (2.83)$$

where $|p_{avg}|$ is the magnitude of the rms value of the sound pressure averaged in *both* space and time. Note that Eq. (2.83) is identical to Eq. (2.73).

Example 2.4. Calculate the average energy density in a plane free-progressive sinusoidal sound wave with a maximum particle displacement of 0.01 cm at a frequency of 100 cps.

Solution. From Eq. (2.30) we find that the rms particle velocity is $u_{rms} = \omega \xi_{rms}$. So

$$u_{rms} = \frac{2\pi \times 100 \times 0.01}{\sqrt{2} \times 100} = 0.0445 \text{ m/sec}$$

The average energy density is given by Eq. (2.75),

$$D_{avg} = (0.0445)^2 \times 1.18 = 2.34 \times 10^{-3} \text{ watt-sec/m}^3$$

2.9. Energy Density in a Spherical Free-progressive Wave. The energy density in a spherical free-progressive wave can be shown[5] to be equal to

$$D_{avg} = \frac{|p_r|^2}{\rho_0 c^2}\left(1 + \frac{1}{2k^2 r^2}\right) \tag{2.84}$$

where $|p_r|$ is the magnitude of the rms value (in time) of the sound pressure at a point a distance r from the center of the spherical source.

If the product of the distance r and the frequency is large ($2k^2 r^2 \gg 1$), the average energy density is the same as for a plane free-traveling or standing wave, as can be seen from Eqs. (2.73) and (2.83). Near the source, however, the energy density becomes very large. This occurs because the impedance [see Eq. (2.64)] becomes largely reactive and the stored energy becomes high.

2.10. Sound Energy Flow—Intensity. Later in this text we make frequent reference to the flow of sound energy through an acoustic system. Because of the law of the conservation of energy, the total acoustic energy starting from a source must be completely accounted for in the system. At any part of an acoustic system, we should be able to state the amount of energy flowing through that part per unit time, and it should equal the power emanating from the source minus any intervening losses.

In Part II we defined intensity as the average time rate at which energy is flowing through unit area of the acoustic medium. In the mks system, the units of intensity are watts per square meter. The intensity is actually the product of the sound pressure times the in-phase component of the particle velocity.

General Equation for Intensity. We can find the average intensity I in a given direction at a given point in the medium by performing the operation†

$$I = \text{Re } p^* q \cos \phi \tag{2.85}$$

---

[5] L. E. Kinsler and A. R. Frey, "Fundamentals of Acoustics," pp. 167–169, John Wiley & Sons, Inc., 1950.

† The average power supplied by an electrical generator to a circuit equals the voltage times the in-phase component of the current. This power can be shown to equal $Re (E^*I)$, where E and I are the complex rms voltage and current, respectively.

where p^* is the complex conjugate† of the rms sound pressure p, q is the complex rms particle velocity in the direction the wave is traveling, and ϕ is the angle between the direction of travel and the direction in which the intensity is being determined. The symbol Re indicates that the real part of the product is to be taken.

Intensity in a Plane Free-progressive Wave. For a plane free-progressive sound wave the intensity equals

$$I = \text{Re } p_+ e^{jkx} \frac{p_+}{\rho_0 c} e^{-jkx} \cos \phi \qquad (2.86)$$

Another way of looking at the question of intensity for a plane progressive wave is to say that all the energy contained in a column of gas equal in length to c m must pass through unit area in 1 sec. Hence, the intensity is

$$I = c D_{\text{avg}} \cos \phi \qquad (2.87)$$

So, regardless of whether the intensity is determined from (2.86) or (2.87), we get for a plane free-progressive wave that

$$I = \frac{|p_+|^2}{\rho_0 c} \cos \phi = |u_+|^2 \rho_0 c \cos \phi \qquad (2.88)$$

Intensity in a Plane Standing Wave. In a plane standing wave the pressure and particle velocity are 90° out of phase in time [see Eqs. (2.76) and (2.77)] so that the real part of the product $p^* u$ is zero. Hence, for a plane standing wave,

$$I = 0 \qquad (2.89)$$

Physically, this means that as much sound energy returns to the source as travels away from it.

Intensity in a Spherical Free-progressive Wave. For a spherical progressive wave, we get the pressure p from Eq. (2.62). By definition, let

$$p_r \equiv \frac{A_+}{r} e^{-jkr} \qquad (2.90)$$

Then,

$$p(r,t) = \sqrt{2}\, p_r e^{j\omega t} \qquad (2.91)$$

The quantity p_r is equal to the complex rms pressure at any point a distance r from the center of the source. Hence, the particle velocity $u(r,t)$ at any point a distance r is

$$u(r,t) = \frac{\sqrt{2}\, p_r}{\rho_0 c}\left(1 + \frac{1}{jkr}\right) e^{j\omega t} \qquad (2.92)$$

† If p is represented by $|p|e^{j\theta}$, then p^* is $|p|e^{-j\theta}$. Similarly, if p is represented by $p_R + jp_I$, then p^* is represented by $p_R - jp_I$.

or the complex rms particle velocity is

$$u_r = \frac{p_r}{\rho_0 c}\left(1 - j\frac{1}{kr}\right) \qquad (2.93)$$

Substitution of the sound pressure at r, p_r, and Eq. (2.93) into Eq. (2.85) yields

$$I = \text{Re } p_r^*\left[\frac{p_r}{\rho_0 c}\left(1 - \frac{j}{kr}\right)\right]\cos\phi = \frac{|p_r|^2}{\rho_0 c}\cos\phi \qquad (2.94)$$

where, as before, ϕ is the angle between the direction of travel of the wave and the direction in which the intensity is being determined.

We can derive these results in a different way. Equation (2.93) states that, for kr large, p and u for a spherical wave are nearly in time phase and $p(r) = \rho_0 c u(r)$ as shown by Eq. (2.65). Hence, for kr large, we see from Eq. (2.88) that in a spherical wave for large distances $I = |u_r|^2\rho_0 c \cos\phi$.

The total power at *any* radius r is equal to $W = 4\pi r^2 I = 4\pi r^2 |p_r|^2/\rho_0 c$. Hence, for a spherical wave,

$$I = \frac{W}{4\pi r^2} \qquad \text{for } \phi = 0 \qquad (2.95)$$

By the law of conservation of energy, W is independent of r if there are no losses in the gas so that the intensity varies inversely as the square of the distance r.

From Eq. (2.90) we see also that the square of the rms magnitude of the sound pressure at any point varies inversely with the square of the distance r. Hence, because the intensity I at any point varies similarly, it is directly proportional to the square of the sound pressure at that point. This result agrees with that shown in Eq. (2.94).

Example 2.5. A spherical sound source is radiating sinusoidally into free space 1 watt of acoustic power at 1000 cps. Calculate (a) intensity in the direction the wave is traveling; (b) sound pressure; (c) particle velocity; (d) phase angle between (b) and (c); (e) energy density; and (f) sound pressure level at a point 30 cm from the center of the source. (Assume 22°F and 0.751 m Hg.)

Solution. a. The intensity may be found from Eq. (2.95).

$$I = \frac{W}{4\pi r^2} = \frac{1}{4\pi(0.3)^2} = 0.885 \text{ watt/m}^2$$

b. The rms sound pressure comes from Eq. (2.94).

$$|p_r| = \sqrt{I\rho_0 c} = \sqrt{0.885 \times 407} = 18.97 \text{ newtons/m}^2$$

c. The rms particle velocity is given by Eq. (2.93).

$$kr = (2\pi \times 1000/344.8)(0.3) = 5.46$$

$$|u_r| = \frac{p_r}{\rho_0 c}\frac{\sqrt{1 + k^2 r^2}}{kr} = \frac{18.97}{407}\frac{\sqrt{1 + 29.8}}{5.46} = 0.0474 \text{ m/sec}$$

d. The phase angle θ between p_r and u_r may be found from Eq. (2.64).

$$\theta = 90° - \tan^{-1} kr = 90° - 79.6° = 10.4°$$

e. The energy density is given by Eq. (2.84).

$$D_{\text{avg}} = \frac{|p_r|^2}{\rho_0 c^2} \left(1 + \frac{1}{2k^2 r^2} \right) = \frac{360}{1.4 \times 10^5} \left(1 + \frac{1}{2 \times 29.8} \right)$$
$$\doteq 2.62 \times 10^{-3} \text{ watt-sec/m}^3$$

f. The sound pressure level is found from Eq. (1.18).

$$\text{SPL} = 20 \log_{10} \frac{18.97}{2 \times 10^{-5}}$$
$$= 119.5 \text{ db } re \text{ } 2 \times 10^{-5} \text{ newton/m}^2 \text{ } (re \text{ } 2 \times 10^{-4} \text{ microbar})$$

This sound pressure level is about 15 db higher than the highest level that is measured at 25 ft above a full symphony orchestra.   In other words, 1 watt of acoustic power creates a very high sound pressure level at 1 ft from the source.

CHAPTER 3

ELECTRO-MECHANO-ACOUSTICAL CIRCUITS

PART VI *Mechanical Circuits*

3.1. Introduction. The subject of electro-mechano-acoustics (sometimes called dynamical analogies) is the application of electrical-circuit theory to the solution of mechanical and acoustical problems. In classical mechanics, vibrational phenomena are represented entirely by differential equations. This situation existed also early in the history of telephony and radio. As telephone and radio communication developed, it became obvious that a schematic representation of the elements and their interconnections was valuable. These schematic diagrams made it possible for engineers to visualize the performance of a circuit without laboriously solving its equations. The performance of radio and television systems can be studied from a single sheet of paper when such schematic diagrams are used. Such a study would have been hopelessly difficult if only the equations of the system were available.

There is another important advantage of a schematic diagram besides its usefulness in visualizing the system. Often one has a piece of equipment for which he desires the differential equations. The schematic diagram may then be drawn from visual inspection of the equipment. Following this, the differential equations may be formed directly from the schematic diagrams. Most engineers are trained to follow this procedure rather than to attempt to formulate the differential equations directly.

Schematic diagrams have their simplest applications in circuits that contain lumped elements, *i.e.*, where the only independent variable is time. In distributed systems, which are common in acoustics, there may be as many as three space variables and a time variable. Here, a schematic diagram becomes more complicated to visualize than the differential equations, and the classical theory comes into its own again. There are many problems in acoustics, however, in which the elements are lumped and the schematic diagram may be used to good advantage.

Four principal requirements are fulfilled by the methods used in this text to establish schematic representations for acoustic and mechanical devices. They are:

1. The methods must permit the formation of schematic diagrams from visual inspection of devices.

2. They must be capable of such manipulation as will make possible the combination of electrical, mechanical, and acoustical elements into one schematic diagram.

3. They must preserve the identity of each element in combined circuits so that one can recognize immediately a force, voltage, mass, inductance, and so on.

4. They must use the familiar symbols and the rules of manipulation for electrical circuits.

Several methods that have been devised fulfill one or two of the above four requirements, but not all four. A purpose of this chapter is to present a new method for handling combined electrical, mechanical, and acoustic systems. It incorporates the good features of previous theories and also fulfills the above four requirements. The symbols used conform with those of earlier texts wherever possible.[1-5]

3.2. Physical and Mathematical Meanings of Circuit Elements. The circuit elements we shall use in forming a schematic diagram are those of electrical-circuit theory. These elements and their mathematical meaning are tabulated in Table 3.1 and should be learned at this time. There are generators of two types. There are four types of circuit elements: resistance, capacitance, inductance, and transformation. There are three generic quantities: (a) the drop across the circuit element; (b) the flow through the circuit element; and (c) the magnitude of the circuit element.[†]

Attention should be paid to the fact that the quantity a is not restricted to voltage e, nor b to electrical current i. In some problems a will represent force f, or velocity u, or pressure p, or volume velocity U. In those cases b will represent, respectively, velocity u, or force f, or volume

[1] B. Gehlshoj, "Electromechanical and Electroacoustical Analogies," Academy of Technical Sciences, Copenhagen, 1947.

[2] F. A. Firestone, A New Analogy between Mechanical and Electrical Systems, *J. Acoust. Soc. Amer.*, **4**: 249–267 (1933); The Mobility Method of Computing the Vibrations of Linear Mechanical and Acoustical Systems: Mechanical-electrical Analogies, *J. Appl. Phys.*, **9**: 373–387 (1938).

[3] H. F. Olson, "Dynamical Analogies," D. Van Nostrand Company, Inc., New York, 1943.

[4] W. P. Mason, Electrical and Mechanical Analogies, *Bell System Tech. J.*, **20**: 405–414 (1941).

[5] A. Bloch, Electro-mechanical Analogies and Their Use for the Analysis of Mechanical and Electro-mechanical Systems, *J. Inst. Elec. Eng.*, **92**: 157–169 (1945).

[†] Among the four circuit elements, the first three are two-poles. This list is exhaustive. The transformation element is a four-pole. There are other lossless four-poles which one might have chosen in addition, e.g., the ideal gyrator.

velocity U, or pressure p. Similarly, the quantity c might be any appropriate quantity such as mass, compliance, inductance, resistance, etc. The physical meaning of the circuit elements c depends on the way in which the quantities a and b are chosen, with the restriction that ab has the dimension of power in all cases. The complete array of alternatives is shown in Table 3.2.

TABLE 3.1. Mathematical and Physical Significance of Symbols

Symbol	Name	Meaning	
		Transient	Steady-state
	Constant-drop generator	The quantity a is independent of what is connected to the generator. The arrow points to the positive terminal of the generator	
	Constant-flow generator	The quantity b is independent of what is connected to the generator. The arrow points in the direction of positive flow	
	Resistance-type element	$a = bc$ $v = iR$	$a = bc$
	Capacitance-type element	$v = \frac{1}{c}\int i\,dt$ $a = \frac{1}{c}\int b\,dt$	$a = \dfrac{b}{j\omega c}$ $v = \dfrac{i}{j\omega c}$
	Inductance-type element	$v = L\dfrac{di}{dt}$ $a = c\dfrac{db}{dt}$	$a = j\omega cb$
	Transformation-type element	$a = cg$ $cb = d$ $\dfrac{a}{b} = c^2\dfrac{g}{d}$	$a = cg$ $cb = d$ $\dfrac{a}{b} = c^2\dfrac{g}{d}$

An important idea to fix in your mind is that the *mathematical operations associated with a given symbol are invariant*. If the element is of the inductance type, for example, the drop a across it is equal to the time derivative of the flow b through it multiplied by its size c. Note that this rule is not always followed in electrical-circuit theory because there conductance and resistance are often indiscriminately written beside the symbol for a resistance-type element. The invariant operations to be associated with each symbol are shown in columns 3 and 4 of Table 3.1.

TABLE 3.2. Values for a, b, and c in Electrical, Mechanical, and Acoustical Circuits

Element	Electrical	Mechanical		Acoustical	
		Mobility analogy†	Impedance analogy	Impedance analogy†	Mobility analogy
a	e	u	f	p	U
b	i	f	u	U	p
	$c = R_E$	$c = \dfrac{1}{R_M} = r_M$	$c = R_M$	$c = R_A$	$c = \dfrac{1}{R_A} = r_A$
	$c = C_E$	$c = M_M$	$c = C_M$	$c = C_A$	$c = M_A$
	$c = L$	$c = C_M$	$c = M_M$	$c = M_A$	$c = C_A$
	$c = Z_E = \dfrac{e}{i}$	$c = z_M = \dfrac{u}{f} = \dfrac{1}{Z_M}$	$c = Z_M = \dfrac{f}{u} = \dfrac{1}{z_M}$	$c = Z_A = \dfrac{p}{U} = \dfrac{1}{z_A}$	$c = z_A = \dfrac{U}{p} = \dfrac{1}{Z_A}$

† Preferred analogies.

3.3. Mechanical Circuits. Mechanical-circuit elements need not always be represented by electrical symbols. Since one frequently draws a mechanical circuit directly from inspection of the mechanical device, more obvious forms of mechanical elements are sometimes useful, at least until the student is thoroughly familiar with the analogous circuit. We shall accordingly devise a set of "mechanical" elements to be used as an introduction to the elements of Table 3.1.

TABLE 3.3.　Conversion from Mobility-type Analogy to Impedance-type Analogy, or Vice Versa

Element	MECHANICAL ANALOGIES		ACOUSTICAL ANALOGIES	
	Mobility type	Impedance type	Mobility type	Impedance type
Infinite mechanical or acoustic impedance generator (zero mobility)	u	u	U	U
Zero mechanical or acoustic impedance generator (infinite mobility)	f	f	p	p
Dissipative element (resistance and responsiveness)	r_M f u	R_M u f	r_A p U	R_A U p
Mass element	M_M f u	M_M u f	M_A p U	M_A U p
Compliant element	C_M f u	C_M u f	C_A p U	C_A U p
Impedance element	z_M f u	Z_M u f	z_A p U	Z_A U p
Transformation element	Mech. to acous. (mobility type) f $1:S$ p u U		Mech. to acous. (impedance type) u $S:1$ U f p	

In electrical circuits, a voltage measurement is made by attaching the leads from a voltmeter *across* the two terminals of the element. Voltage is a quantity that we can measure without breaking into the circuit. To measure electric current, however, we must break into the circuit because this quantity acts *through* the element. In mechanical devices, on the other hand, we can measure the velocity (or the displacement) without disturbing the machine by using a capacitive or inertially operated vibration pickup to determine the quantity at any point on the machine. It is not velocity but force that is analogous to electric current. Force cannot be measured unless one breaks into the device.

MOBILITY = $\dfrac{1}{IMPEDANCE}$

MOBILITY ANALOGY

It becomes apparent then that if a mechanical element is strictly analogous to an electrical element it must have a velocity difference appearing between (or across) its two terminals and a force acting through it. Analogously, also, the product of the rms force f in newtons and the in-phase component of the rms velocity u in meters per second is the power in watts. We shall call this type of analogy, in which a velocity corresponds to a voltage and a force to a current, the *mobility-type analogy*. It is also known as the "inverse" analogy.

IMPEDANCE ANALOGY Many texts teach in addition a "direct" analogy. It is the opposite of the mobility analogy in that force is made to correspond to voltage and velocity to current. In this text we shall call this kind of analogy an *impedance-type analogy*. To familiarize the student with both concepts, all examples will be given here both in mobility-type and impedance-type analogies.

Mechanical Impedance Z_M, and Mechanical Mobility z_M. The mechanical impedance is the complex ratio of force to velocity at a given point in a mechanical device. We commonly use the symbol Z_M for mechanical impedance, where the subscript M stands for "mechanical." The units are newton-seconds per meter, or mks mechanical ohms.

The mechanical mobility is the inverse of the mechanical impedance. It is the complex ratio of velocity to force at a given point in a mechanical device. We commonly use the symbol z_M for mechanical mobility. The units are meters per second per newton, or mks mechanical mohms.†

FIG. 3.1. Mechanical symbol for a mass.

Mass M_M. Mass is that physical quantity which when acted on by a force is accelerated in direct proportion to that force. The unit is the kilogram. At first sight, mass appears to be a one-terminal quantity because only one connection is needed to set it in motion. However, the force acting on a mass and the resultant acceleration are reckoned with respect to the earth (inertial frame) so that in reality the second terminal of mass is the earth.

The mechanical symbol used to represent mass is shown in Fig. 3.1. The upper end of the mass moves with a velocity u with respect to the ground. The ⊥-shaped configuration represents the "second" terminal of the mass and has zero velocity. The force can be measured by a suitable device inserted between the point 1 and the next element or generator connecting to it.

Mass M_M obeys Newton's second law that

$$f(t) = M_M \frac{du(t)}{dt} \tag{3.1}$$

† The word "mohm" stands for mobility ohm. The units are meters per second per newton.

where $f(t)$ is the instantaneous force in newtons, M_M is the mass in kilograms, and $u(t)$ is the instantaneous velocity in meters per second.

In the steady state [see Eqs. (2.33) to (2.35)], with an angular frequency ω equal to 2π times the frequency of vibration, we have the special case of Newton's second law,

$$f = j\omega M_M u \tag{3.2}$$

where $j = \sqrt{-1}$ as usual and f and u are rms complex quantities.

The mobility-type analogous symbol that we use as a replacement for the mechanical symbol in our circuits is a capacitance type. It is shown in Fig. 3.2a. The mathematical operation invariant for this symbol is found from Table 3.1. In the steady state we have

$$a = \frac{b}{j\omega c} \quad \text{or} \quad u = \frac{f}{j\omega M_M} \tag{3.3}$$

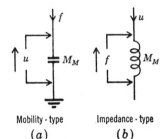

Mobility - type   Impedance - type
(a)   (b)

Fig. 3.2. (a) Mobility-type and (b) impedance-type symbols for a mass.

This equation is seen to satisfy the physical law given in Eq. (3.2). Note the similarity in appearance of the mechanical and analogous symbols in Figs. 3.1 and 3.2a. In electrical circuits the time integral of the current through a capacitor is charge. The analogous quantity here is the time integral of force, which is momentum.

The impedance-type analogous symbol for a mass is an inductance. It is shown in Fig. 3.2b. The invariant operation for steady state is $a = j\omega cb$ or $f = j\omega M_M u$. It also satisfies Eq. (3.2). Note, however, that in this analogy one side of the mass element is not necessarily grounded; this often leads to confusion. In electrical circuits the time integral of the voltage across an inductance is flux-turns. The analogous quantity here is momentum. ⟳ displacement α force

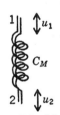

Fig. 3.3. Mechanical symbol for a mechanical compliance.

Mechanical Compliance C_M. A physical structure is said to be a mechanical compliance C_M if, when it is acted on by a force, it is displaced in direct proportion to the force. The unit is the meter per newton. Compliant elements usually have two apparent terminals.

The mechanical symbol used to represent a mechanical compliance is a spring. It is shown in Fig. 3.3. The upper end of the element moves with a velocity u_1 and the lower end with a velocity u_2. The force required to produce the difference between the velocities u_1 and u_2 may be measured by breaking into the machine at either point 1 or point 2. Just as the same current would be measured at either end of an element in an electrical circuit, so the same force will be found here at either end of the compliant element.

Mechanical compliance C_M obeys the following physical law,

$$a = \frac{1}{c} \int b\, dt \quad \text{or} \quad f(t) = \frac{1}{C_M} \int u(t)\, dt \qquad (3.4)$$

where C_M is the mechanical compliance in meters per newton and $u(t)$ is the instantaneous velocity in meters per second equal to $u_1 - u_2$, the difference in velocity of the two ends.

In the steady state, with an angular frequency ω equal to 2π times the frequency of vibration, we have,

$$f = \frac{u}{j\omega C_M} \qquad (3.5)$$

where f and u are taken to be rms complex quantities.

The mobility-type analogous symbol used as a replacement for the mechanical symbol in our circuits is an inductance.   It is shown in Fig. 3.4a.   The invariant mathematical operation that this symbol represents is given in Table 3.1.   In the steady state we have

$$u = j\omega C_M f \qquad (3.6)$$

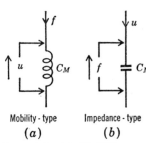

Mobility - type   Impedance - type
(a)   (b)

FIG. 3.4. (a) Mobility-type and (b) impedance-type symbols for a mechanical compliance.

In electrical circuits the time integral of the voltage across an inductance is flux-turns. The analogous quantity here is the time integral of velocity, which is displacement. This equation satisfies the physical law given in Eq. (3.5).   Note the similarity in appearance of the mechanical and analogous symbols in Figs. 3.3 and 3.4a.

The impedance-type analogous symbol for a mechanical compliance is a capacitance.   It is shown in Fig. 3.4b.   The invariant operation for steady state is $a = b/j\omega c$, or $f = u/j\omega C_M$.   It also satisfies Eq. (3.5).   In electrical circuits the time integral of the current through a capacitor is the charge.   The analogous quantity here is the displacement.

Mechanical Resistance R_M, and Mechanical Responsiveness r_M.   A physical structure is said to be a mechanical resistance R_M if, when it is acted on by a force, it moves with a velocity directly proportional to the force.   The unit is the mks mechanical ohm.

We also define here a quantity r_M, the mechanical *responsiveness*, that is the reciprocal of R_M.   The unit of responsiveness is the mks mechanical mohm.

The above representation for mechanical resistance is usually limited to viscous resistance.   Frictional resistance is excluded because, for it, the ratio of force to velocity is not a constant.   Both terminals of resistive elements can usually be located by visual inspection.

MECHANICAL RESPONSIVENESS $= \dfrac{1}{\text{MECH. RESISTANCE}}$

MECHANICAL RESISTANCE IS USUALLY LIMITED TO
VISCOUS RESISTANCE. R ∝ V

The mechanical element used to represent viscous resistance is the fluid dashpot shown schematically in Fig. 3.5. The upper end of the element moves with a velocity u_1 and the lower with a velocity u_2. The force required to produce the difference between the two velocities u_1 and u_2 may be measured by breaking into the machine at either point 1 or point 2.

Mechanical resistance R_M obeys the following physical law,

$$f = R_M u = \frac{1}{r_M} u \tag{3.7}$$

where f is the force in newtons, u is the difference between the velocities u_1 and u_2 of the two ends, R_M is the mechanical resistance in mechanical ohms, i.e., newtons/(meter per second), and r_M is the mechanical responsiveness in mks mechanical mohms, i.e., meters per second per newton.

The mobility-type analogous symbol used to replace the mechanical symbol in our circuits is a resistance. It is shown in Fig. 3.6a. The

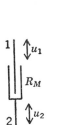

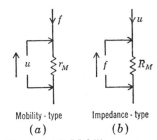

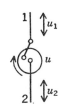

FIG. 3.5. Mechanical symbol for mechanical (viscous) resistance.

FIG. 3.6. (a) Mobility-type and (b) impedance-type symbols for a mechanical resistance.

FIG. 3.7. Mechanical symbol for a constant-velocity generator.

invariant mathematical operation that this symbol represents is given in Table 3.1. In either the steady or transient state we have

$$u = r_M f = \frac{1}{R_M} f \tag{3.8}$$

In the steady state u and f are taken to be rms complex quantities. This equation satisfies the physical law given in Eq. (3.7).

The impedance-type analogous symbol for a mechanical resistance is shown in Fig. 3.6b. It also satisfies Eq. (3.7).

Mechanical Generators. The mechanical generators considered will be one of two types, constant-velocity or constant-force. A _constant-velocity generator_ is represented as a very strong motor attached to a shuttle mechanism in the manner shown in Fig. 3.7. The opposite ends of the generator have velocities u_1 and u_2. One of these velocities, either u_1 or u_2, is determined by factors external to the generator. The difference between the velocities u_1 and u_2, however, is a velocity u that is independent of the external load connected to the generator.

The symbols that we used in the two analogies to replace the mechanical symbol for a constant-velocity generator are shown in Fig. 3.8. The invariant mathematical operations that these symbols represent are also given in Table 3.1. The tips of the arrows point to the "positive" terminals of the generators. The double circles in Fig. 3.8a indicate that the internal mobility of the generator is zero. The dashed line in Fig. 3.8b indicates that the internal impedance of the generator is infinite.

A *constant-force generator* is represented here by an electromagnetic transducer (*e.g.*, a moving-coil loudspeaker) in the primary of which an electric current of constant amplitude is maintained. Such a generator produces a force equal to the product of the current i, the flux density B, and the effective length of the wire l cutting the flux ($f = Bli$). This device is shown schematically in Fig. 3.9. The opposite ends of the

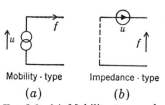 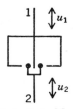

| Mobility - type | Impedance - type | | Mobility - type | Impedance - type |
| (a) | (b) | | (a) | (b) |

FIG. 3.8. (a) Mobility-type and (b) impedance-type symbols for a constant-velocity generator.

FIG. 3.9. Mechanical symbol for a constant-force generator.

FIG. 3.10. (a) Mobility-type and (b) impedance-type symbols for a constant-force generator.

generator have velocities u_1 and u_2 that are determined by factors external to the generator. The force that the generator produces and that may be measured by breaking into the device at either point 1 or point 2 is a constant force, independent of what is connected to the generator.

The symbols used in the two analogies to replace the mechanical symbol for a constant-force generator are given in Fig. 3.10. The invariant mathematical operations that these symbols represent are also given in Table 3.1. The arrows point in the direction of positive flow. Here, the dashed line indicates infinite mobility, and the double circles indicate zero impedance.

Levers. SIMPLE LEVER. It is apparent that the lever is a device closely analogous to a transformer. The lever in its simplest form consists of a weightless bar resting on an immovable fulcrum, so arranged that a downward force on one end causes an upward force on the other end (see Fig. 3.11). From elementary physics we may write the equation of balance of moments around the fulcrum,

$$f_1 l_1 = f_2 l_2$$

or, if not balanced, assuming small displacements,

$$u_1 l_2 = u_2 l_1 \qquad \frac{u_1}{f_1} = \frac{u_2}{l_2} \qquad (3.9)$$

$$= \frac{f_2}{l_1} \frac{l_2}{} = $$

Also,

MECH. MOBILITY: $\qquad z_{M1} = \dfrac{u_1}{f_1} = \left(\dfrac{l_1}{l_2}\right)^2 z_{M2}$

MECH. IMPEDANCE: $\qquad Z_{M1} = \dfrac{f_1}{u_1} = \left(\dfrac{l_2}{l_1}\right)^2 Z_{M2}$

$$(3.10)$$

The above equations may be represented by the ideal transformers of Fig. 3.12, having a transformation ratio of $\left(\dfrac{l_1}{l_2}\right):1$ for the mobility type and $\left(\dfrac{l_2}{l_1}\right):1$ for the impedance type.

FLOATING LEVER. As an example of a simple floating lever, consider a weightless bar resting on a fulcrum that yields under force. The bar is

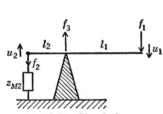

FIG. 3.11. Simple lever.

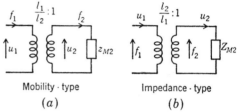

Mobility - type
(a)

Impedance - type
(b)

FIG. 3.12. (a) Mobility-type and (b) impedance-type symbols for a simple lever.

so arranged that a downward force on one end tends to produce an upward force on the other end. An example is shown in Fig. 3.13.

To solve this type of problem, we first write the equations of moments. Summing the moments about the center support gives

$$l_1 f_1 = l_2 f_2$$

and summing the moments about the end support gives

$$(l_1 + l_2)f_1 = l_2 f_3 \qquad (3.11)$$

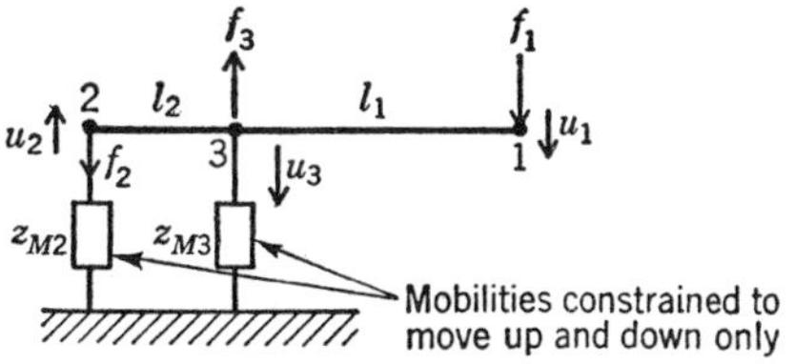

FIG. 3.13. Floating lever.

Mobilities constrained to move up and down only

When the forces are not balanced, and if we assume infinitesimal displacements, the velocities are related to the forces through the mobilities, so that

$$u_3 = z_{M3} f_3 = z_{M3} \frac{l_1 + l_2}{l_2} f_1$$

$$u_2 = z_{M2} f_2 = z_{M2} \frac{l_1}{l_2} f_1$$

$$(3.12)$$

Also, by superposition, it is seen from simple geometry that

$$u_1' = u_3 \frac{l_1 + l_2}{l_2} \quad \text{for } u_2 = 0$$

$$u_1'' = u_2 \frac{l_1}{l_2} \quad \text{for } u_3 = 0$$

so that

$$u_1 = u_1' + u_1'' = \frac{l_1 + l_2}{l_2} u_3 + \frac{l_1}{l_2} u_2 \tag{3.13}$$

and, finally,

$$\frac{u_1}{f_1} = z_{M1} = z_{M3} \left(\frac{l_1 + l_2}{l_2}\right)^2 + z_{M2} \left(\frac{l_1}{l_2}\right)^2 \tag{3.14}$$

This equation may be represented by the analogous circuit of Fig. 3.14. The lever loads the generator with two mobilities connected in series, each of which behaves as a simple lever when the other is equal to zero. It will be seen that this is a way of obtaining the equivalent of two series masses without a common zero-velocity (ground) point. This will be illustrated in Example 3.3.

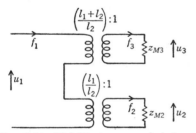

FIG. 3.14. Mobility-type symbol for a floating lever.

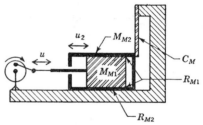

FIG. 3.15. Six-element mechanical device.

Example 3.1. The mechanical device of Fig. 3.15 consists of a piston of mass M_{M1} sliding on an oil surface inside a cylinder of mass M_{M2}. This cylinder in turn slides in an oiled groove cut in a rigid body. The sliding (viscous) resistances are R_{M1} and R_{M2}, respectively. The cylinder is held by a spring of compliance C_M. The mechanical generator maintains a constant sinusoidal velocity of angular frequency ω, whose rms magnitude is u m/sec. Solve for the force f produced by the generator.

Solution. Although the force will be determined ultimately from an analysis of the mobility-type analogous circuit for this mechanical device, it is frequently useful to draw a mechanical-circuit diagram. This interim step to the desired circuit will be especially helpful to the student who is inexperienced in the use of analogies. Its use virtually eliminates errors from the final circuit.

To draw the mechanical circuit, note first the junction points of two or more elements. This locates all element terminals which move with the same velocity. There are in this example two velocities, u and u_2, in addition to "ground," or zero velocity. These two velocities are represented in the mechanical-circuit diagram by the velocities of two imaginary rigid bars, 1 and 2 of Fig. 3.16, which oscillate in a vertical direction. The circuit drawing is made by attaching all element terminals with velocity u to the first bar and all terminals with velocity u_2 to the second bar. All terminals with zero velocity are drawn to a ground bar. Note that a mass always has one terminal on ground.† Three elements of Fig. 3.15 have one terminal with the velocity u: the generator, the mass M_{M1}, and the viscous resistance R_{M1}. These are attached to bar 1. Four elements have one terminal with the velocity u_2: the viscous resistances R_{M1} and R_{M2}, the mass M_{M2}, and the compliance C_M. These are attached to bar 2. Five elements have one terminal with zero velocity: the generator, both masses, the viscous resistance R_{M2}, and the compliance C_M.

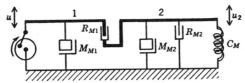

FIG. 3.16. Mechanical circuit for the device of Fig. 3.15.

We are now in a position to transform the mechanical circuit into a mobility-type analogous circuit. This is accomplished simply by replacing the mechanical elements with the analogous mobility-type elements. The circuit becomes that shown in Fig. 3.17. Remember that, in the mobility-type analogy, force "flows" through the elements and velocity is the drop across them. The resistors must have lower case r's written alongside them. As defined above, $r_M = 1/R_M$, and the unit is the mks mechanical mohm.

The equations for this circuit are found in the usual manner, using the rules of Table 3.1. Let us determine $z_M = u/f$, the mechanical mobility presented to the

$r_m \equiv$ MECHANICAL RESPONSIVENESS

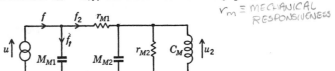

FIG. 3.17. Mobility-type analogous circuit for the device of Fig. 3.15.

generator. The mechanical mobility of the three elements in parallel on the right-hand side of the schematic diagram is

$$\frac{u_2}{f_2} = \frac{1}{\dfrac{1}{1/j\omega M_{M2}} + \dfrac{1}{r_{M2}} + \dfrac{1}{j\omega C_M}}$$

$$= \frac{1}{j\omega M_{M2} + R_{M2} + \dfrac{1}{j\omega C_M}}$$

Including the element r_{M1} the mechanical mobility for that part of the circuit through which f_2 flows is, then,

$$\frac{u}{f_2} = r_{M1} + \frac{1}{j\omega M_{M2} + R_{M2} + \dfrac{1}{j\omega C_M}}$$

Note that the input mechanical mobility z_M is given by

$$z_M = \frac{u}{f} = \frac{u}{f_1 + f_2}$$

† An exception to this rule may occur when the mechanical device embodies one or more floating levers, as we just learned.

and

$$f_1 = \frac{u}{1/j\omega M_{M1}} = j\omega M_{M1} u$$

Substituting f_1 and f_2 into the second equation preceding gives us the input mobility.

$$z_M = \frac{u}{f} = \cfrac{1}{j\omega M_{M1} + \cfrac{1}{r_{M1} + \cfrac{1}{j\omega M_{M2} + R_{M2} + \cfrac{1}{j\omega C_M}}}} \qquad (3.15a)$$

The mechanical impedance is the reciprocal of Eq. (3.15a).

$$Z_M = \frac{f}{u} = j\omega M_{M1} + \cfrac{1}{r_{M1} + \cfrac{1}{j\omega M_{M2} + R_{M2} + \cfrac{1}{j\omega C_M}}} \qquad (3.15b)$$

The result is

$$f = Z_M u \qquad \text{newtons} \qquad (3.16)$$

Example 3.2. As a further example of a mechanical circuit, let us consider the two masses of 2 and 4 kg shown in Fig. 3.18. They are assumed to rest on a frictionless

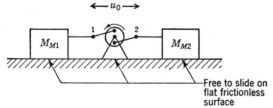

FIG. 3.18. Three-element mechanical device.

plane surface and to be connected together through a generator of constant velocity that is also free to slide on the frictionless plane surface. Let its velocity be

$$u_0 = 2 \cos 1000t \qquad \text{cm/sec}$$

Draw the mobility-type analogous circuit, and determine the force f produced by the generator. Also, determine the mobility presented to the generator.

Solution. The masses do not have the same velocity with respect to ground. The difference between the velocities of the two masses is u_0. The element representing a mass is that shown in Fig. 3.2a with one end grounded and the other moving at the velocity of the mass.

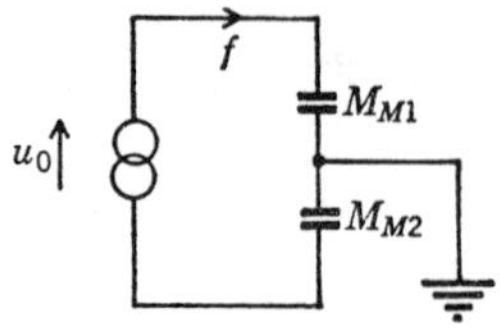

The mobility-type circuit for this example is shown in Fig. 3.19. The velocity u_0 equals $u_1 + u_2$, where u_1 is the velocity *with respect to ground* of M_{M1}, and u_2 is that for M_{M2}. The force f is

FIG. 3.19. Mobility-type analogous circuit for the device of Fig. 3.18.

$$\begin{aligned} f_{rms} &= \frac{(u_0)_{rms}}{(1/j\omega M_{M1}) + (1/j\omega M_{M2})} \\ &= \frac{j\omega M_{M1} M_{M2}}{M_{M1} + M_{M2}} u_0 \\ &= \frac{j1000 \times 2 \times 4 \times 0.02}{(2+4)\sqrt{2}} = j18.9 \text{ newtons} \quad (3.17) \end{aligned}$$

The j indicates that the time phase of the force is 90° leading with respect to that of the velocity of the generator.

Obviously, when one mass is large compared with the other, the force is that necessary to move the smaller one alone. This example reveals the only type of case in which masses can be in series without the introduction of floating levers. At most, only two masses can be in series because a common ground is necessary.

The mobility presented to the generator is

$$z_M = \left(\frac{u_0}{f}\right)_{rms} = \frac{M_{M1} + M_{M2}}{j\omega M_{M1}M_{M2}}$$

$$= \frac{6}{j1000 \times 8} = -j7.5 \times 10^{-4} \text{ mohms} \qquad (3.18)$$

Example 3.3. An example of a mechanical device embodying a floating lever is shown in Fig. 3.20. The masses attached at points 2 and 3 may be assumed to be

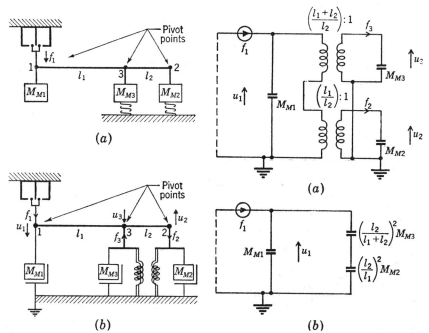

FIG. 3.20. (a) Mechanical device embodying a floating lever. (b) Mechanical diagram of (a). The compliances of the springs are very large so that all of f_2 and f_3 go to move M_{M2} and M_{M3}.

FIG. 3.21. (a) Mobility-type analogous circuit for the device of Fig. 3.20. (b) Same as (a) but with transformers removed.

resting on very compliant springs. The driving force f_1 will be assumed to have a frequency well above the resonance frequencies of the masses and their spring supports so that

$$z_{M2} \doteq \frac{1}{j\omega M_{M2}}$$

$$z_{M3} \doteq \frac{1}{j\omega M_{M3}}$$

Also, assume that a mass is attached to the weightless lever bar at point 1, with a mobility

$$z_{M1} = \frac{1}{j\omega M_{M1}}$$

Solve for the total mobility presented to the constant-force generator f_1.

Solution. By inspection, the mobility-type analogous circuit is drawn as shown in Fig. 3.21a and b. Solving for $z_M = u_1/f_1$, we get

$$z_M = \frac{1}{j\omega \left[\dfrac{M_{M2}M_{M3}l_2{}^2}{M_{M3}l_1{}^2 + M_{M2}(l_1 + l_2)^2} + M_{M1} \right]} \tag{3.19}$$

Note that if $l_2 \to 0$, the mobility is simply that of the mass M_{M1}. Also, if $l_1 \to 0$, the mobility is that of M_{M1} and M_{M3}, that is,

$$z_M = \frac{1}{j\omega(M_{M3} + M_{M1})} \tag{3.20}$$

It is possible with one or more floating levers to have one or more masses with no ground terminal.

PART VII *Acoustical Circuits*

3.4. Acoustical Elements. Acoustical circuits are frequently more difficult to draw than mechanical ones because the elements are less easy to identify. As was the case for mechanical circuits, the more obvious forms of the elements will be useful as an intermediate step toward drawing the analogous circuit diagram. When the student is more familiar with acoustical circuits, he will be able to pass directly from the acoustic device to the final form of the equivalent circuit.

In acoustic devices, the quantity we are able to measure most easily without modification of the device is sound pressure. Such a measurement is made by inserting a small hollow probe tube into the sound field at the desired point. This probe tube leads to one side of a microphone diaphragm. The other side of the diaphragm is exposed to atmospheric pressure. A movement of the diaphragm takes place when there is a difference in pressure across it. This difference between atmospheric pressure and the pressure with the sound field is the sound pressure p.

Because we can measure sound pressure by such a probe-tube arrangement without disturbing the device, it seems that sound pressure is analogous to voltage in electrical circuits. Such a choice requires us to consider current as being analogous to some quantity which is proportional to velocity. As we shall show shortly, a good choice is to make current analogous to volume velocity, the volume of gas displaced per second.

A strong argument can be made for this choice of analogy when one considers the relations governing the flow of air inside such acoustic devices as loudspeakers, microphones, and noise filters. Inside a certain type of microphone, for example, there is an air cavity that connects to the outside air through a small tube (see Fig. 3.22). Assume, now, that the outer end of this tube is placed in a sound wave. The wave will cause a movement of the air particles in the tube. Obviously, there is a junction between the tube and the cavity at the inner end of the tube at point A. Let us ask ourselves the question, What physical quantities are continuous at this junction point?

First, the sound pressure just inside the tube at A is the same as that in the cavity just outside A. That is to say, we have continuity of sound pressure. Second, the quantity of air leaving the inner end of the small tube in a given interval of time is the quantity that enters the cavity in the same interval of time. That is, the mass per second of gas leaving the small tube equals the mass per second of gas entering the volume.

Because the pressure is the same at both places, the density of the gas must also be the same, and it follows that there is continuity of volume velocity (cubic meters per second) at this junction. Analogously, in the case of electricity, there is continuity of electric current at a junction. Continuity of volume velocity must exist even if there are several tubes or cavities joining near one point. A violation of the law of conservation of mass otherwise would occur.

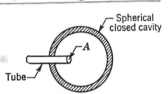

Fig. 3.22. Closed cavity connecting to the outside air through a tube of cross-sectional area S. The junction plane between the tube and the cavity occurs at A.

We conclude that the quantity that flows *through* our acoustical elements must be the volume velocity U in cubic meters per second and the drop across our acoustical elements must be the pressure p in newtons per square meter. This conclusion indicates that the impedance type of analogy is the preferred analogy for acoustical circuits. The product of the effective sound pressure p times the in-phase component of the effective volume velocity U gives the acoustic power in watts.

In this part, we shall discuss the more general aspects of acoustical circuits. In Chap. 5 of this book, we explain fully the approximations involved and the rules for using the concepts enunciated here in practical problems.

Acoustic Mass M_A. Acoustic mass is a quantity proportional to mass but having the dimensions of kilograms per meter⁴. It is associated with a mass of air accelerated by a net force which acts to displace the gas without appreciably compressing it. (The concept of acceleration without compression is an important one to remember.) It will assist you in distinguishing acoustic masses from other elements.

The acoustical element that is used to represent an acoustic mass is a tube filled with the gas as shown in Fig. 3.23.

The physical law governing the motion of a mass that is acted on by a force is Newton's second law, $f(t) = M_M \, du(t)/dt$. This law may be expressed in acoustical terms as follows,

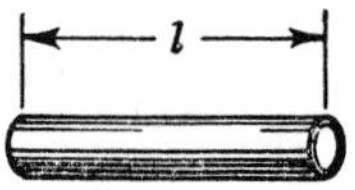

$$\frac{f(t)}{S} = \frac{M_M}{S} \frac{d[u(t)S]}{dt\,S} = p(t) = \frac{M_M}{S^2} \frac{dU(t)}{dt}$$

FIG. 3.23. Tube of length l and cross-sectional area S.

or

$$p(t) = M_A \frac{dU(t)}{dt} \qquad (3.21)$$

where $p(t) =$ instantaneous difference between pressures in newtons per square meter existing at each end of a mass of gas of M_M kg undergoing acceleration.

$M_A = M_M/S^2 =$ acoustic mass in kilograms per meter^4 of the gas undergoing acceleration. This quantity is nearly equal to the mass of the gas inside the containing tube divided by the square of the cross-sectional area. To be more exact we must note that the gas in the immediate vicinity of the ends of the tube also adds to the mass. Hence, there are "end corrections" which must be considered. These corrections are discussed in Chap. 5 (pages 132 to 139).

$U(t) =$ instantaneous volume velocity, of the gas in cubic meters per second across any cross-sectional plane in the tube. The volume velocity $U(t)$ is equal to the linear velocity $u(t)$ multiplied by the cross-sectional area S.

In the steady state, with an angular frequency ω, we have

$$p = j\omega M_A U \qquad (3.22)$$

where p and U are taken to be rms complex quantities.

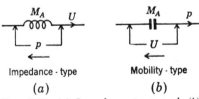

Impedance - type

(a)

Mobility - type

(b)

FIG. 3.24. (a) Impedance-type and (b) mobility-type symbols for an acoustic mass.

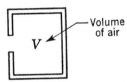

FIG. 3.25. Enclosed volume of air V with opening for entrance of pressure variations.

The impedance-type analogous symbol for acoustic mass is shown in Fig. 3.24a, and the mobility-type is given in Fig. 3.24b. In the steady state, for either, we get Eq. (3.22). The arrows point in the direction of positive flow or positive drop.

ACOUSTIC COMPLIANCE, C_A.

Acoustic Compliance C_A. Acoustic compliance is a constant quantity having the dimensions of meter⁵ per newton. It is associated with a volume of air that is compressed by a net force without an appreciable average displacement of the center of gravity of air in the volume. In other words, compression without acceleration identifies an acoustic compliance.

The acoustical element that is used to represent an acoustic compliance is a volume of air drawn as shown in Fig. 3.25.

The physical law governing the compression of a volume of air being acted on by a net force was given as $f(t) = (1/C_M)\int u(t)\, dt$. Converting from mechanical to acoustical terms,

$$p(t) = \frac{f(t)}{S} = \frac{1}{C_M S}\int u(t)\,\frac{S}{S}\,dt \quad \text{or} \quad p(t) = \frac{1}{C_M S^2}\int U(t)\,dt$$

or

$$p(t) = \frac{1}{C_A}\int U(t)\,dt \tag{3.23}$$

where $p(t)$ = instantaneous pressure in newtons per square meter acting to compress the volume V of the air.

$C_A = C_M S^2$ = acoustic compliance in meters⁵ per newton of the volume of the air undergoing compression. The acoustic compliance is nearly equal to the volume of air divided by γP_0, as we shall see in Chap. 5 (pages 128 to 131).

$U(t)$ = instantaneous volume velocity in cubic meters per second of the air flowing into the volume that is undergoing compression. The volume velocity $U(t)$ is equal to the linear velocity $u(t)$ multiplied by the cross-sectional area S.

In the steady state with an angular frequency ω, we have

$$p = \frac{U}{j\omega C_A} \tag{3.24}$$

where p and U are taken to be rms complex quantities.

The impedance-type analogous element for acoustic compliance is shown in Fig. 3.26a and the mobility-type in Fig. 3.26b. In the steady state for either, Eq. (3.24) applies.

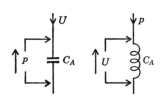

Impedance-type (a)   Mobility-type (b)

Fig. 3.26. (a) Impedance-type and (b) mobility-type symbols for an acoustic compliance.

Acoustic Resistance R_A, and Acoustic Responsiveness r_A. Acoustic resistance R_A is associated with the dissipative losses occurring when there is a viscous movement of a quantity of gas through a fine-mesh screen or through a capillary tube. It is a constant quantity having the dimensions newton-seconds per meter⁵. The unit is the mks acoustic ohm.

ACOUSTIC RESISTANCE R_A.

FIG. 3.27. Fine-mesh screen which serves as an acoustical symbol for acoustic resistance.

The acoustic element used to represent an acoustic resistance is a fine-mesh screen drawn as shown in Fig. 3.27.

The reciprocal of acoustic resistance is the acoustic responsiveness r_A. The unit is the mks acoustic mohm with dimensions meter^5 per second per newton.

The physical law governing dissipative effects in a mechanical system was given by $f(t) = R_M u(t)$, or, in terms of acoustical quantities,

$$p(t) = R_A U(t) = \frac{1}{r_A} U(t) \qquad (3.25)$$

where $p(t)$ = difference between instantaneous pressures in newtons per square meter across the dissipative element. In the steady state p is an rms complex quantity.

$R_A = R_M/S^2$ = acoustic resistance in acoustic ohms, i.e., newton-seconds per meter^5.

$r_A = r_M S^2$ = acoustic responsiveness in acoustic mohms, i.e., meter^5 per newton-seconds.

$U(t)$ = instantaneous volume velocity in cubic meters per second of the gas through the cross-sectional area of resistance. In the steady state U is an rms quantity.

The impedance-type analogous symbol for acoustic resistance is shown in Fig. 3.28a and the mobility-type in Fig. 3.28b.

Acoustic Generators. Acoustic generators can be of either the constant-volume velocity or the constant-pressure type. The prime movers in our

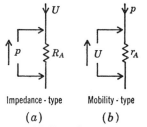

Impedance - type   Mobility - type

(a)   (b)

FIG. 3.28. (a) Impedance-type symbol for acoustic resistance and (b) mobility-type symbol for acoustic responsiveness.

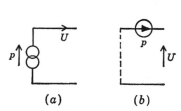

(a)   (b)

FIG. 3.29. (a) Impedance-type and (b) mobility-type symbols for a constant-pressure generator.

acoustical circuits will be exactly like those shown in Figs. 3.7 and 3.9 except that u_2 often will be zero and u_1 will be the velocity of a small piston of area S. Remembering that $u = u_1 - u_2$, we see that the generator of Fig. 3.7 has a constant-volume velocity $U = uS$ and that of Fig. 3.9 a constant pressure of $p = f/S$.

The two types of analogous symbols for acoustic generators are given in Figs. 3.29 and 3.30. The arrows point in the direction of the positive terminal or the positive flow.   As before, the double circles indicate zero impedance or mobility and a dashed line infinite impedance or mobility.

Mechanical Rotational Systems.   Mechanical rotational systems are handled in the same manner as mechanical rectilineal systems.   The following quantities are analogous in the two systems.

Rectilineal systems	Rotational systems
f = force, newtons	T = torque, newton-m
v = velocity, m/sec	θ = angular velocity, radians/sec
ξ = displacement, m	ϕ = angular displacement, radians
$Z_M = f/v$ = mechanical impedance, mks mechanical ohms	$Z_R = T/\theta$ = rotational impedance, mks rotational ohms
$z_M = v/f$ = mechanical mobility, mks mechanical mohms	$z_R = \theta/T$ = rotational mobility, mks rotational mohms
R_M = mechanical resistance, mks mechanical ohms	R_R = rotational resistance, mks rotational ohms
r_M = mechanical responsiveness, mks mechanical mohms	r_R = rotational responsiveness, mks rotational mohms
M_M = mass, kg	I_R = moment of inertia, kg-m^2
C_M = mechanical compliance, m/newton	C_R = rotational compliance, radians/ newton-m
W_M = mechanical power, watts	W_R = rotational power, watts

Example 3.4.   The acoustic device of Fig. 3.31 consists of three cavities V_1, V_2, and V_3, two fine-mesh screens R_{A1} and R_{A2}, four short lengths of tube T_1, T_2, T_3, and T_4, and a constant-pressure generator.   Because the air in the tubes is not confined, it

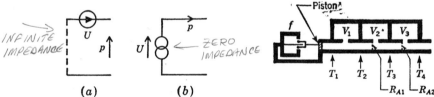

Fig. 3.30. (a) Impedance-type and (b) mobility-type symbols for a constant-volume velocity generator.

Fig. 3.31. Acoustic device consisting of four tubes, three cavities, and two screens driven by a constant-pressure generator.

experiences negligible compression.   Because the air in each of the cavities is confined, it experiences little average movement.   Let the force of the generator be

$$f = 10^{-5} \cos 1000t \qquad \text{newtons}$$

the radius of the tube $a = 0.5$ cm; the length of each of the four tubes $l = 5$ cm; the volume of each of the three cavities $V = 10$ cm^3; and the magnitude of the two acoustic resistances $R_A = 10$ mks acoustic ohms.   Neglecting end corrections, solve for the volume velocity U_0 at the end of the tube T_4.

Solution.   Remembering that there is continuity of volume velocity and pressure at the junctions, we can draw the impedance-type analogous circuit from inspection. It is shown in Fig. 3.32.   The bottom line of the schematic diagram represents atmospheric pressure, which means that here the variational pressure p is equal to zero.   At each of the junctions of the elements 1 to 4, a different variational pressure can be observed.   The end of the fourth tube (T_4) opens to the atmosphere, which requires that M_{A4} be connected directly to the bottom line of Fig. 3.32.

Note that the volume velocity of the gas leaving the tube T_1 is equal to the sum of the volume velocities of the gas entering V_1 and T_2. The volume velocity of the gas leaving T_2 is the same as that flowing through the screen R_{A1} and is equal to the sum of the volume velocities of the gas entering V_2 and T_3.

One test of the validity of an analogous circuit is its behavior for direct current. If one removes the piston and blows into the end of the tube T_1 (Fig. 3.31), a steady flow of air from T_4 is observed. Some resistance to this flow will be offered by the two screens R_{A1} and R_{A2}. Similarly in the schematic diagram of Fig. 3.32, a steady pressure p will produce a steady flow U through M_{A4}, resisted only by R_{A1} and R_{A2}.

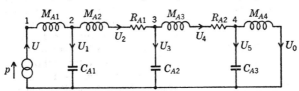

FIG. 3.32. Impedance-type analogous circuit for the acoustic device of Fig. 3.31.

As an aside, let us note that an acoustic compliance can occur in a circuit without one of the terminals being at ground potential only if it is produced by an elastic diaphragm. For example, if the resistance R_{A1} in Fig. 3.31 were replaced by an impervious but elastic diaphragm, the element R_{A1} in Fig. 3.32 would be replaced by a compliance-type element with both terminals above ground potential. In this case a steady flow of air could not be maintained through the device of Fig. 3.31, as can also be seen from the circuit of Fig. 3.32, with R_{A1} replaced by a compliance.

Determine the element sizes of Fig. 3.32.

$$p = \frac{f}{S} = \frac{10^{-5}\cos 1000t}{\pi(5\times 10^{-3})^2} = 0.1273\cos 1000t \qquad \text{newton/m}^2$$

$$M_{A1} = M_{A2} = M_{A3} = M_{A4} = \frac{\rho_0 l}{S} = \frac{1.18\times 0.05}{7.85\times 10^{-5}} = 750 \text{ kg/m}^4$$

$$C_{A1} = C_{A2} = C_{A3} = \frac{V}{\gamma P_0} = \frac{10^{-5}}{1.4\times 10^5} = 7.15\times 10^{-11} \text{ m}^5/\text{newton}$$

$$R_{A1} = R_{A2} = 10 \text{ mks acoustic ohms}$$

As is customary in electric-circuit theory, we solve for U_0 indirectly. First, arbitrarily let $U_0 = 1$ m³/sec, and determine the ratio p/U_0.

$P = M_A \frac{dU}{dt} = j\omega M_A U$

$P = \frac{1}{C_A}\int U\,dt$

$P = \frac{U}{j\omega C_A}$

$P = R_A U$

$p_4 = j\omega M_{A4}U_0 = j7.5\times 10^5$ newtons/m²

$U_5 = j\omega C_{A3}p_4 = -5.36\times 10^{-2}$ m³/sec

$U_4 = U_5 + U_0 = 0.946$

$p_3 = (R_{A2} + j\omega M_{A3})U_4 + p_4 = 9.46 + j14.6\times 10^5$

$U_3 = j\omega C_{A2}p_3 = -0.1043 + j6.77\times 10^{-7}$

$U_2 = U_3 + U_4 = 0.842 + j6.77\times 10^{-7}$

$p_2 = (R_{A1} + j\omega M_{A2})U_2 + p_3 = 17.37 + j2.091\times 10^6$

$U_1 = j\omega C_{A1}p_2 = -0.1496 + j1.242\times 10^{-6}$

$U = U_2 + U_1 = 0.692 + j1.919\times 10^{-6}$

$p = j\omega M_{A1}U + p_2 = 15.93 + j2.61\times 10^6 = \dfrac{p}{U_0}$ for $U_0 = 1$

The desired value of U_0 is

$$U_0 = p\,\frac{U_0}{p} = \frac{0.1273\cos 1000t}{15.93 + j2.61\times 10^6}$$
$$\doteq 4.88\times 10^{-8}\cos(1000t - 90°)$$
$$\doteq 4.88\times 10^{-8}\sin 1000t$$

In other words, the impedance is principally that of the four acoustic masses in series so that U_0 lags p by nearly 90°.

Example 3.5. A Helmholtz resonator is frequently used as a means for eliminating an undesired frequency component from an acoustic system. An example is given in Fig. 3.33a. A constant-force generator G produces a series of tones, among which is one that is not wanted. These tones actuate a microphone M whose acoustic impedance is 500 mks acoustic ohms. If the tube T has a cross-sectional area of 5 cm², $l_1 = l_2 = 5$ cm, $l_3 = 1$ cm, $V = 1000$ cm³, and the cross-sectional area of l_3 is 2 cm², what frequency is eliminated from the system?

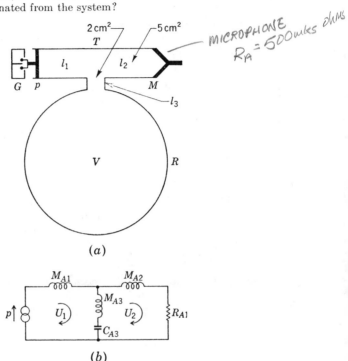

(a)

(b)

FIG. 3.33. (a) Acoustic device consisting of a constant-force generator G, piston P, tube T with length $l_1 + l_2$, microphone M, and Helmholtz resonator R with volume V and connecting tube as shown. (b) Impedance-type analogous circuit for the device of (a).

Solution. By inspection we may draw the impedance-type analogous circuit of Fig. 3.33b. The element sizes are

$$M_{A1} = M_{A2} = \frac{\rho_0 l_1}{S_T} = \frac{1.18 \times 0.05}{5 \times 10^{-4}} = 118 \text{ kg/m}^4$$

$$M_{A3} = \frac{\rho_0 l_3}{S_R} = \frac{1.18 \times 0.01}{2 \times 10^{-4}} = 59 \text{ kg/m}^4$$

$$C_{A3} = \frac{V}{\gamma P_0} = \frac{10^{-3}}{1.4 \times 10^5} = 7.15 \times 10^{-9} \text{ m}^5/\text{newton}$$

$$R_{A1} = 500 \text{ mks acoustic ohms}$$

It is obvious that the volume velocity U_2 of the transducer M will be zero when the shunt branch is at resonance. Hence,

$$\omega = \frac{1}{\sqrt{M_{A3}C_{A3}}} = \frac{10^4}{\sqrt{42.2}} = 1540 \text{ radians/sec}$$

$$f = 245 \text{ cps}$$

PART VIII *Transducers*

A transducer is defined as a device for converting energy from one form to another. Of importance in this text is the electromechanical transducer for converting electrical energy into mechanical energy, and vice versa. There are many types of such transducers. In acoustics we are concerned with microphones, earphones, loudspeakers, and vibration pickups and vibration producers which are generally linear passive reversible networks.

The type of electromechanical transducer chosen for each of these instruments depends upon such factors as the desired electrical and mechanical impedances, durability, and cost. It will not be possible here to discuss all means for electromechanical transduction. Instead we shall limit the discussion to electromagnetic and electrostatic types. Also, we shall deal with mechano-acoustic transducers for converting mechanical energy into acoustic energy.

3.5. Electromechanical Transducers. Two types of electromechanical transducers, electromagnetic and electrostatic, are commonly employed in loudspeakers and microphones. Both may be represented by transformers with properties that permit the joining of mechanical and electrical circuits into one schematic diagram.

Electromagnetic-mechanical Transducer. This type of transducer can be characterized by four terminals. Two have voltage and current associated with them. The other two have velocity and force as the measurable properties. Familiar examples are the moving-coil loudspeaker or microphone and the variable-reluctance earphone or microphone.

The simplest type of moving-coil transducer is a single length of wire in a uniform magnetic field as shown in Fig. 3.34. When a wire is moved upward with a velocity u as shown in Fig. 3.34a, a potential difference e will be produced in the wire such that terminal 2 is positive. If, on the other hand, the wire is fixed in the magnetic field (Fig. 3.34b) and a current i is caused to flow into terminal 2 (therefore, 2 is positive), a force will be produced that acts on the wire upward in the same direction as that indicated previously for the velocity.

The basic equations applicable to the moving-coil type of transducer are

$$f = Bli \qquad (3.26a)$$
$$e = Blu \qquad (3.26b)$$

where i = electrical current in amperes
  f = "open-circuit" force in newtons produced on the mechanical circuit by the current i
  B = magnetic-flux density in webers per square meter

l = effective length in meters of the electrical conductor that moves at right angles across the magnetic lines of force of flux density B

u = velocity in meters per second

e = "open-circuit" electrical voltage in volts produced by a velocity u

The right-hand sides of Eqs. (3.26) have the same sign because when u and f are in the same direction the electrical terminals have the same sign.

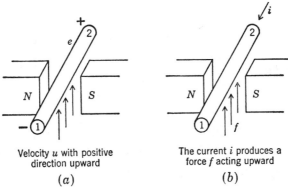

Velocity u with positive direction upward

(a)

The current i produces a force f acting upward

(b)

Fig. 3.34. Simplified form of moving-coil transducer consisting of a single length of wire cutting a magnetic field of flux density B. (a) The conductor is moving vertically at constant velocity so as to generate an open-circuit voltage across terminals 1 and 2. (b) A constant current is entering terminal 2 to produce a force on the conductor in a vertical direction.

The analogous symbol for this type of transducer is the "ideal" transformer given in Fig. 3.35. The "windings" on this ideal transformer have infinite impedance, and the transformer obeys Eqs. (3.26) at all frequencies, including steady flow. The mechanical side of this symbol necessarily is of the mobility type if current flows in the primary. The invariant mathematical operations which this symbol represents are given in Table 3.1. They lead directly to Eqs. (3.26). The arrows point in the directions of positive flow or positive potential.

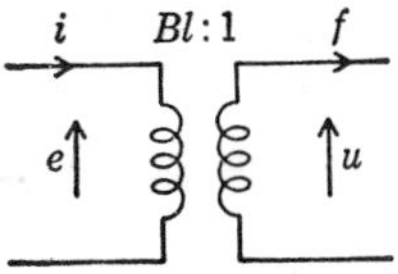

Electrostatic-mechanical Transducer. This type of transducer may also be characterized by four terminals. At two of them, voltages and currents can be measured. At the other two, forces and velocities can be measured. This transducer is satisfactorily described by the following mathematical relations,

Fig. 3.35. Analogous symbol for the electromagnetic-mechanical transducer of Fig. 3.34. The mechanical side is of the mobility type.

~Voltage α–displacement

$$e = -\tau\xi \tag{3.27a}$$
$$f = \tau q \tag{3.27b}$$

force α charge

where e = "open-circuit" electrical voltage in volts produced by a displacement ξ.

ξ = displacement in meters of a dimension of the piezoelectric device.

q = electrical charge in coulombs stored in the dielectric of the piezoelectric device.

f = "open-circuit" force in newtons produced by an electrical charge q.

τ = coupling coefficient† with dimensions of newtons per coulomb or volts per meter. It is a real number when the network is linear, passive, and reversible.

An example is a piezoelectric crystal microphone such as is shown in Fig. 3.36. A force applied uniformly over the face of the crystal causes an inward displacement of magnitude ξ. As a result of this displacement, a voltage e appears across the electrical terminals 1 and 2. Let us assume that a positive displacement (inward) of the crystal causes terminal 1 to become positive. For small displacements, the induced voltage is proportional to displacement. The inverse of this effect occurs when no external force acts on the crystal face but an electrical generator is connected to the terminals 1 and 2. If the external generator is connected so that terminal 1 is positive, an internal force f is produced which acts to expand the crystal. For small displacements, the developed force f is proportional to the electric charge q stored in the dielectric of the crystal.

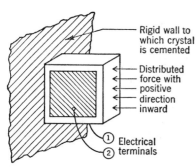

Rigid wall to which crystal is cemented

Distributed force with positive direction inward

① ② Electrical terminals

Fig. 3.36. Piezoelectric crystal transducer mounted on a rigid wall.

Equations (3.27) are often inconvenient to use because they contain charge and displacement. One prefers to deal with current and velocity, which appear directly in the equation for power. Conversion to current and velocity may be made by the relations

$$u = \frac{d\xi}{dt} = j\omega\xi \qquad (3.28a)$$

$$i = \frac{dq}{dt} = j\omega q \qquad (3.28b)$$

† The coupling coefficient is frequently defined differently in advanced texts on electrostatic-mechanical transducers. For example, in some texts it is defined as the square root of $C'_M C'_E \tau^2 \pi^2 / 8$, where C'_M and C'_E are defined after Eq. (3.37). The author does not intend that the definition for coupling coefficient in this text should be adopted as standard; rather, the term is used simply for convenience in the discussion.

so that Eqs. (3.27) become

$$e = -\frac{\tau}{j\omega}u \qquad (3.29a)$$

PIEZOELECTRIC DEVICES

$$f = \frac{\tau}{j\omega}i \qquad (3.29b)$$

Unfortunately, the usual analogous symbol for this type of transformer is not as simple as that for the electromagnetic-mechanical type. Two possible forms are shown in Fig. 3.37. The mechanical sides are of the impedance-type analogy. Let us discuss Fig. 3.37a first.

The element C'_M is the mechanical compliance of the transducer. In order to measure C'_M, a sinusoidal driving force f is applied to the transducer terminals 3 and 4, and the resulting sinusoidal displacement is measured. During this measurement the electrical terminals are short-circuited ($e = 0$). A very low driving frequency is used so that the mass reactance and mechanical resistance can be neglected.

The element C'_E is the electrical capacitance of the crystal measured at low frequencies with the mechanical terminals open-circuited ($u = 0$). Application of a current i to the primary will produce a voltage across the condenser C'_E of $i/j\omega C'_E$.

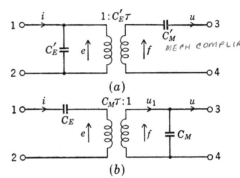

(a)

(b)

Fig. 3.37. Two forms of analogous symbols for piezoelectric transducers. The mechanical sides are of the impedance type.

This in turn produces an open-circuit force

$$f = \frac{i}{j\omega C'_E}\tau C'_E = \frac{\tau i}{j\omega} \qquad (3.30)$$

A velocity u applied at the secondary of the transducer by an external generator produces a current through the condenser C'_E equal to $-uC'_E\tau$. This in turn generates an open-circuit voltage

$$e = -uC'_E\tau\frac{1}{j\omega C'_E} = -\frac{\tau u}{j\omega} \qquad (3.31)$$

Equations (3.30) and (3.31) are seen to equal Eqs. (3.29).

In Fig. 3.37b, C_M is also the mechanical compliance of the transducer, but measured in a different way. A sinusoidal driving force is applied to terminals 3 and 4 of the transducer at a very low frequency so that mass reactance and mechanical resistance can be neglected, and the resulting sinusoidal displacement is measured. During this measurement the electrical terminals 1 and 2 are open-circuited ($i = 0$). The element C_E is the electrical capacitance measured at low frequencies with the mechanical terminals *short-circuited* ($f = 0$).

Application of a current i to the primary will produce a velocity across the compliance C_M equal to $C_M \tau i$. This velocity will produce an open-circuit force

$$f = \frac{\tau i}{j\omega} \tag{3.32}$$

A velocity u applied at the secondary of the transformer by an external generator produces a force across the compliance C_M equal to $-u/j\omega C_M$. This force will in turn generate a voltage across C_M equal to

$$e = -\frac{\tau u}{j\omega} \tag{3.33}$$

Equations (3.32) and (3.33) are seen to equal Eqs. (3.29). The transducers of Fig. 3.37 are identical. The elements in Fig. 3.37 are related by the equations

$$C'_E = \frac{C_E}{1 + C_M C_E \tau^2} \tag{3.34}$$

$$C_M = \frac{C'_M}{1 + C'_M C'_E \tau^2} \tag{3.35}$$

$$C_E = C'_E(1 + C'_M C'_E \tau^2) \tag{3.36}$$

$$C'_M = C_M(1 + C_E C_M \tau^2) \tag{3.37}$$

where C'_E = electrical capacitance measured with the mechanical "terminals" blocked so that no motion occurs ($u = 0$)

C_E = electrical capacitance measured with the mechanical "terminals" operating into zero mechanical impedance so that no force is built up ($f = 0$)

C_M = mechanical compliance measured with the electrical terminals open-circuited ($i = 0$)

C'_M = mechanical compliance measured with the electrical terminals short-circuited ($e = 0$)

The choice between the alternative analogous symbols of Fig. 3.37 is usually made on the basis of the use to which the transducer will be put. If the electrostatic transducer is a microphone, it usually is operated into the grid of a vacuum tube so that the electrical terminals are essentially open-circuited. In this case the circuit of Fig. 3.37b is the better one to use, because C_E can be neglected in the analysis when $i = 0$. On the other hand, if the transducer is a loudspeaker, it usually is operated from a low-impedance amplifier so that the electrical terminals are essentially short-circuited. In this case the circuit of Fig. 3.37a is the one to use, because $C'_E \omega$ is small in comparison with the output admittance of the amplifier.

The circuit of Fig. 3.37a corresponds more closely to the physical facts than does that of Fig. 3.37b. If the crystal could be held motionless ($u = 0$) when a voltage was impressed across terminals 1 and 2, there would be no stored mechanical energy. All the stored energy would be electrical. This is the case for circuit (a), but not for (b). In other respects the two circuits are identical.

At higher frequencies, the mass M_M and the resistance R_M of the crystal must be considered in the circuit. These elements can be added in series with terminal 3 of Fig. 3.37.

These analogous symbols indicate an important difference between electromagnetic and electrostatic types of coupling. For the electromagnetic case, we ordinarily use a mobility-type analogy, but for the electrostatic case we usually employ the impedance-type analogy.

In the next part we shall introduce a different method for handling electrostatic transducers. It involves the use of the mobility-type analog in place of the impedance-type analog. The simplification in analysis that results will be immediately apparent. By this new method it will also be possible to use the impedance-type analog for the electromagnetic case.

3.6. Mechano-acoustic Transducer. This type of transducer occurs at a junction point between the mechanical and acoustical parts of an analogous circuit. An example is the plane at which a loudspeaker diaphragm acts against the air. This transducer may also be characterized by four terminals. At two of the terminals, forces and velocities can be measured. At the other two, pressures and volume velocities can be measured. The basic equations applicable to the mechano-acoustic transducer are

$$f = Sp \qquad (3.38a)$$
$$U = Su \qquad (3.38b)$$

where f = force in newtons

$\quad p$ = pressure in newtons per square meter

$\quad U$ = volume velocity in cubic meters per second

$\quad u$ = velocity in meters per second

$\quad S$ = area in square meters

The analogous symbols for this type of transducer are given at the bottom of Table 3.3 (page 51). They are seen to lead directly to Eqs. (3.38).

3.7. Examples of Transducer Calculations

Example 3.6. An ideal moving-coil loudspeaker produces 2 watts of acoustic power into an acoustic load of 4×10^4 mks acoustic ohms when driven from an amplifier with a constant-voltage output of 1.0 volts. The area of the diaphragm is 100 cm². What open-circuit voltage will it produce when operated as a microphone with an rms diaphragm velocity of 10 cm/sec?

Solution. From Fig. 3.35 we see that, always,

$$e = Blu$$

The power dissipated W gives us the rms volume velocity of the diaphragm U.

$$U = \sqrt{\frac{W}{R_A}} = \sqrt{\frac{2}{4 \times 10^4}} = 7.07 \times 10^{-3} \text{ m}^3/\text{sec}$$

or

$$u_{\not B} = 0.707 \text{ m/sec}$$

$$Bl = \frac{e}{u} = \frac{1}{0.707} = 1.4 \text{ volts/m-sec}$$

Hence, the open-circuit voltage for an rms velocity of 0.1 m/sec is

$\underline{?\ H\!O\!\omega\ ?}$

$$e = 1.316 \times 0.1 = 0.1316 \text{ volt}$$

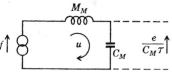

FIG. 3.38. Analogous circuit of the impedance type for a crystal microphone.

Example 3.7. An ammonium dihydrogen phosphate (ADP) crystal of the Z-cut expander-bar type (discussed in Chap. 6) has the following mechanical and electrical properties:

$$\tau = 2 \times 10^8 \text{ newtons/coulomb, or volts/m}$$
$$C_M = 9.5 \times 10^{-8} \text{ m/newton}$$
$$M_M = 1.47 \times 10^{-4} \text{ kg}$$
$$C_E = 26 \times 10^{-12} \text{ farad}$$
$$R_M = \text{negligibly small}$$

This crystal is to be used in a microphone with a circular (weightless) diaphragm. Determine the diameter of the diaphragm if the microphone is to yield an open-circuit voltage of -70 db re 1 volt for a sound pressure level of 74 db re 0.0002 microbar at 10,000 cps.

Solution. The circuit for this transducer with the transformer removed is shown in Fig. 3.38. Because only the open-circuit voltage is desired, C_E may be neglected in the calculations. f is the total force applied to the crystal by the diaphragm. Solving for e yields

$\underline{?}$
$\underline{H\!O\!\omega\ T\!O}$
$\underline{G\!E\!T\ H\!E\!R\!E\ ?}$ $e = \dfrac{f C_{MT}}{1 - \omega^2 C_M M_M}$

The force f equals the area of the diaphragm S times the sound pressure p. Solving for p,

$$p = 0.0002 \text{ antilog } {}^{74}\!/_{20}$$
$$= 1 \text{ dyne/cm}^2 = 0.1 \text{ newton/m}^2$$

Solving for e,

$$\frac{1}{e} = \text{antilog } \frac{70}{20} = 3.16 \times 10^3$$

or

$$e = 3.16 \times 10^{-4} \text{ volts}$$

Hence,

$$S = \frac{f}{p} = \frac{3.16 \times 10^{-4}(1 - 9.5 \times 1.47 \times 10^{-12} \times 6.28^2 \times 10^8)}{0.1 \times 19}$$
$$= 1.57 \times 10^{-4} \text{ m}^2$$
$$S = 1.57 \text{ cm}^2$$

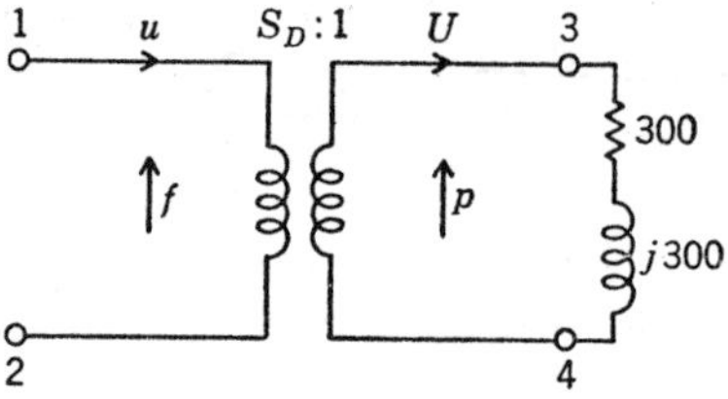

FIG. 3.39. Example of a mechano-acoustic transducer. The acoustic impedance of a horn (at terminals 3 and 4) loads the diaphragm with a mechanical impedance $S_D^2(300 + j300)$ mks mechanical ohms.

This corresponds to a diaphragm with a diameter of about 1.41 cm.

Example 3.8. A loudspeaker diaphragm couples to the throat of an exponential horn that has an acoustic impedance of $300 + j300$ mks acoustic ohms. If the area of the loudspeaker diaphragm S_D is 0.08 m², determine the mechanical-impedance load on the diaphragm due to the horn.

Solution. The analogous circuit is shown in Fig. 3.39. The mechanical impedance at terminals 1 and 2 represent the load on the diaphragm.

$$Z_M = \frac{f}{u} = S_D{}^2 \,(300 + j300)$$
$$= 6.4 \times 10^{-3}(300 + j300)$$
$$= 1.92 + j1.92 \text{ mks mechanical ohms}$$

PART IX *Circuit Theorems, Energy, and Power*

In this part we discuss conversions from one type of analogy to the other, Thévenin's theorem, energy and power relations, transducer impedances, and combinations of transducers.

3.8. Conversion from Mobility-type Analogies to Impedance-type Analogies. In the preceding parts we showed that electromagnetic and electrostatic transducers require two different types of analogy if they are to be represented by the networks shown in Table 3.1. A further need for two types of analogy is apparent from the standpoint of ease of drawing an analogous circuit by inspection. The mobility type of analogy is better for mechanical systems and the impedance type for acoustic systems.   The circuits we shall use, however, will frequently contain electrical, mechanical, and acoustical elements. Since analogies cannot be mixed in a given circuit, we must have a simple means for converting from one to the other.

We may readily derive one analogy from the other if we recognize that:

1. Elements in series in the circuit of one analogy correspond to elements in parallel on the other.

2. Resistance-type elements become responsiveness-type elements, capacitance-type elements become inductance-type elements, and inductance-type elements become capacitance-type elements.

3. The sum of the drops across the series elements in a mesh of one analogy corresponds to the sum of the currents at a branch point of the other analogy.

This is equivalent to saying that one analogy is the *dual* of the other. In electrical-circuit theory one learns that the quantities that "flow" in one circuit are the same as the "drops" in the dual of that circuit.   This is also true here.

To facilitate the conversion from one type of analogy to another, a method that we shall dub the "dot" method is used.[6] Assume that we

---

[6] M. F. Gardner and J. L. Barnes, "Transients in Linear Systems," pp. 46–49, John Wiley & Sons, Inc., New York, 1942.

have the mobility-type analog of Fig. 3.17 and that we wish to convert it to an impedance-type analog.   The procedure is as follows (see Fig. 3.40):

1. Place a dot at the center of each mesh of the circuit and one dot outside all meshes.   Number these dots consecutively.

2. Connect the dots together with lines so that there is a line through each element and so that no line passes through more than one element.

3. Draw a new circuit such that each line connecting two dots now contains an element that is the inverse of that in the original circuit.   The inverse of any given element may be seen by comparing corresponding

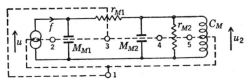

FIG. 3.40. Preparation by the "dot" method for taking the dual of Fig. 3.17.

columns for mobility-type analogies and impedance-type analogies of Table 3.3.   The complete inversion (dual) of Fig. 3.40 is shown in Fig. 3.41.

4. Solving for the velocities or the forces in the two circuits using the rules of Table 3.1 will readily reveal that they give the same results.

After completing the formation of an analogous circuit, it is always profitable to ask concerning each element, If this element becomes very small or very large, does the circuit behave in the same way the device itself would behave?   If the circuit behaves properly in the extremes, it is probably correct.

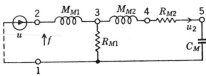

FIG. 3.41. Dual of the circuit of Fig. 3.40.   Solving for the forces or velocities in this circuit using the rules of Table 3.1 yields the same values as solving for the forces or velocities in Fig. 3.40.

3.9. Thévenin's Theorem.

It appears possible, from the foregoing discussions, to represent the operation of a transducer as a combination of electrical, mechanical, and acoustical elements.   The connection between the electrical and mechanical circuit takes place through an electromechanical transducer. Similarly, the connection between the mechanical and acoustical circuit takes place through a mechano-acoustic transducer.   A Thévenin's theorem may be written for the combined circuits, just as is written for electrical circuits only.

The requirements which must be satisfied in the proper statement and use of Thévenin's theorem are that all the elements be linear and there be no hysteresis effects.

In the next few paragraphs we shall demonstrate the application of Thévenin's theorem to a loudspeaker problem.   The mechanical-radiation impedance presented by the air to the vibrating diaphragm of a loudspeaker or microphone will be represented simply as Z_{MR} in the impedance-type analogy or $z_{MR} = 1/Z_{MR}$ in the mobility-type analogy. The exact physical nature of Z_{MR} will be discussed in Chap. 5.

Assume a simple electromagnetic (moving-coil) loudspeaker with a diaphragm that has only mass and a voice coil that has only electrical resistance (see Fig. 3.42a). Let this loudspeaker be driven by a constant-voltage generator. By making use of Thévenin's theorem, we wish to find the equivalent mechanical generator u_0 and the equivalent mechanical mobility z_{MS} of the loudspeaker, as seen in the interface between the diaphragm and the air. The circuit of Fig. 3.42a with the transformer removed is shown in Fig. 3.42b. The Thévenin equivalent circuit is shown in Fig. 3.42c. We arrive at the values of u_0 and z_{MS} in two steps.

1. Determine the open-circuit velocity u_0 by terminating the loudspeaker in an infinite mobility, $z_{MA} = \infty$ (that is, $Z_{MA} = 0$) and then measuring the velocity of the diaphragm u_0. As we discussed in Part II, $Z_{MA} = 0$ can be obtained by acoustically connecting the diaphragm to a tube whose length is equal to one-fourth wavelength. This is possible at low frequencies.

Inspection of Fig. 3.42b shows that

$$u_0 = \frac{eBl}{j\omega M_{MD}R_E + (Bl)^2} \quad (3.39)$$

2. Short-circuit the generator e without changing the mesh impedance in that part of the electrical circuit. Then determine the mobility z_{MS} looking back into the out-

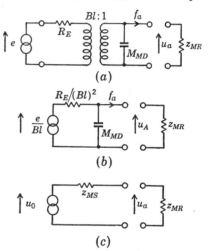

Fig. 3.42. Analogous circuits for a simplified moving-coil loudspeaker radiating sound into air. (a) Analogous circuit. (b) Same, with transformer removed. (c) Same, reduced to its Thévenin's equivalent.

put terminals of the loudspeaker. For example, z_{MS} for the circuit of Fig. 3.42b is equal to the parallel combination of $1/j\omega M_{MD}$ and $R_E/(Bl)^2$, that is,

$$z_{MS} = \frac{R_E}{j\omega M_{MD}R_E + (Bl)^2} \quad (3.40)$$

The Thévenin's equivalent circuit for the loudspeaker (looking into the diaphragm) is shown schematically in Fig. 3.42c, where u_0 and the mobility z_{MS} are given by Eqs. (3.39) and (3.40), respectively.

The application of Thévenin's theorem as discussed above is an example of how general theorems originally applying to linear passive electrical networks can be applied to great advantage to the analogs of mechanical and acoustic systems including transducers.

3.10. Energy and Power Relations. *Electrical Elements.* The expressions for energy stored in the inductance and capacitance elements of electrical circuits are familiar to most students from fundamental studies in physics and circuit theory. They are

$$\text{Instantaneous magnetic stored energy} = T(t) = \tfrac{1}{2}L[i(t)^2] \qquad (3.41)$$
$$\text{Instantaneous electric stored energy} = V(t) = \tfrac{1}{2}C[e(t)^2] \qquad (3.42)$$

where $i(t)$ and $e(t)$ are, respectively, the instantaneous current in the inductance and the instantaneous voltage across the capacitance.

In the steady state for an angular frequency ω, these equations become

$$T(t) = \tfrac{1}{2}L|i|^2(1 + \cos 2\omega t) \qquad (3.43)$$
$$V(t) = \tfrac{1}{2}C|e|^2(1 + \cos 2\omega t) \qquad (3.44)$$

where $|i|$ and $|e|$ are the rms magnitude of the complex current through and voltage across the elements L and C, respectively. The quantities T and V have a constant component, and a component which alternates at double frequency. On the average, the alternating component is zero, so that

$$T_{\text{avg}} = \tfrac{1}{2}L|i|^2 \qquad \text{watt-sec} \qquad (3.45)$$

and

$$V_{\text{avg}} = \tfrac{1}{2}C|e|^2 \qquad \text{watt-sec} \qquad (3.46)$$

On the average, power is dissipated only in resistive elements. The instantaneous power $W(t)$ dissipated in a resistance R by a sinusoidal current is

$$W(t) = R|i|^2(1 + \cos 2\omega t) \qquad (3.47)$$

On the average, the alternating component is zero, so that

$$W_{\text{avg}} = R|i|^2 \qquad \text{watts} \qquad (3.48)$$

Mechanical and Acoustical Elements. From basic mechanics, the student has also become acquainted with the energy and power relations for the mechanical elements: mass, compliance, and resistance. A summary of the steady-state values of W_{avg}, T_{avg}, and V_{avg} for electrical, mechanical, and acoustical elements is given in Table 3.4. It is interesting to see that T_{avg} for the mobility-type analogy has the same form as V_{avg} for the impedance-type analogy, and vice versa. We are now ready to use these energy and power concepts in the analysis of complete electro-mechano-acoustical circuits.

Energy and Power Functions for Combined Circuits. In a complete electro-mechano-acoustical circuit for a device such as a loudspeaker, the energy and power relations for the individual elements may be used to determine the total active and reactive power supplied by the generator.

TABLE 3.4.   Average Values of Power Dissipated and Energy Stored†

Element (see Table 3.2)		Electrical	Type of analogy	Mechanical	Acoustical
W_{avg}	⟿	$R\lvert i\rvert^2$	Mobility	$r_M\lvert f\rvert^2$	$r_A\lvert p\rvert^2$
			Impedance	$R_M\lvert u\rvert^2$	$R_A\lvert U\rvert^2$
T_{avg}	⟾⟾⟾	$\tfrac{1}{2}L\lvert i\rvert^2$	Mobility	$\tfrac{1}{2}C_M\lvert f\rvert^2$	$\tfrac{1}{2}C_A\lvert p\rvert^2$
			Impedance	$\tfrac{1}{2}M_M\lvert u\rvert^2$	$\tfrac{1}{2}M_A\lvert U\rvert^2$
V_{avg}	⊣⊢	$\tfrac{1}{2}C_E\lvert e\rvert^2$	Mobility	$\tfrac{1}{2}M_M\lvert u\rvert^2$	$\tfrac{1}{2}M_A\lvert U\rvert^2$
			Impedance	$\tfrac{1}{2}C_M\lvert f\rvert^2$	$\tfrac{1}{2}C_A\lvert p\rvert^2$

† Rms values are used for time-varying quantities.

As an example, let us consider in detail the loudspeaker circuit shown in Fig. 3.43. The transducer is of the electromagnetic-mechanical type, which requires the use of the mobility analogy in the mechanical and therefore in the acoustical parts of the circuit.

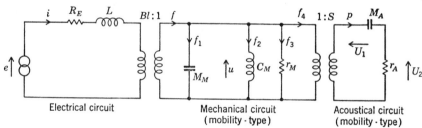

FIG. 3.43. Complete analogous circuit for a direct-radiator loudspeaker of the moving-coil type.

Logically, the total power supplied by the generator e to the circuit must be equal to the sums of the powers dissipated in the resistive elements because the circuit is passive. The power delivered by the loudspeaker to the air is given by the power dissipated in the acoustic responsiveness r_A. From Table 3.4 we find for the mobility case that the total power for the circuit of Fig. 3.43 is equal to

$$W_{avg} = R_E\lvert i\rvert^2 + r_M\lvert f_3\rvert^2 + r_A\lvert p\rvert^2 \tag{3.49}$$

We know that the total amount of inductive energy T_{avg} stored in the circuit must be the sum of the energies stored in the individual inductance-type elements. From Table 3.4 we see for the mobility analogy that the total T_{avg} of Fig. 3.43 is equal to

$$T_{avg} = \tfrac{1}{2}L\lvert i\rvert^2 + \tfrac{1}{2}C_M\lvert f_2\rvert^2 \tag{3.50}$$

Similarly, the capacitive energy V_{avg} stored in the capacitative-type elements is found to be

$$V_{avg} = \frac{1}{2} M_M |u|^2 + \frac{1}{2} M_A |U_1|^2 \tag{3.51}$$

By way of justification for the choice of mks units made early in this text, it should be noted that direct addition of electrical, mechanical, and acoustic powers and energies is possible in Eqs. (3.49) to (3.51) provided all quantities are given in the mks system. By contrast, if volts and amperes were used in the electrical circuit and cgs units in the mechanical and acoustical circuits, energy would be in joules in the electrical circuit and in ergs in the mechanical and acoustical circuits. Conversion of ergs to joules by multiplying the number of ergs by 10^{-7} would be necessary before addition were possible.

Active and Reactive Power: Vector Power. It is common in electrical-circuit theory to speak of vector power: the quadrature combination of active and reactive power. By definition the vector power supplied by the generator is the product of the current and the complex conjugate of the voltage at the source.

$$\text{Vector power} = W_{avg} + jQ_{avg} = e^*i \tag{3.52}$$

where e^* is the complex conjugate of the rms voltage e at the source and i is the rms current at the source. Note that if $e = |e| \angle \phi$, then $e^* = |e| \angle -\phi$.

Since $i = e/Z$, one may alternatively write

$$W_{avg} + jQ_{avg} = \frac{|e|^2}{Z} \tag{3.53}$$

where $|e|$ is the rms magnitude of the voltage across the complex impedance Z. Clearly the active power W_{avg} in a linear passive circuit must always be positive. The reactive power, however, may have either sign, being *positive in a predominantly capacitative-type circuit and negative in a predominantly inductive one.* The units of W_{avg} are watts, and the units of Q_{avg} are called "vars"—a contraction of "volt-ampere reactive."

Instead of Eq. (3.53) it is also possible to write

$$W_{avg} + jQ_{avg} = |i|^2 Z^* \tag{3.54}$$

where Z^* is the complex conjugate of Z. For example, if $Z = R + jX$, then $Z^* = R - jX$.

In a complex circuit, such as that given in Fig. 3.43, the total vector power supplied by the generator is found by summing the vector powers supplied to each element in the circuit. For a purely electrical circuit, we write

$$W_{avg} + jQ_{avg} = \sum e_k^* i_k = \sum |i_k|^2 Z_k^* = \sum \frac{|e_k|^2}{Z_k} \tag{3.55}$$

where i_k is the complex current in the kth element of impedance Z_k. The voltage across the kth element is e_k. A similar equation holds for electro-mechano-acoustical circuits. For example, from Fig. 3.43, remembering that we have the mobility analogy, we find that

$$W_{\text{avg}} + jQ_{\text{avg}} = |i|^2(R_E - jL\omega) - \frac{|f_1|^2}{jM_M\omega} - |f_2|^2 jC_M\omega$$
$$+ |f_3|^2 r_M + |p|^2 \left(\frac{j}{M_A\omega} + r_A \right)$$

or

$$W_{\text{avg}} + jQ_{\text{avg}} = (|i|^2 R_E + |f_3|^2 r_M + |p|^2 r_A)$$
$$+ 2j\omega \left(-\frac{1}{2} L|i|^2 + \frac{1}{2} \frac{|f_1|^2}{M_M\omega^2} - \frac{1}{2} |f_2|^2 C_M + \frac{1}{2} \frac{|p|^2}{M_A\omega^2} \right) \quad (3.56)$$

We see that the quantities contained within the first parentheses are exactly the same as those found in Eq. (3.49). Consider the terms in the second parentheses. From Eq. (3.2) we see that $|f_1|^2 = \omega^2 M_M{}^2 |u|^2$ and from Eq. (3.22) that $|p|^2 = \omega^2 M_A{}^2 |U_1|^2$. Hence, comparison of the second parentheses of Eq. (3.56) with Eqs. (3.50) and (3.51) shows that the value in these second parentheses is equal to $V_{\text{avg}} - T_{\text{avg}}$.

In general, the reactive power is

$$Q_{\text{avg}} = 2\omega(V_{\text{avg}} - T_{\text{avg}}) \quad (3.57)$$

The total stored energies V_{avg} and T_{avg} are found by the procedures described in connection with Eqs. (3.50) and (3.51). It should be noted that *at resonance* the average energy stored in the inductance-type elements T_{avg} is just equal to the average energy stored in the capacitance-type elements V_{avg} so that the total reactive power supplied by the generator is zero.

Finally, it is seen from Eqs. (3.54) and (3.57) that the impedance presented to the generator is equal to

$$Z = \frac{W_{\text{avg}} - jQ_{\text{avg}}}{|i|^2} = \frac{W_{\text{avg}} + j2\omega(T_{\text{avg}} - V_{\text{avg}})}{|i|^2} \quad (3.58)$$

3.11. Transducer Impedances. Let us look a little closer at the impedances at the terminals of electromechanical transducers. It has become popular in recent years for electrical-circuit specialists to express the equations for their circuits in matrix form. The matrix notation is a condensed manner of writing systems of linear equations.[7] We shall express the properties of transducers in matrix form for those who are familiar with this concept. An explanation of the various mathematical operations to be performed with matrices is beyond the scope of this book.

[7] P. LeCorbeiller, "Matrix Analysis of Electric Networks," Harvard University Press, Cambridge, Mass., and John Wiley & Sons, Inc., New York, 1950.

The student not familiar with matrix theory is advised to deal directly with the simultaneous equations from which the matrix is derived. A knowledge of matrix theory is not necessary, however, for an understanding of any material in this text.

Let us determine the impedance matrix for the electromagnetic-mechanical transducer of Fig. 3.44a. In that circuit Z_E is the electrical impedance measured with the mechanical terminals "blocked," that is, $u = 0$; z_M is the mechanical mobility of the mechanical elements in the transducer measured with the electrical circuit "open-circuited"; and z_L is the mechanical mobility of the acoustic load on the diaphragm. The

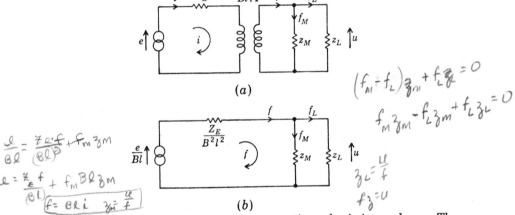

FIG. 3.44. Analogous circuits for an electromagnetic-mechanical transducer. The mechanical side is of the mobility type.

quantity Bl is the product of the flux density times the effective length of the wire cutting the lines of force perpendicularly.

Removing the transformer from Fig. 3.44a yields the two-mesh circuit of Fig. 3.44b. The equations for this circuit are

$$e = iZ_E + uBl \tag{3.59}$$
$$0 = -iBl + u(Z_M + Z_L) \tag{3.60}$$

where $Z_M = 1/z_M$, $Z_L = 1/z_L$, and $f = Bli$. The impedance matrix is drawn from these equations and is a square array of four impedances.

$$Z = \begin{bmatrix} Z_E & Bl \\ -Bl & Z_M + Z_L \end{bmatrix} \tag{3.61}$$

The total electrical impedance Z_{ET} as viewed from the voltage generator is found from the matrix or from Eqs. (3.59) and (3.60) to be

$$Z_{ET} = Z_E + \frac{(Bl)^2}{Z_M + Z_L} \tag{3.62}$$

The second term on the right-hand side is usually called the *motional impedance* because, if the mechanical side is blocked so there is no movement, then $Z_{ET} = Z_E$. This equation illustrates a striking fact, *viz.*, that the electromagnetic transducer is an *impedance* inverter. By an inverter we mean that a mass reactance on the mechanical side becomes a capacitance reactance when referred to the electrical side of the transformer, and vice versa. Similarly, an inductance on the electrical side reflects through the transformer as a mechanical compliance. These statements are well illustrated by the circuit of Fig. 3.47.

For an electrostatic-mechanical transducer of the type shown in Fig. 3.45, the circuit equations are

$$e_0 = iZ_E - u\frac{\tau}{j\omega} \tag{3.63}$$

$$0 = -i\frac{\tau}{j\omega} + u(Z_M + Z_L) \tag{3.64}$$

where

$Z_E \equiv Z'_E + \dfrac{1}{j\omega C'_E}$

    = the electrical impedance with the mechanical motion blocked

Z_L = mechanical impedance of the acoustical load on the diaphragm

$Z_M \equiv R_M + j\omega M_M + \dfrac{1}{j\omega C_M}$ = mechanical impedance of the mechanical

    elements in the transducer measured with $i = 0$

$C_M = \dfrac{C'_M}{1 + C'_M C'_E \tau^2}$ = mechanical compliance in the transducer with $i = 0$

The impedance matrix is

$$Z = \begin{bmatrix} Z_E & -\tau/j\omega \\ -\tau/j\omega & Z_M + Z_L \end{bmatrix} \tag{3.65}$$

This matrix is symmetrical about the main diagonal, as for any ordinary electrical passive network. By contrast matrix (3.61) is skew-symmetrical. For transient problems, replace $j\omega$ by the operator $s = d/dt$.[6]

The impedance matrix for the electrostatic transducer is almost identical in form to that for the electromagnetic transducer, the difference being that the mutual terms have the same sign, as contrasted to opposite signs for the electromagnetic case.

For the electrostatic transducer the total impedance is

$$Z_{ET} = Z_E - \frac{\tau^2/j^2\omega^2}{Z_M + Z_L} = Z_E + \frac{\tau^2/\omega^2}{Z_M + Z_L} \tag{3.66}$$

The second term on the right-hand side is called the motional impedance as before.

Again we see that the transducer acts as a sort of *impedance* inverter. An added positive mechanical reactance $(+X_M)$ comes through the transducer as a negative electrical reactance.

Some interesting facts can be illustrated by assuming that we have an electrostatic and an electromagnetic transducer, each stiffness controlled on the mechanical side so that

$$Z_M + Z_L = \frac{1}{j\omega C_{M1}} \tag{3.67}$$

Substitution of Eq. (3.67) into (3.62) yields

$$Z_{ET} = Z_E + j\omega(B^2 l^2 C_{M1}) \tag{3.68}$$

The mechanical compliance C_M appears from the electrical side to be an inductance with a magnitude $B^2 l^2 C_{M1}$. Substitute now Eq. (3.67) into (3.66).

$$Z_{ET} = Z_E + j\frac{\tau^2 C_{M1}}{\omega} \tag{3.69}$$

The mechanical compliance C_M of this transducer appears from the electrical side to be a negative capacitance, that is to say, C_{M1} appears to be

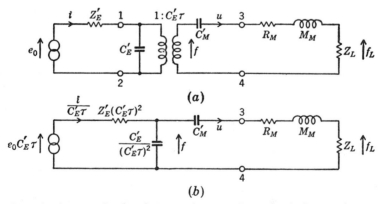

Fig. 3.45. Analogous circuits for an electrostatic-mechanical transducer. The mechanical side is of the impedance type.

an inductance with a magnitude that varies inversely with ω^2. The effect of this is simply to reduce the value of C_M. Another way of looking at this is to note from Fig. 3.45 that with $R_M = M_M = 0$ and $Z_L = 1/j\omega C_{ML}$, the total compliance is less than C_M because of the added compliance C_{ML}.

3.12. Combinations of Electrostatic and Electromagnetic Transducers. The engineer sometimes is called upon to join an electromagnetic and an electrostatic transducer together both electrically and mechanically, say, by a feedback loop.   Inspection of Figs. 3.44a and 3.45a reveals that it is not possible to make such a connection directly, because we have the mobility-type analogy on the secondary side in one circuit and the impedance-type analogy on that side in the other.

One method has been advanced for overcoming this difficulty.[8]   Inspection of the impedance matrices [Eqs. (3.61) and (3.65)] reveals that the same circuit could be used for both types of devices provided some means were introduced for changing one of the signs of the mutual terms.   To do this, the "β operator" will be introduced.

The Operator β.   The method for transforming one of the impedance matrices into the other is to multiply the mutual terms of one matrix by an arbitrary operator β.   This operator is a 90° "direction rotator," and

$$\beta^2 = -1$$
$$\beta^4 = 1 \qquad\qquad (3.70)$$
$$j\beta = -\beta j$$

For example, the circuit for an electrostatic transducer can be drawn like that shown in Fig. 3.46.   The impedance matrix for that circuit is

$$Z = \begin{bmatrix} Z_E & \tau\beta/j\omega \\ -(\tau\beta/j\omega) & Z_M + Z_L \end{bmatrix} \qquad (3.71)$$

The total impedance of that circuit, with the application of Eq. (3.71) above, is that given in Eq. (3.66).   As a word of caution, the quantity β^2

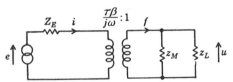

FIG. 3.46. Analogous circuit of the mobility type, using the β operator, for handling electrostatic-mechanical transducers.   For use with transient problems, replace $j\omega$ by the operator s.

should not be replaced by (-1) until the equations have been reduced to their final form.   This avoids the problem of having to decide whether a negative number should be replaced by j or β when a square root is taken. In Fig. 3.46, going from left to right, $u = iz\ \tau/j\omega/\beta$, and, going from right to left, the open-circuit voltage $e_0 = f_0\tau z/j\omega/-\beta$, so that the transfer impedances in the two directions are the negative of each other.

When both β and j appear in the transformation ratio, the values of the stored energy components T_{avg} and V_{avg} are the same as those found when

---

[8] F. V. Hunt, "Symmetry in the Equations for Electromechanical Coupling," *Paper* B1, presented at the thirty-ninth meeting of the Acoustical Society of America.   Professor Hunt has used this concept of a β operator in a variety of useful ways.

neither appear in the transformation ratio, because

$$j^2\beta^2 = \frac{j^2}{\beta^2} = 1 \tag{3.72}$$

Analytically the results for power and energy are the same as were found in connection with Eqs. (3.50), (3.51), (3.56), and (3.57).

In a similar manner, the circuit for an electromagnetic transducer may be drawn like that of Fig. 3.37b if the transducer ratio is $jBl\beta C_M\omega : 1$. In this case, the impedance matrix would be

$$Z = \begin{bmatrix} Z_E & -Bl\beta \\ -Bl\beta & Z_M + Z_L \end{bmatrix} \tag{3.73}$$

The total electrical impedance is found from this matrix and is the same as that given by Eq. (3.62).

Example 3.9. A moving-coil earphone actuated by frequencies above its first resonance frequency may be represented by the circuit of Fig. 3.42a. Its mechanical and electrical characteristics are

$R_E = 10$ ohms
$B = 10^4$ gauss (1 weber/m²)
$l = 3$ m
$M_{MD} = 2$ g
$z_{MR} = j\omega 2.7 \times 10^{-4}$ m/newton-sec

where z_{MR} is the mobility that the diaphragm sees when the earphone is on the ear, M_{MD} is the mass of the diaphragm, R_E and l are the resistance and the length of wire

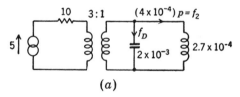

(a)

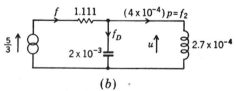

(b)

FIG. 3.47. Analogous circuits for Example 3.9.

wound on the voice coil, and B is the flux density cut by the moving coil. Determine the sound pressure level produced at the ear at 1000 cps when the earphone is operated from a very low impedance amplifier with an output voltage of 5 volts. Assume that the area of the diaphragm is 4 cm².

Solution. The circuit diagram for the earphone with the element sizes given in mks units is shown in Fig. 3.47a. Eliminating the transformer gives the circuit of

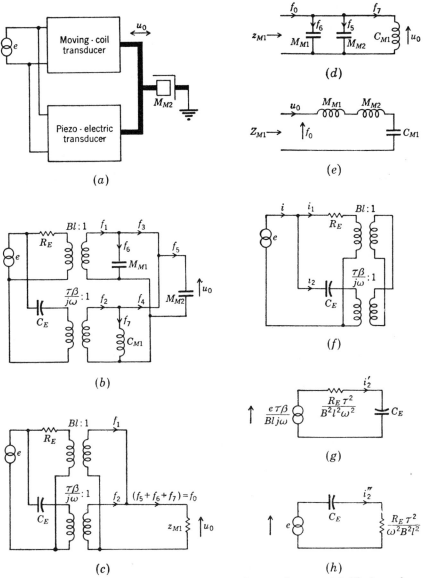

FIG. 3.48. Combined electrostatic-electromagnetic transducers. (*a*) Block mechanical diagram of the device. (*b*) Analogous circuit with mobilities on mechanical side. The β operator is used for the piezoelectric transducer. (*c*) Same as (*b*), except that z_{M1} replaces the three parallel mobilities as shown by (*d*). (*e*) Dual of (*d*). (*f*) Because the circuit of (*d*) has infinite mobility, (*b*) simplifies to this form. (*g*) and (*h*) Solution of (*f*) by superposition.

Fig. 3.47b.   Solving, we get

$$u = f_2 z_{MR} = (4 \times 10^{-4}p)j6280(2.7 \times 10^{-4})$$
$$= j6.78 \times 10^{-4}p$$
$$f_D = j\omega M_{MD}u = -8.54 \times 10^{-3}p$$
$$f = f_D + f_2 = -8.14 \times 10^{-3}p$$
$$\tfrac{5}{3} = u + 1.111f = p(j6.78 \times 10^{-4} - 9 \times 10^{-3})$$
$$|p| = \frac{1.667 \times 10^2}{0.9} \doteq 185 \text{ newton/m}^2$$

$$\text{SPL} = 20 \log \frac{185}{2 \times 10^{-5}} = 139.3 \text{ db } re \text{ } 0.0002 \text{ microbar}$$

Example 3.10.   Two transducers, one a piezoelectric crystal and the other a moving coil in a magnetic field, are connected to a mass M_{M2} of 1.533 kg as shown in Fig. 3.48a. Determine the stored electrical energy in the condenser C_E at 100 cps for the following constants:

$$e = 1 \text{ volt}$$
$$R_E = 10 \text{ ohms}$$
$$B = 1 \text{ weber/m}^2$$
$$l = 30 \text{ m}$$
$$C_E = 4 \times 10^{-9} \text{ farad}$$
$$M_{M1} = 1.0 \text{ kg}$$
$$C_{M1} = 10^{-6} \text{ m/newton}$$
$$\tau = 1.28 \times 10^7$$
$$\omega = 628 \text{ radians/sec}$$

Solution.   The transducers are shown schematically in (b) of Fig. 3.48.   A further simplification of this diagram is shown in (c).   Let us determine the value of z_{M1} first. We note that the dual of (d) is given by (e).

$$\frac{1}{z_{M1}} = Z_{M1} = j\omega(M_{M1} + M_{M2}) - j\frac{1}{C_M\omega}$$
$$= j(629 + 964) - j1593 = 0$$

In other words, the mobility is infinite at 100 cps.   Hence, circuit (c) simplifies to that shown in (f).   By superposition, i_2 can be broken into two parts i_2' and i_2'' given by the two circuits (g) and (h), so that $i_2 = i_2'' - i_2'$.

$$\frac{\tau}{Bl\omega} = \frac{1.28 \times 10^7}{1 \times 30 \times 628} = 680$$
$$\frac{1}{\omega C_E} = 4 \times 10^5$$
$$i_2' = \frac{e\tau\beta/Blj\omega}{(R_{ET}^2/B^2l^2\omega^2) + (1/j\omega C_E)} = \frac{eBl\omega}{jR_{ET}} = \frac{1}{j6800}$$
$$i_2'' = \frac{e\omega^2 B^2l^2}{R_{ET}^2} = \frac{1}{(10)(680)^2} = \frac{1}{4.64 \times 10^6}$$
$$i_2 = i_2'' - i_2' = j1.47 \times 10^{-4}$$
$$|i_2| = 1.47 \times 10^{-4} \text{ amp}$$

The voltage drop across the capacitor is

$$|e_c| = 1.47 \times 10^{-4} \times 4 \times 10^5 = 59 \text{ volts}$$

The electric stored energy on the capacitor C_E is

$$\tfrac{1}{2}C_E|e_c|^2 = 2 \times 10^{-9} \times 3.48 \times 10^3 = 7 \times 10^{-6} \text{ watt-sec}$$

CHAPTER 4

RADIATION OF SOUND

In order fully to specify a source of sound, we need to know, in addition to other properties, its directivity characteristics at all frequencies of interest. Some sources are nondirective, that is to say, they radiate sound equally in all directions and as such are called spherical radiators. Others may be highly directional, either because their size is naturally large compared to a wavelength or because of special design.

The most elementary radiator of sound is a spherical source whose radius is small compared to one-sixth of a wavelength. Such a radiator is called a *simple source* or a *point source*. Its properties are specified by the magnitude of the velocity of its surface and by its phase relative to some reference. More complicated sources such as plane or curved radiators may be treated analytically as a combination of simple sources, each with its own surface velocity and phase.

A particularly important consideration in the design of loudspeakers and horns is their directivity characteristics. This chapter serves as an important basis for later chapters dealing with loudspeakers, baffles, horns, and noise sources.

The basic concepts governing radiation of sound must be grasped thoroughly at the outset. It is then possible to reason from those concepts in deducing the performance of any particular equipment or in planning new systems. Examples of measured radiation patterns for common loudspeakers are given here as evidence of the applicability of the basic concepts.

PART X *Directivity Patterns*

The *directivity pattern* of a transducer used for the emission or for reception of sound is a description, usually presented graphically, of the response of the transducer as a function of the direction of the transmitted or incident sound waves in a specified plane and at a specified frequency.

The *beam width* of a directivity pattern is used in this text as the angular distance between the two points on either side of the principal axis where the sound pressure level is down 6 db from its value at $\theta = 0$.

4.1. Spherical Sources.[1,2] A spherical source is the simplest to consider because it radiates sound uniformly in all directions. As we saw from Eq. (2.62), the sound pressure at a point a distance r in any direction from the center of a spherical source of any radius in free space is equal to

$$p(r,t) = \frac{\sqrt{2}\,A_+}{r}\,e^{j(\omega t - kr)} \tag{4.1}$$

where A_+ is the magnitude of rms sound pressure at unit distance from the center of the sphere.

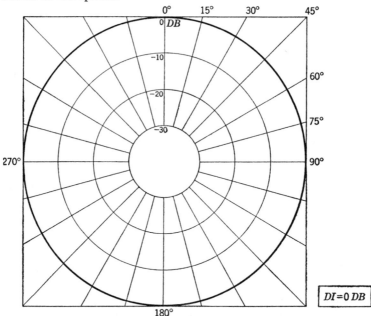

Fig. 4.1. Directivity pattern for a nondirectional source. Such a pattern is drawn on a particular plane intersecting the center of the source. The directivity index DI (defined in Part XI) equals 0 db at all angles.

Directivity Pattern. On a polar diagram, the directivity pattern on any plane surface intersecting the center of such a spherical source is given in Fig. 4.1. It is obviously a nondirectional source.

Simple Source (Point Source).[3] For the special case of a very small source, whose radius a is small compared with one-sixth wavelength (that

[1] P. M. Morse, "Vibration and Sound," 2d ed., pp. 311–326, McGraw-Hill Book Company, Inc., New York, 1948.

[2] L. E. Kinsler and A. R. Frey, "Fundamentals of Acoustics," pp. 163–173, John Wiley & Sons, Inc., New York, 1950.

[3] Morse, *op. cit.*, pp. 312–313.

is, $ka \ll 1$), the velocity at the surface of the sphere is [see Eq. (2.63)]

$$u(a,t) \doteq \frac{\sqrt{2}\,A_+}{\rho_0 ca}\; \frac{c}{j2\pi fa}\; e^{j(\omega t - ka)} \qquad (4.2)$$

A source for which this formula is valid is called a *simple source*.

Substitution of Eq. (4.2) into Eq. (4.1) yields

$$p \doteq j\,\frac{U_0 f \rho_0}{2r}\; e^{-jk(r-a)} \qquad (4.3)$$

where U_0 = rms volume velocity in cubic meters per second of the very small source and is equal to $(4\pi a^2) u_{rms}$

p = rms sound pressure in newtons per square meter at a distance r from the simple source

Strength of a Simple Source.[3] The rms magnitude of the total air flow at the surface of a simple source in cubic meters per second (or cubic centimeters per second in the cgs system) is given by U_0 and is called the *strength of a simple source.*†

Intensity at Distance r. At a distance r from the center of a simple source the intensity is given by

$$I = \frac{|p|^2}{\rho_0 c} = \frac{U_0^2 f^2 \rho_0}{4r^2 c} \qquad \text{watts/m}^2 \qquad (4.4)$$

When the dimensions of a source are *much smaller* than a wavelength, the radiation from it will be much the same no matter what shape the radiator has, as long as all parts of the radiator vibrate substantially in phase. The intensity at any distance is directly proportional to the square of the volume velocity and the frequency.

4.2. Combination of Simple Sources.[4] The basic principles governing the directivity patterns from loudspeakers can be learned by studying combinations of simple sources. This approach is very similar to the consideration of Huygens wavelets in optics. Basically, our problem is to add, vectorially, at the desired point in space, the sound pressures arriving at that point from all the simple sources. Let us see how this method of analysis is applied.

Two Simple Sources in Phase. The geometric situation is shown in Fig. 4.2. It is assumed that the distance r from the two point sources to the point A at which the pressure p is being measured is large compared with the separation b between the two sources.

The spherical sound wave arriving at the point p from source 1 will have

† In some texts the peak magnitude of the total air flow instead of the rms magnitude is used. In these texts, the "strength of a simple source" is $\sqrt{2}\,(4\pi a^2) u_{rms}$.

[4] H. F. Olson, "Elements of Acoustical Engineering," 2d ed., pp. 31–34, D. Van Nostrand Company, Inc., New York, 1947.

traveled a distance $r - (b/2) \sin \theta$, and the sound pressure will be

$$p_1(r_1;t) = \frac{\sqrt{2}\,A_+}{r}\,e^{j\omega t}e^{-j(2\pi/\lambda)\,[r-(b/2)\sin\theta]} \qquad (4.5a)$$

The wave from source 2 will have traveled a distance $r + (b/2) \sin \theta$, so that

$$p_2(r_2;t) = \frac{\sqrt{2}\,A_+}{r}\,e^{j\omega t}e^{-j(2\pi/\lambda)\,[r+(b/2)\sin\theta]} \qquad (4.5b)$$

The sum of $p_1 + p_2$, assuming $r \gg b$, gives

$$p(r,t) = \frac{\sqrt{2}\,A_+}{r}\,e^{j\omega t}e^{-j(2\pi/\lambda)r}\big(e^{j(\pi b/\lambda)\sin\theta} + e^{-j(\pi b/\lambda)\sin\theta}\big) \qquad (4.6)$$

Multiplication of the numerator and the denominator of Eq. (4.6) by

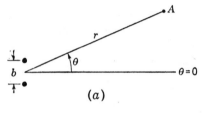

(a)

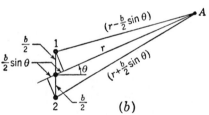

(b)

Fig. 4.2. Two simple (point) sources vibrating in phase located a distance b apart and at distance r and angle θ with respect to the point of measurement A.

$\exp\,(j\pi b \sin \theta/\lambda) - \exp\,(-j\pi b \sin \theta/\lambda)$ and replacement of the exponentials by sines, yields

$$p(r,t) = \frac{\sqrt{2}\,A_+}{r}\,e^{j\omega t}e^{-j(2\pi r/\lambda)}\,\frac{\sin\,[(2\pi b/\lambda)\sin\theta]}{\sin\,[(\pi b/\lambda)\sin\theta]} \qquad (4.7)$$

The equation for the magnitude of the rms sound pressure $|p|$ is

$$|p| = \frac{2A_+}{r}\left|\frac{\sin\,[(2\pi b/\lambda)\sin\theta]}{2\sin\,[(\pi b/\lambda)\sin\theta]}\right| \qquad (4.8)$$

The portion of this equation within the straight lines yields the directivity pattern.

Referring to Fig. 4.2, we see that if b is very small compared with a wavelength, the two sources essentially coalesce and the pressure at a distance r at any angle θ is double that for one source acting alone. The directivity pattern will be that of Fig. 4.1.

As b gets larger, however, the pressures arriving from the two sources will be different in phase and the directivity pattern will not be a circle.

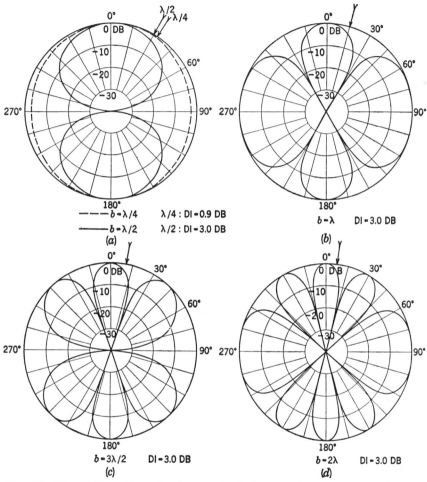

Fig. 4.3. Directivity patterns for the two simple in-phase sources of Fig. 4.2. Symmetry of the directivity patterns occurs about the axis passing through the two sources. Hence, only a single plane is necessary to describe the directivity characteristics at any particular frequency. The boxes give the directivity index at $\theta = 0°$. One angle of zero directivity index is also indicated. (The directivity index is discussed in Part XI.)

In other words, the sources will radiate sound in some directions better than in others. As a specific example, let $b = \lambda/2$. For $\theta = 0$ or $180°$ it is clear that the pressure arriving at a point A will be double that from either source. However, for $\theta = \pm 90°$ the time of travel between the two simple sources is just right so that the radiation from one source completely cancels the radiation from the other. Hence, the pressure at all points along the $\pm 90°$ axis is zero. Remember, we have limited our discussion to $r \gg b$.

Directivity patterns, expressed in decibels relative to the pressure at $\theta = 0$, are given in Fig. 4.3 for the two in-phase sources with $b = \lambda/4$; $\lambda/2$; λ; $3\lambda/2$; and 2λ.

A very important observation can be made from the directivity patterns for this simple type of radiator that applies to all types of radiation. The longer the extent of the radiator (*i.e.*, here, the greater b is), the sharper will be the principal lobe along the $\theta = 0$ axis at any given frequency and the greater the number of side lobes. As we shall see in the next paragraph, it is possible to suppress the side lobes, that is to say, those other than the principal lobes at 0 and 180°, by simply increasing the number of elements.

Linear Array of Simple Sources.[4] The geometric situation for this type of radiating array is shown in Fig. 4.4. The rms sound pressure produced at a point A by n identical simple in-phase sources, lying in a straight line, the sources a distance b apart and with the extent

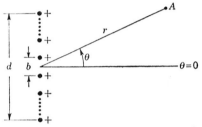

Fig. 4.4. A linear array of n simple sources, vibrating in phase, located a distance b apart. The center of the array is at distance r and angle θ with respect to the point of measurement A.

$d = (n - 1)b$ small compared with the distance r, is

$$p = \frac{nA_+}{r} \left| \frac{\sin\left[(n\pi b/\lambda)\sin\theta\right]}{n\sin\left[(\pi b/\lambda)\sin\theta\right]} \right| \qquad (4.9)$$

As a special case, let us assume that the number of points n becomes very large and that the separation b becomes very small. Then, as before,

$$d = (n - 1)b \doteq nb \qquad (4.10)$$

and

$$p = p_0 \left| \frac{\sin\left[(\pi d/\lambda)\sin\theta\right]}{(\pi d/\lambda)\sin\theta} \right| \qquad (4.11)$$

where p_0 is the magnitude of the rms sound pressure at a distance r from the array at an angle $\theta = 0$. As before, it is assumed that the extent of the array d is small compared with the distance r.

Plots of Eq. (4.9) for $n = 4$ and $d = \lambda/4$, $\lambda/2$, λ, $3\lambda/2$, and 2λ are shown in Fig. 4.5. Similar plots for $n \to \infty$ and $b \to 0$, that is, Eq. (4.11), are given in Fig. 4.6.

The principal difference among Figs. 4.3, 4.5, and 4.6 for a given ratio of array length to wavelength is in the suppression of the "side lobes."

That is, sound is radiated well in the $\theta = 0°$ and $\theta = 180°$ directions for all three arrays. However, as the array becomes longer and the number of elements becomes greater, the radiation becomes less in other directions than at $\theta = 0°$ and $\theta = 180°$.

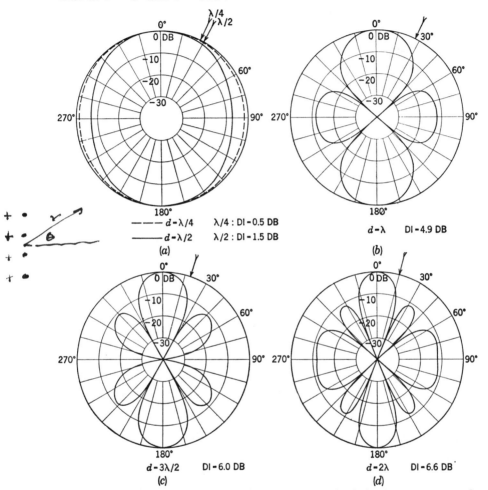

Fig. 4.5. Directivity patterns for a linear array of four simple in-phase sources evenly spaced over a length d. The boxes give the directivity index at $\theta = 0°$. One angle of zero directivity index is also indicated by the arrow.

Doublet Sound Source. A doublet sound source is a pair of simple sound sources, separated a very small distance b apart and vibrating in opposing phase. The geometric situation is shown in Fig. 4.7. The distance r to the point A is assumed to be large compared with the separation b between the two sources.

It can be clearly seen that the sound pressure at $\theta = 90°$ and $\theta = 270°$ will be zero, because the contribution at those points will be equal from the two sources and 180° out of phase. The pressures at $\theta = 0°$ and $\theta = 180°$ will depend upon the ratio of b to the wavelength λ. For example, if $b = \lambda$, we shall have zero sound pressure at those angles just as we did for $b = \lambda/2$ in the case of two in-phase sources. In the present case, we have a maximum pressure at $\theta = 0°$ and $\theta = 180°$ for $b = \lambda/2$.

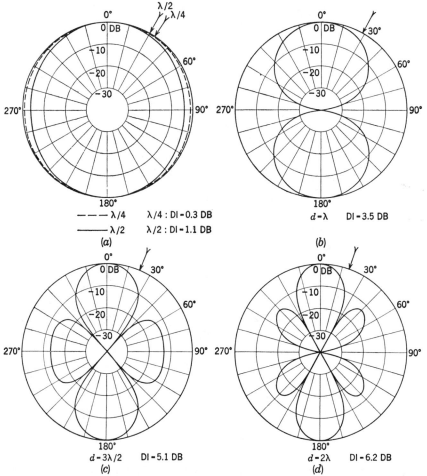

FIG. 4.6. Directivity patterns for a linear line array radiating uniformly along its length d. The boxes give the directivity index at $\theta = 0°$. One angle of zero directivity index is also indicated by the arrow.

The usual case of interest, however, is the one for

$$b \ll \lambda \tag{4.12}$$

In this case, the complex rms pressure p_d at a point A can be shown to equal[5]

[5] Kinsler and Frey, *op. cit.*, pp. 280–285.

$$p_d(r,\theta) = \frac{\rho_0 f U_0 b}{2r}\left(+k - j\frac{1}{r}\right)\cos\theta \, e^{-jkr} \qquad (4.13)$$

where U_0 = rms strength in cubic meters per second of each simple source.

The ratio of the complex rms sound pressure p_d produced by the doublet to the complex rms sound pressure p_s produced by a simple source is found by dividing Eq. (4.13) by Eq. (4.3). This division yields

$$\frac{p_d}{p_s} = -\frac{b}{r}(1 + jkr)\cos\theta \qquad (4.14)$$

When the square of the distance r from the acoustic doublet is large

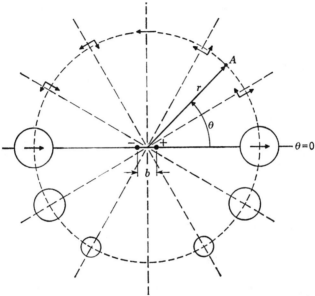

$\theta=0$

FIG. 4.7. Doublet sound source. This type of source consists of two simple (point) sources vibrating 180° out of phase. They are located a distance b apart and are at an angle θ and a distance r with respect to the point of measurement A. The lower half of the graph shows by the area of the circles the magnitude of the sound pressure as a function of angle θ. The upper half of the graph shows the variation of the radial and tangential components of the particle velocity as a function of angle θ.

compared with $\lambda^2/36$ ($k^2r^2 \gg 1$), Eq. (4.13) reduces to

$$p_d = \frac{\rho_0 \omega^2 U_0 b}{4\pi rc}\cos\theta \, e^{-jkr} \qquad (4.15)$$

For this case the pressure varies with θ as shown in Figs. 4.7 and 4.8. It changes inversely with distance r in exactly the same manner as for the simple source.

Near the acoustic doublet, for $r^2 \ll \lambda^2/36$, Eq. (4.13) reduces to

$$p_d = -\frac{\rho_0 f U_0 b}{2r^2}\cos\theta \, e^{i(\pi/2 - kr)} \qquad (4.16)$$

For this case, the pressure also varies with cos θ as shown in Fig. 4.8, but it changes inversely with the square of the distance r. We are still assuming that $r \gg b$.

Near-field and Far-field. The difference between *near-field* and *far-field* behaviors of sources must always be borne in mind. When the directivity pattern of a loudspeaker or some other sound source is presented in a technical publication, it is always understood that the data were taken at a distance r sufficiently large so that the sound pressure was decreasing linearly with distance along a radial line connecting with the source, as was the case for Eq. (4.15). This is the *far-field* case. For this to be true, two conditions usually have to be met. First, the extent b of the

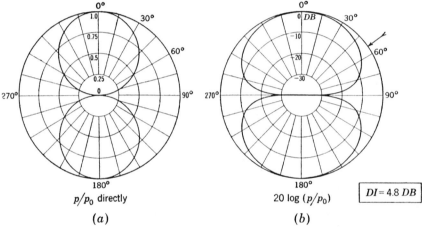

FIG. 4.8. Directivity pattern for a doublet sound source. (a) Sound-pressure ratio p/p_0 vs. θ. (b) $20 \log_{10} p/p_0$ vs. θ. The boxes give the directivity index at $\theta = 0°$. One angle of zero directivity index is also indicated by the arrow.

radiating array must be small compared with r, and r^2 must be large compared with $\lambda^2/36$. In acoustics the size factor indicated is usually taken to be larger than 3 to 10.

One more item is of interest in connection with the acoustic doublet. The particle velocity is composed of two components, one radially directed, and the other perpendicular to that direction. At $\theta = 0$ and 180° the particle velocity is directed radially entirely (see Fig. 4.7). At $\theta = 90$ and 270° the particle velocity is entirely perpendicular to the radial line. In between, the radial component varies as the cos θ and the perpendicular component as the sin θ.

An interesting fact is that at $\theta = 90°$ and 270° a doublet sound source appears to propagate a transversely polarized sound wave. To demonstrate this, take two unbaffled small loudspeakers into an anechoic chamber. Unbaffled loudspeakers (transducers) are equivalent to doublets because the pressure increases on one side of the diaphragm whenever it decreases on the other. Hold the two transducers about 0.5 m apart with *both* diaphragms facing the floor (not facing each other). Let one transducer radiate a low-frequency sound and the other act as a microphone connected to the input of an audio amplifier. As we see from Fig. 4.7, no sound pressure will be produced at the diaphragm of the microphone, but there will be transverse particle velocity. A particle velocity is always the result of a pressure gradient in the direction of the velocity. Therefore, the diaphragm of the microphone will be caused to move when the two transducers are held as described above. When one of the transducers is rotated through 90° about the axis joining the units, the diaphragm of the microphone will not move because the pressure gradient will be in the plane of the diaphragm. Hence, the sound wave appears to be plane polarized.

You have now learned the elementary principles governing the directional characteristics of sound sources. We shall be able to use these

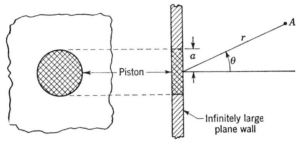

Fig. 4.9. Rigid circular piston in a rigid baffle. The point of measurement A is located at distance r and angle θ with respect to the center of the piston.

principles in understanding the measured or calculated behavior of some of the more complicated sound sources found in acoustics.

4.3. Plane Piston Sources. *Rigid Circular Piston in Infinite Baffle.* Many radiating sources can be represented by the simple concept of a vibrating piston located in an infinitely large rigid wall. The piston is assumed to be rigid so that all parts of its surface vibrate in phase and its velocity amplitude is independent of the mechanical or acoustic loading on its radiating surface. The rigid wall surrounding the piston is usually called a baffle, which, by definition, is a shielding structure or partition used to increase the effective length of the external transmission path between the front and back of the radiating surface.

The geometry of the problem is shown in Fig. 4.9. We wish to know the sound pressure at a point A located at a distance r and an angle θ from the center of the piston. To do this, we divide the surface of the piston into a number of small elements, each of which is a simple source vibrating in phase with all the other elements. The pressure at A is, then, the sum in magnitude and phase of the pressures from these elementary elements.

This summation appears in many texts[6] and, for the case of r large compared with the radius of the piston a, leads to the equation

$$p(r,t) = \frac{\sqrt{2}\, jf\rho_0 u_0 \pi a^2}{r} \left[\frac{2J_1(ka \sin \theta)}{ka \sin \theta} \right] e^{j\omega(t-r/c)} \qquad (4.17)$$

where u_0 = rms velocity of the piston

$J_1(\ )$ = Bessel function of the first order for cylindrical coordinates[6]

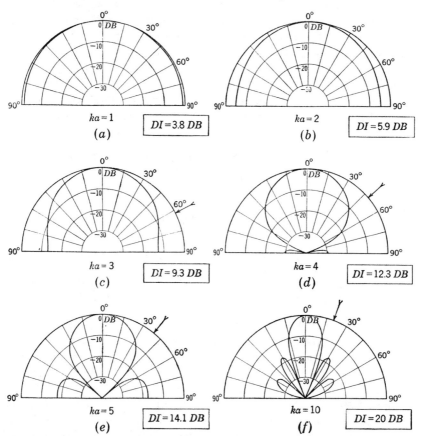

Fig. 4.10. Directivity patterns for a rigid circular piston in an infinite baffle as a function of $ka = 2\pi a/\lambda$, where a is the radius of the piston. The boxes give the directivity index at $\theta = 0°$. One angle of zero directivity index is also indicated. The DI never becomes less than 3 db because the piston radiates only into half-space.

The portion of Eq. (4.17) within the square brackets yields the directivity pattern and is plotted in decibels as a function of θ in Fig. 4.10 for six values of $ka = 2\pi a/\lambda$, that is, for six values of the ratio of the circumference of the piston to the wavelength.

When the circumference of the piston ($2\pi a$) is less than one-half wave-

[3] Morse, op. cit., pp. 326–346. A table of Bessel functions is given on page 444.

length, that is, $ka < 0.5$, the piston behaves essentially like a point source. When ka becomes greater than 3, the piston is highly directional.   We see from Fig. 4.24 that an ordinary loudspeaker also becomes quite directive at higher frequencies in much the same manner as does the vibrating piston.

Rigid Circular Piston in End of a Long Tube.[7]   In many instances, sound is radiated from a diaphragm whose rear side is shielded from the front side by a box or a tube.   If the box does not extend appreciably beyond the edges of the diaphragm, its performance may be estimated by comparison with that of a rigid piston placed in the end of a long tube.

The geometrical situation is shown in Fig. 4.11.   The pressure at point A is again found by summing the pressures from a number of small elements on the surface of the piston, each acting as a simple source.   The

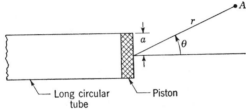

FIG. 4.11. Rigid circular piston in the end of a long tube.   The point of measurement is located at distance r and angle θ with respect to the center of the piston.

solution of this problem is complex, however, because radiation can take place in all directions and the sound must diffract around the edge of the tube to get to the left-hand part of space (Fig. 4.11).   Hence, a theory that includes the effects of diffraction must be used in solving the problem analytically.

The results from such a theory are shown in Fig. 4.12 for six values of ka.   It is assumed here also that the distance r is large compared with a, so that the directivity pattern applies to the far-field.

Rigid Circular Piston without Baffle.[8]   To complete the cases wherein pistons are commonly used, we present the results of theoretical studies on the directivity pattern of a rigid piston of radius a without any baffle, radiating into free space.   These results are shown graphically in Fig. 4.13 for four values of ka.   It is interesting to note the resemblance between these curves and those for an acoustic doublet.   In fact, to a first approximation, an unbaffled thin piston is simply a doublet, because an axial movement in one direction compresses the air on one side of it and causes a rarefaction of the air on the other side.

[7] H. Levine and J. Schwinger, On the Radiation of Sound from an Unflanged Circular Pipe, *Phys. Rev.*, **73**: 383–406 (Feb. 15, 1948).

[8] F. M. Wiener, On the Relation between the Sound Fields Radiated and Diffracted by Plane Obstacles, *J. Acoust. Soc. Amer.*, **23**: 697–700 (1951).

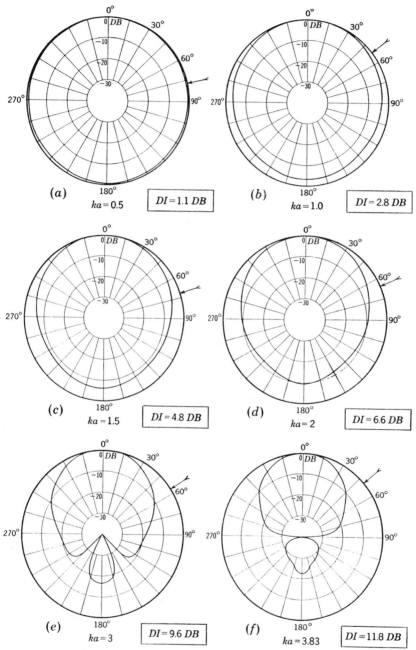

Fig. 4.12. Directivity patterns for a rigid circular piston in the end of a long tube as a function of $ka = 2\pi a/\lambda$, where a is the radius of the piston. The boxes give the directivity index at $\theta = 0°$. One angle of zero directivity index is also indicated.

Square or Rectangular Piston Sources.[9] RIGID SQUARE PISTON IN INFI-
NITE BAFFLE. This type of radiating source is not very common in
acoustics. It suffices to say here that the directivity pattern for such a
piston in a plane perpendicular to the piston and parallel to a side is
identical to that for a linear array of simple sources as given by Eq. (4.11).

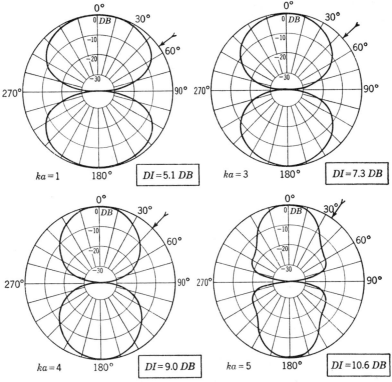

FIG. 4.13. Directivity patterns for an unbaffled rigid circular piston of radius a
located in free space at an angle θ a large distance r from the point of measurement A.
For $ka < 1$, the directivity pattern is the same as that for the doublet. The boxes
give the directivity index at $\theta = 0°$. One angle of zero directivity index is also
indicated by the arrow.

The pattern in a plane that is parallel to either of the two diagonals differs
very little from that in a plane parallel to a side.

RIGID RECTANGULAR PISTON IN INFINITE BAFFLE. The directivity patterns for this type of radiating source with dimensions d_1 and d_2 are given
by the formula

$$\text{Directivity pattern} = \left| \frac{\sin\left[(\pi d_1/\lambda)\sin\theta_1\right]}{(\pi d_1/\lambda)\sin\theta_1} \frac{\sin\left[(\pi d_2/\lambda)\sin\theta_2\right]}{(\pi d_2/\lambda)\sin\theta_2} \right| \qquad (4.18)$$

[9] Olson, *op. cit.*, pp. 39–40.

where θ_1 = angle between the normal to the surface of the piston and the projection of the line joining the middle of the surface and the observation point on the plane normal to the surface and parallel to d_1

θ_2 = same as θ_1, with d_2 substituted for d_1

Note that the directivity pattern is equal to the product of the directivity patterns for two line arrays at right angles to each other [see Eq. (4.11)].

4.4. Curved Sources.[10] In the preceding paragraphs of this part we have dealt with the radiation of sound from straight-line and plane-surface arrays. These types of arrays are closely resembled by open-ended organ pipes, direct-radiator loudspeakers, simple horns, and other devices. One characteristic common to all these arrays, except the doublet array, is that they become more directional as the ratio of their length to the wavelength becomes greater. This is usually an undesirable trait for a loudspeaker to exhibit because it means that the spectrum

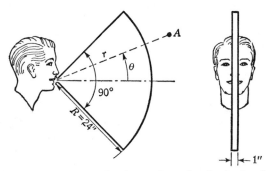

FIG. 4.14. Parabolic megaphone suitable for use by a cheerleader in a football stadium.

of music or speech as reproduced will vary from one position to another around the loudspeaker.

To overcome, in part, this increase in directivity with increasing frequency, curved surfaces are commonly employed as sound radiators. These surfaces can be made up of a number of small loudspeakers or small horns or as a megaphone with a curved front.

Curved-line Source (Parabolic Megaphone). An example of a curved-line source is the parabolic megaphone of Fig. 4.14. The megaphone opening is thin enough (1 in.) to be roughly equivalent to a simple line source for frequencies below 4000 cps. The horn is parabolic because the sectional area is proportional to the distance from the apex.

The sound pressure at a point A a distance r from the apex is found by summing the pressures in amplitude and phase originating from an assumed curved line composed of simple sources. When the distance r is

[10] *Ibid.*, pp. 40–47.

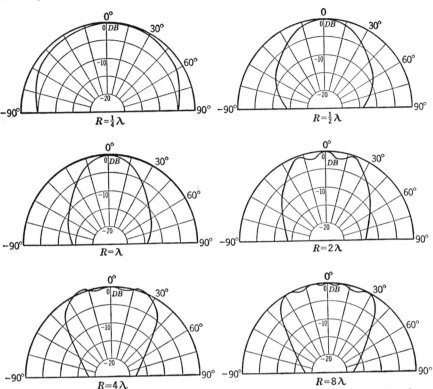

FIG. 4.15. Directivity patterns for the parabolic megaphone of Fig. 4.14 in the plane containing the arc of the opening.

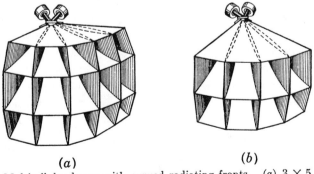

FIG. 4.16. Multicellular horns with curved radiating fronts. (a) $3 \times 5 = 15$ cells. (b) $2 \times 4 = 8$ cells.

large compared with the radius R of the horn, we obtain the directivity patterns shown in Fig. 4.15.

This simple case illustrates basic principles applicable to all curved radiating sources. At low frequencies the curved source is largely nondirectional. As the frequency is increased, the source becomes more directional, achieving its minimum angle of spread when the chord of the curved source is approximately equal to one wavelength. At high frequencies, the directivity pattern becomes broader, reaching its maximum width when it becomes equal to the width of the arc expressed in degrees, for example, 90° in Fig. 4.14.

Curved-surface Sources (Multicellular Horns). The most common curved-surface sources found in acoustics at the present time are multicellular horns. Two typical examples of such curved-surface horns are shown in Fig. 4.16.

The beam width of the directivity pattern was defined as the angular distance between the two points on either side of the principal axis where

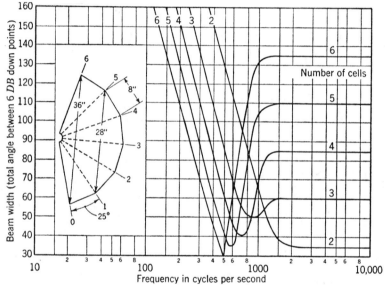

Fig. 4.17. Beam widths of multicellular horns constructed as shown in the insert and as sketched in Fig. 4.16.

the sound pressure level is down 6 db† from its value at $\theta = 0$. The beam widths of the directivity patterns for two, three, four, and five cell widths of multicellular horns were measured on commercial units and are shown in Fig. 4.17. These data are useful in the design of sound systems.

It should be noted that the minimum beam width occurs when the arc of the multicellular horn about equals λ. Also, at high frequencies the beam reaches a width of about $n \cdot 25° - 15°$ for the size of cell shown in Fig. 4.17. The theoretical directivity indexes for these maximum widths of beams are found from the nomogram in Fig. 4.22, which is based on the sketch of Fig. 4.21.

† No standard value has been chosen for the number of decibels down from the $\theta = 0$ value of the sound pressure level in determining beam width. Values of 3, 6, and 10 db are often encountered in the literature.

PART XI *Directivity Index and Directivity Factor*

Charts of the directivity patterns of sound sources are sufficient in many cases, such as when the source is located outdoors at a distance from reflecting surfaces. Indoors, it is necessary in addition to know something about the total power radiated in order to calculate the reinforcing effect of the reverberation in the room on the output of the sound source. A number is calculated at each frequency that tells the degree of directivity without the necessity for showing the entire directivity pattern. This number is the directivity factor or, when expressed in decibels, the directivity index.

4.5. Directivity Factor $[Q(f)]$. The directivity factor is the ratio of the intensity† on a designated axis of a sound radiator at a stated distance r to the intensity that would be produced at the same position by a point source if it were radiating the *same* total acoustic power as the radiator. Free space is assumed for the measurements. Usually, the designated axis is taken as the axis of maximum radiation, in which case $Q(f)$ always exceeds unity. In some cases, the directivity factor is desired for other axes where $Q(f)$ may assume any value equal to or greater than zero.

4.6. Directivity Index $[\mathrm{DI}(f)]$. The directivity index is 10 times the logarithm to the base 10 of the directivity factor.

$$\mathrm{DI}(f) = 10 \log_{10} Q(f) \tag{4.19}$$

4.7. Calculation of $Q(f)$ **and** $\mathrm{DI}(f)$. The intensity I at a point removed a distance r from the acoustical center of a source of sound located in free space is determined by first measuring the effective sound pressure p and letting $I = |p|^2/\rho_0 c$. If the source is a point source so that I is not a function of θ and is located in free space, the total acoustic power radiated is

$$W_p = 4\pi r^2 I$$

If the source is not a point source, the total acoustic power radiated is determined by summing the intensities over the surface of a sphere of radius r. That is, the total radiated power is

$$W = \frac{r^2}{\rho_0 c} \int_0^{2\pi} \int_0^{\pi} p^2(\theta,\phi,r) \sin\theta \, d\theta \, d\phi \tag{4.20}$$

where the coordinate of any point in space is given by the angles θ and ϕ and the radius r (see Fig. 4.18) and $p^2(\theta,\phi,r)$ equals the mean-square sound pressure at the point designated by θ, ϕ, and r.

---

† See the definition for intensity on page 11. The intensity equals the sound pressure squared, divided by $\rho_0 c$ for a plane wave in free space or for a spherical wave.

Usually an analytical expression for $p(\theta,\phi)$ does not exist. In practice, therefore, data are taken at the centers of a number of areas, approximately equal in magnitude, on the surface of a sphere of radius r surrounding the source. As an example, we show in Fig. 4.19a a spherical

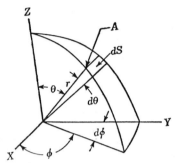

FIG. 4.18. Coordinate system defining the angle θ and ϕ and the length r of a line connecting a point A to the center of a sphere. The area of the incremental surface $dS = r^2 \sin\theta\, d\theta\, d\phi$.

Coordinates of mid-points of sectors			
Sector numbers	Coordinate		
	X	Y	Z
1-8	$\pm 1/\sqrt{3}$	$\pm 1/\sqrt{3}$	$\pm 1/\sqrt{3}$
9-12	± 0.934	± 0.357	0
13-16	0	± 0.934	± 0.357
17-20	± 0.357	0	± 0.934

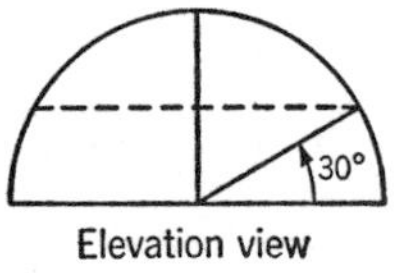

Elevation view

(b)

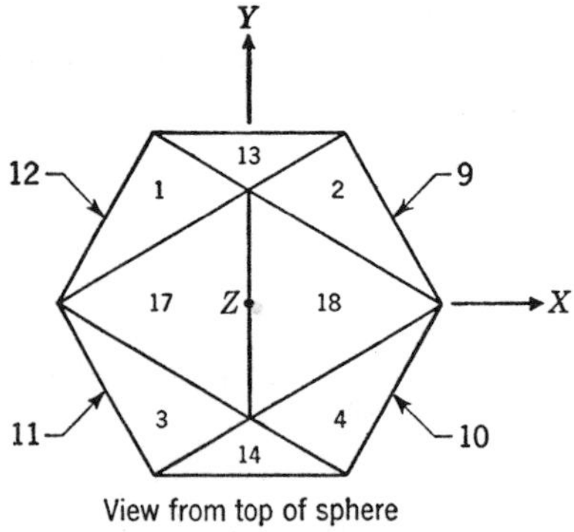

View from top of sphere

(a)

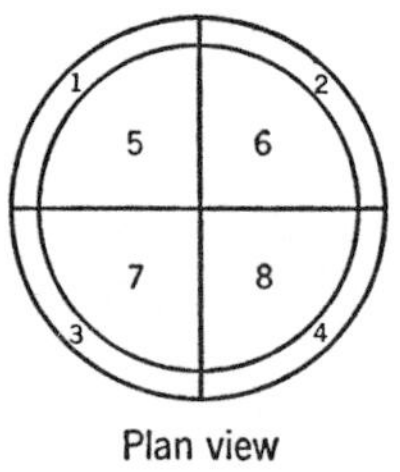

Plan view

(c)

FIG. 4.19. (a) Division of a spherical surface into 20 equal areas of identical shape. (b) and (c) Division of hemisphere into 8 parts of equal area but unequal shape.

surface divided into 20 equal parts of the same shape. The measured intensities on each of these parts may be called I_1, I_2, I_3, etc. The total power radiated W is found from

$$W = I_1 S_1 + I_2 S_2 + \cdots + I_{20} S_{20} \qquad (4.21)$$

where $S_1, S_2, \ldots, S_{20}$ are the areas of the 20 parts of the spherical surface. If, as in Fig. 4.19a, the surface is divided into 20 equal parts, then $S_1 = S_2 = S_1 = \cdots = S_{20}$. For less critical cases, it is possible to divide the spherical surface into 16 parts of equal area but of different shape, as we can see from Fig. 4.19b and c.

By definition, the directivity factor $Q(f)$ is

$$Q(f) = \frac{|p_{ax}|^2}{\rho_0 c} \frac{4\pi r^2}{W} = \frac{4\pi |p_{ax}|^2}{\int_0^{2\pi} \int_0^{\pi} |p(\theta,\phi)|^2 \sin\theta \, d\theta \, d\phi} \qquad (4.22)$$

where $|p_{ax}|^2$ is the magnitude of the mean-square sound pressure on the designated axis of the sound source at a certain distance r (see Fig. 4.23, 0° axis, as an example).

For the special case where, for any particular value of θ, the sound pressure produced by the sound source is independent of the value of ϕ, that is to say, there is an axis of symmetry, Eq. (4.22) simplifies to

$$Q(f) = \frac{4\pi p_{ax}^2}{2\pi \int_0^{\pi} p^2(\theta) \sin\theta \, d\theta} \qquad (4.23)$$

The magnitude signs are left off for convenience.

Many sources, such as loudspeakers, are fairly symmetrical about the principal axes so that Eq. (4.23) is valid. In this case, data are generally taken at a number of points with the angles θ_n in a horizontal plane around the source so that

$$Q(f) = \frac{(4\pi p_{ax}^2)(57.3)}{2\pi \sum_{n=1}^{180°/\Delta\theta} p^2(\theta_n) \sin\theta_n \, \Delta\theta} \qquad (4.24)$$

where $\Delta\theta$ = separation in degrees of the successive points around the sound source at which measurement of $p(\theta_n)$ was made (see Fig. 4.23 as an example).

180°/$\Delta\theta$ = number of measurements that were made in passing from a point directly in front of the source to one directly behind the source (0 to 180°). The sound source is assumed to be symmetrical so that the variation between 360 and 180° is the same as that between 0 and 180°.

If the source is mounted in an infinite baffle, measurement is possible only in a hemisphere. Hence, the value of n in Eq. (4.23) varies from 1 to 90°/$\Delta\theta$. If the source in an infinite baffle is nondirectional in the hemisphere, which is usually the case for $ka < 0.5$, then the directivity factor $Q = Q_h = 2$, that is, DI = 3 db.

If the directivity pattern is not quite symmetrical, then the factor of 4 in the numerator of Eq. (4.23) becomes 8 and the value of n varies from 1 to $360°/\Delta\theta$. This, in effect, averages the two sides of the directivity pattern.

The directivity index at $\theta = 0°$ for each directivity pattern shown in Part X and in this part is written alongside each directivity pattern.

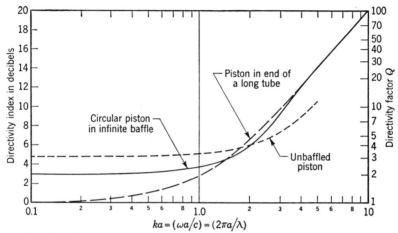

Fig. 4.20. Directivity indexes for the radiation from (1) *one side only* of a piston in an infinite plane baffle; (2) a piston in the end of a long tube; and (3) a piston in free space without any baffle.

The reference axis is the principal axis at $\theta = 0$ in every case. An angle θ at which the directivity index equals 0 db is also marked on these graphs. Hence, the directivity index at any other angle θ can be found by subtracting the decibel value for that axis from the decibel value at the axis where DI = 0.

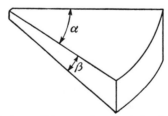

Fig. 4.21. Radiation into a solid cone of space defined by the angles α and β.

For easy reference, the directivity indexes for a piston in (1) an infinite plane baffle, (2) a long tube, and (3) free space are plotted as a function of ka in Fig. 4.20.

Many horn loudspeakers at high frequencies (above 1500 cps) radiate sound uniformly into a solid rectangular cone of space as shown in Fig. 4.21. These horns are of the type discussed in Par. 4.4 (page 106). The directivity indexes in the frequency range above 1500 cps as a function of

α and β may be estimated with the aid of the nomogram given in Fig. 4.22.[11]

Detailed calculations are shown in Table 4.1 for a box-enclosed loud-speaker having the directivity pattern at a frequency of 1500 cps shown

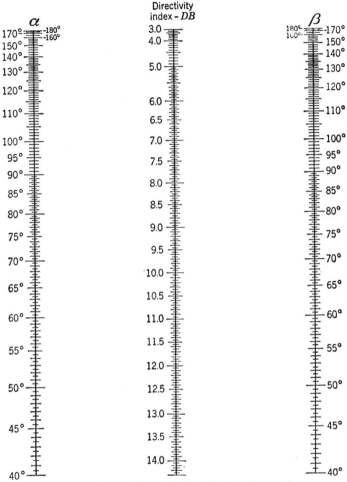

FIG. 4.22. Nomogram for determining the directivity indexes of a source of sound radiating uniformly into a solid cone of space of the type shown by Fig. 4.21. [*After Molloy, Calculation of the Directivity Index for Various Types of Radiators, J. Acoust. Soc. Amer.*, **20**: 387–405 (1948).]

in Fig. 4.23. The left (L) and right (R) sides of the directivity characteristics are not alike, so that the averaging process of the previous paragraph is used.

[11] C. T. Molloy, Calculation of the Directivity Index for Various Types of Radiators, *J. Acoust. Soc. Amer.*, **20**: 387–405 (1948).

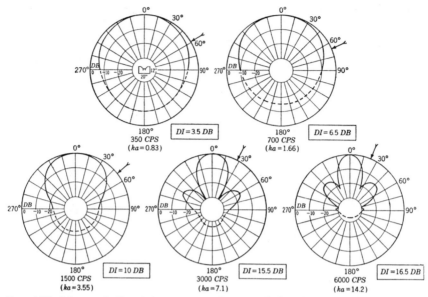

FIG. 4.23. Measured directivity patterns for a typical 12-in. direct-radiator loud-speaker in a 27- by 20- by 12-in. rectangular box. The squares give the directivity index at $\theta = 0°$. One angle of zero directivity index is also indicated.

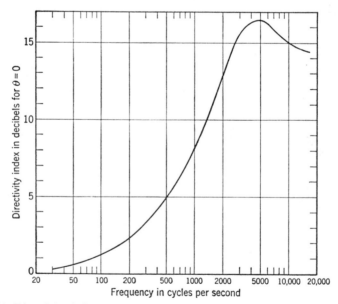

FIG. 4.24. Directivity indexes for 0° axes of the directivity patterns of Fig. 4.23 computed as though the source were symmetrical about the 0° axis. The data apply to a typical 12-in. direct-radiator loudspeaker mounted in a 27- by 20- by 12-in. rectangular box.

After a directivity factor has been calculated at each frequency, a plot of directivity index $DI(f)$ in decibels is made with the aid of Eq. (4.19). For the loudspeaker with the directivity patterns of Fig. 4.23, the directivity index as a function of frequency is shown in Fig. 4.24.

TABLE 4.1.   Calculation of Directivity Index $DI(f)$†

θ_n		$\sin \theta_n$	Directivity		$\left\lvert\dfrac{p(\theta_n)}{p_{ax}}\right\rvert^2$		$\left\lvert\dfrac{p(\theta_n)}{p_{ax}}\right\rvert^2 \sin \theta_n$	
L	R		L, db	R, db	L	R	L	R
355	5	0.087	0	−0.2	1.00	0.95	0.09	0.08
345	15	0.259	0	−1.0	1.00	0.79	0.26	0.21
335	25	0.423	−1.5	−3.0	0.71	0.50	0.30	0.21
325	35	0.574	−3.5	−6.0	0.45	0.25	0.26	0.14
315	45	0.707	−5.7	−10.0	0.27	0.10	0.19	0.07
305	55	0.819	−8.6	−14.5	0.14	0.04	0.11	0.03
295	65	0.906	−11.7	−18.5	0.07	0.01	0.06	0.01
285	75	0.966	−15.4	−22.3	0.03	0.01	0.03	0.01
275	85	0.996	−18.4	−23	0.01	0.01	0.01	0.01
265	95	0.996	−20	−20	0.01	0.01	0.01	0.01
255	105	0.966	−20	−20	0.01	0.01	0.01	0.01
245	115	0.906	−20	−20	0.01	0.01	0.01	0.01
235	125	0.819	−20	−20	0.01	0.01	0.01	0.01
225	135	0.707	−20	−20	0.01	0.01	0.01	0.01
215	145	0.574	−20	−20	0.01	0.01	0.01	0.01
205	155	0.423	−20	−20	0.01	0.01	0.00	0.00
195	165	0.259	−20	−20	0.01	0.01	0.00	0.00
185	175	0.087	−20	−20	0.01	0.01	0.00	0.00
							1.37	0.83

$$Q(f) = \frac{8\pi \times 57.3°}{2\pi \displaystyle\sum_{1}^{36} \left\lvert\frac{p(\theta_n)}{p_{ax}}\right\rvert^2 \sin \theta_n \times 10°} = \frac{23}{1.37 + 0.83} = 10.4$$

$$DI(f) = 10 \log 10.4 = 10.2 \text{ db}$$

† At $f = 1500$ cps for a commercially available loudspeaker having the directivity patterns shown in Fig. 4.23.   The quantity $\Delta\theta = 10° = \pi/18$ radians.

CHAPTER 5

ACOUSTIC COMPONENTS

PART XII *Radiation Impedances*

The fields of radio and television have advanced rapidly because of the availability of electrical components with well-known physical properties that are simple to assemble into a completed mechanism. With such components (resistors, capacitors, and inductors) the advanced research engineer and the high-school student alike are able to experiment with new circuits. Such complicated devices as electric-wave filters often can be designed by selecting from among readily available parts until a desired performance characteristic is achieved—a feat that otherwise might require a lengthy mathematical analysis.

No such satisfactory situation exists in the field of acoustics. Acoustical elements have not been available commercially. Advanced textbooks have often side-stepped the theoretical treatment of their performance. Even those texts which deal primarily with acoustic devices give limited information on how to predict the performance of cavities, holes, tubes, screens, slots, and diaphragms—the elements of acoustical circuits. This text does not pretend to advance the science of acoustical-circuit theory to anything approaching a state of completion. Much basic research remains to be done. It does attempt to interpret the latest theories in such a way that the reader can construct and understand the performance of the usual types of acoustic devices.

Loudspeakers, microphones, and acoustic filters are the most common devices composed of mechanical and acoustical elements. One obvious acoustical element is the air into which the sound is radiated. Others are air cavities, tubes, slots, and porous screens both behind and in front of actively vibrating diaphragms. These various elements have acoustic impedances associated with them, which can, in some frequency ranges, be represented as simple lumped elements. In other frequency ranges, distributed elements, analogous to electric lines, must be used in explaining the performance of the devices.

The first acoustical element that we shall deal with is the radiation impedance of the air itself. Radiation impedance is a quantitative statement of the manner in which the medium reacts against the motion of a vibrating surface.

Sound is produced by vibrating surfaces such as the diaphragm of a loudspeaker. In addition to the energy required to move the vibrating surface itself, energy is radiated into the air by the diaphragm. Part of this radiated energy is useful and represents the power output of the loudspeaker. The remainder is stored (reactive) energy that is returned to the generator. Consequently, the acoustic impedance has a real part, accounting for the radiated power, and an imaginary part, accounting for the reactive power.

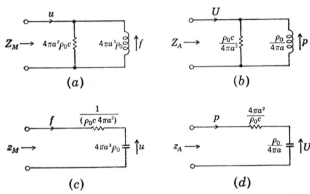

FIG. 5.1. Exact radiation impedances and mobilities for all values of ka for a sphere with a surface that vibrates radially. (a) Mechanical-impedance analogy; (b) acoustic-impedance analogy; (c) mechanical-mobility analogy; (d) acoustic-mobility analogy. The quantity a is the radius of the sphere.

The four simplest types of vibrating surface treated here are (1) a pulsating sphere, (2) a plane circular piston mounted in an infinite surface (baffle), (3) a plane circular piston in the end of a long tube, and (4) a plane circular piston without a baffle. We have already derived the radiation impedance for a pulsating sphere. The mathematical solution of the radiation from a circular piston mounted in an infinite baffle appears in many advanced texts so that only the results will be presented here.[1] More complicated problems are to solve analytically for the radiation impedances and directivity patterns of a long tube and a vibrating piston without baffle. Those solutions are now available, and the results are given in this part. Most other types of vibrating surfaces are exceedingly difficult in mathematical treatment, and the results will not be presented here.

[1] P. M. Morse, "Vibration and Sound," 2d ed., pp. 326–346, McGraw-Hill Book Company, Inc., New York, 1948.

5.1. Sphere with Uniformly Pulsating Surface. In Part IV we derived the radiation impedance for a sphere with a uniformly pulsating surface. For the results, refer to Eq. (2.64) and Fig. 2.10.

It is seen from Fig. 2.10 that for $ka < 0.3$, that is, when the diameter is less than one-tenth the wavelength, the impedance load on the surface of

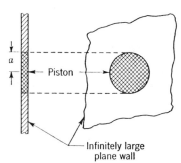

the sphere is that of a mass reactance because the resistive component is negligible compared with the reactive component. This mass loading may be thought of as a layer of air on the outside of the sphere, the thickness of the layer equaling 0.587 of the radius of the sphere.

At all frequencies, the loading shown in Fig. 2.10 may be represented by the equivalent circuits of Fig. 5.1. The element sizes for the mechanical and acoustic mobilities and impedances are given with the circuits.

FIG. 5.2. Plane circular piston vibrating perpendicular to the plane of an infinite wall.

5.2. Plane Circular Piston in Infinite Baffle. The mechanical impedance in mks mechanical ohms (newton-seconds per meter) of the air load upon one side of a plane piston mounted in an infinite baffle (see Fig. 5.2) and vibrating sinusoidally is[2,3]

$$Z_M = \Re_M + jX_M = \pi a^2 \rho_0 c \left[1 - \frac{J_1(2ka)}{ka} \right] + j\frac{\pi \rho_0 c}{2k^2} K_1(2ka) \quad (5.1)$$

where Z_M = mechanical impedance in newton-seconds per meter, *i.e.*, mks mechanical ohms.

a = radius of piston in meters.

ρ_0 = density of gas in kilograms per cubic meter.

c = speed of sound in meters per second.

$\Re_M$ = mechanical resistance in newton-seconds per meter. The German $\Re$ indicates that the resistive component is a function of frequency.

X_M = mechanical reactance in newton-seconds per meter.

$k = \omega/c = 2\pi/\lambda$ = wave number.

J_1, K_1 = two types of Bessel function given by the series.[3,4]

---

[2] L. E. Kinsler and A. R. Frey, "Fundamentals of Acoustics," pp. 187–195, John Wiley & Sons, Inc., New York, 1950.

[3] Morse, *op. cit.*, pp. 332, 333. Morse gives in Table VIII on page 447 a function $M(2ka)$ that equals $K_1(2ka)/2k^2a^2$.

[4] G. N. Watson, "Theory of Bessel Functions," Cambridge University Press, London, 1922.

$$J_1(W) = \frac{W}{2} - \frac{W^3}{2^2 \cdot 4} + \frac{W^5}{2^2 \cdot 4^2 \cdot 6} - \frac{W^7}{2^2 \cdot 4^2 \cdot 6^2 \cdot 8} \cdots \qquad (5.2)$$

$$K_1(W) = \frac{2}{\pi} \left(\frac{W^3}{3} - \frac{W^5}{3^2 \cdot 5} + \frac{W^7}{3^2 \cdot 5^2 \cdot 7} \cdots \right) \qquad (5.3)$$

where $W = 2ka$.

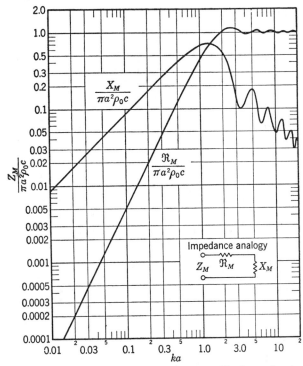

FIG. 5.3. Real and imaginary parts of the normalized mechanical impedance $(Z_M/\pi a^2 \rho_0 c)$ of the air load on one side of a plane piston of radius a mounted in an infinite flat baffle. Frequency is plotted on a normalized scale, where $ka = 2\pi f a/c = 2\pi a/\lambda$. Note also that the ordinate is equal to $Z_A \pi a^2/\rho_0 c$, where Z_A is the acoustic impedance.

Graphs of the real and imaginary parts of

$$\frac{Z_M}{\pi a^2 \rho_0 c} = \frac{\Re_M + j X_M}{\pi a^2 \rho_0 c} \qquad (5.4)$$

are shown in Fig. 5.3 as a function of ka. The German $\Re$ indicates that the quantity varies with frequency.

Similar graphs of the real and imaginary parts of the mechanical mobility

$$z_M \pi a^2 \rho_0 c = \pi a^2 \rho_0 c (\mathfrak{r}_M + j x_M) = \pi a^2 \rho_0 c \left(\frac{\Re_M}{\Re_M{}^2 + X_M{}^2} - j \frac{X_M}{\Re_M{}^2 + X_M{}^2} \right) \qquad (5.5)$$

are shown in Fig. 5.4.   The mechanical mobility is in meters per newton-second, *i.e.*, mks mechanical mohms.

The data of Fig. 5.3 are used in dealing with impedance analogies and the data of Fig. 5.4 in dealing with mobility analogies.

We see from Fig. 5.3 that, for $ka < 0.5$, the reactance varies as the first power of frequency while the resistance varies as the second power of frequency.   At high frequencies, for $ka > 5$, the reactance becomes small

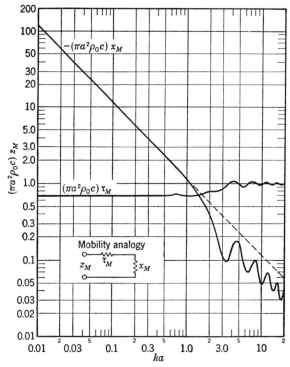

FIG. 5.4. Real and imaginary parts of the normalized mechanical mobility $(\pi a^2 \rho_0 c z_M)$ of the air load upon one side of a plane piston of radius a mounted in an infinite flat baffle.   Frequency is plotted on a normalized scale, where $ka = 2\pi fa/c = 2\pi a/\lambda$. Note also that the ordinate is equal to $z_A \rho_0 c / \pi a^2$, where z_A is the acoustic mobility.

compared with the resistance, and the resistance approaches a constant value.

The mobility, on the other hand, is better behaved.   The responsiveness is constant for $ka < 0.5$, and it is also constant for $ka > 5$ although its value is larger.

Approximate Analogous Circuits.   The behavior just noted suggests that, except for the wiggles in the curves for ka between 1 and 5, the impedance and the mobility for a piston in an infinite baffle can be approximated over the whole frequency range by the analogous circuits of Fig. 5.5. Those circuits give the mechanical and acoustic impedances and mobilities, where

$$R_{M2} = \pi a^2 \rho_0 c \text{ mks mechanical ohms (newton-sec/m)} \tag{5.6}$$

$$R_M = R_{M2} + R_{M1} = 128 a^2 \rho_0 c / 9\pi$$
$$= 4.53 a^2 \rho_0 c \text{ mks mechanical ohms} \tag{5.7}$$

$$R_{M1} = 1.386 a^2 \rho_0 c \text{ mks mechanical ohms} \tag{5.8}$$

$$C_{M1} = 0.6 / a\rho_0 c^2 \text{ m/newton} \tag{5.9}$$

$$M_{M1} = 8 a^3 \rho_0 / 3 = 2.67 a^3 \rho_0 \text{ kg} \tag{5.10}$$

$$r_{M2} = 1/\pi a^2 \rho_0 c = 0.318 / a^2 \rho_0 c \text{ mks mechanical mohms} \tag{5.11}$$

$$r_{M1} = 0.721 a^2 \rho_0 c \text{ mks mechanical mohms} \tag{5.12}$$

$$R_{A2} = \rho_0 c / \pi a^2 = 0.318 \rho_0 c / a^2 \text{ mks acoustic ohms} \tag{5.13}$$

$$R_A = R_{A2} + R_{A1} = 128 \rho_0 c / 9\pi^3 a^2$$
$$= 0.459 \rho_0 c / a^2 \text{ mks acoustic ohms} \tag{5.14}$$

$$R_{A1} = 0.1404 \rho_0 c / a^2 \text{ mks acoustic ohms} \tag{5.15}$$

$$C_{A1} = 5.94 a^3 / \rho_0 c^2 \text{ m}^5/\text{newton} \tag{5.16}$$

$$M_{A1} = 8 \rho_0 / 3\pi^2 a = 0.27 \rho_0 / a \text{ kg/m}^4 \tag{5.17}$$

$$r_{A2} = \pi a^2 / \rho_0 c \text{ mks acoustic mohms} \tag{5.18}$$

$$r_{A1} = 7.12 a^2 / \rho_0 c \text{ mks acoustic mohms} \tag{5.19}$$

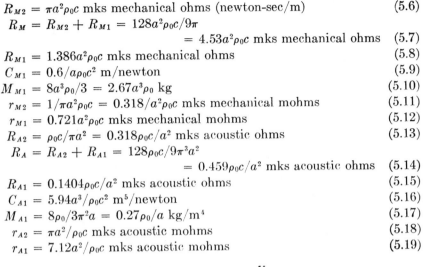

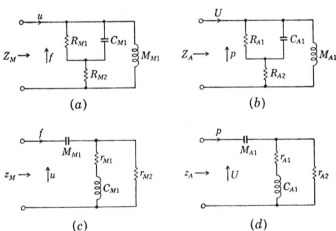

FIG. 5.5. Approximate radiation impedances and mobilities for a piston in an infinite baffle or for a piston in the end of a long tube for all values of ka. (a) Mechanical-impedance analogy; (b) acoustic-impedance analogy; (c) mechanical-mobility analogy; (d) acoustic-mobility analogy.

All constants are dimensionless and were chosen to give the best average fit to the functions of Figs. 5.3 and 5.4.

Low- and High-frequency Approximations. At low and at high frequencies these circuits may be approximated by the simpler circuits given in the last column of Table 5.1.

It is apparent that when $ka < 0.5$, that is, when the circumference of the piston $2\pi a$ is less than one-half wavelength $\lambda/2$, the impedance load presented by the air on the vibrating piston is that of a mass shunted by a very large resistance. In other words $R^2 = (R_1 + R_2)^2$ is large compared with $\omega^2 M_1^2$. In fact, this loading mass may be imagined to be a layer of air equal in area to the area of the piston and equal in thickness to about 0.85 times the radius, because

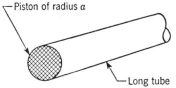

FIG. 5.6. Piston vibrating in the end of a long tube.

$$(\pi a^2)(0.85a)\rho_0 \doteq 2.67a^3\rho_0 = M_{M1}$$

At high frequencies, $ka > 5$, the air load behaves exactly as though it were connected to one end of a tube of the same diameter as the piston, with the other end of the tube perfectly absorbing. As we saw in Eq. (2.60), the input mechanical resistance for such a tube is $\pi a^2 \rho_0 c$. Hence, intuitively one might expect that at high frequencies the vibrating rigid piston beams the sound outward in lines perpendicular to the face of the piston. This

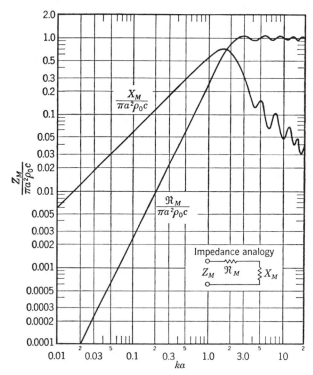

FIG. 5.7. Real and imaginary parts of the normalized mechanical impedance ($Z_M/\pi a^2\rho_0 c$) of the air load upon one side of a plane piston of radius a mounted in the end of a long tube. Frequency is plotted on a normalized scale, where $ka = 2\pi fa/c = 2\pi a/\lambda$. Note also the ordinate is equal to $Z_A\pi a^2/\rho_0 c$, where Z_A is the acoustic impedance.

is actually the case for the near-field.   At a distance, however, the far-field radiation spreads, as we learned in the preceding chapter.

5.3. Plane Circular Piston in End of Long Tube.[5]   The mechanical impedance (newton-seconds per meter) of the air load on one side of a plane piston mounted in the end of a long tube (see Fig. 5.6) and vibrating sinusoidally is given by a complicated mathematical expression that we shall not reproduce here.

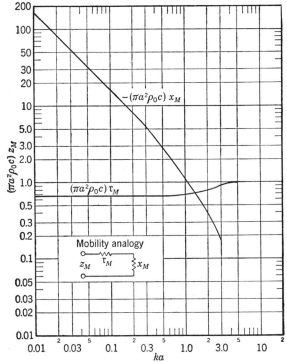

FIG. 5.8. Real and imaginary parts of the normalized mechanical mobility $(\pi a^2 \rho_0 c z_M)$ of the air load upon one side of a plane piston of radius a mounted in the end of a long tube.   Frequency is plotted on a normalized scale, where $ka = 2\pi f a/c = 2\pi a/\lambda$. Note also that the ordinate is equal to $z_A \rho_0 c/\pi a^2$, where z_A is the acoustic mobility.

Graphs of the real and imaginary parts of the normalized mechanical impedance $Z_M/\pi a^2 \rho_0 c$ as a function of ka for a piston so mounted are shown in Fig. 5.7.   Similar graphs of the real and imaginary parts of the normalized mechanical mobility are shown in Fig. 5.8.   The data of Fig. 5.7 are used in dealing with impedance analogies and those of Fig. 5.8 in dealing with mobility analogies.

To a fair approximation, the radiation impedance for a piston in the end of a long tube may be represented over the entire frequency range by

[5] H. Levine and J. Schwinger, On the Radiation of Sound from an Unflanged Circular Pipe, *Phys. Rev.*, **73**: 383–406 (1948).

TABLE 5.1. Radiation Impedance and Mobility for One Side of a Plane Piston in Infinite Baffle†

Impedance	Mechanical	Specific acoustic	Acoustic	Analogous circuits
	f = drop u = flow	p = drop u = flow	p = drop U = flow	
$ka < 0.5$: Series resistance, $\mathfrak{R}$ Shunt resistance, R Mass, M_1	$\mathfrak{R}_M = 1.57\omega^2 a^4 \rho_0/c$ $R_M = 4.53 a^2 \rho_0 c$ $M_{M1} = 2.67 a^3 \rho_0$	$\mathfrak{R}_S = 0.5\omega^2 a^2 \rho_0/c$ $R_S = 1.441 \rho_0 c$ $M_{S1} = 0.849 a \rho_0$	$\mathfrak{R}_A = 0.159\omega^2 \rho_0/c$ $R_A = 0.459 \rho_0 c/a^2$ $M_{A1} = 0.270 \rho_0/a$	
$ka > 5$: Resistance, R_2	$R_{M2} = \pi a^2 \rho_0 c$	$R_{S2} = \rho_0 c$	$R_{A2} = \rho_0 c/\pi a^2$	
Mobility	u = drop f = flow	u = drop p = flow	U = drop p = flow	
$ka < 0.5$: Series responsiveness, r Mass, M_1	$r_M = 0.221/a^2 \rho_0 c$ $M_{M1} = 2.67 a^3 \rho_0$	$r_S = 0.694/\rho_0 c$ $M_{S1} = 0.849 a \rho_0$	$r_A = 2.18 a^2/\rho_0 c$ $M_{A1} = 0.270 \rho_0/a$	
$ka > 5$: Responsiveness, r_2	$r_{M2} = 1/\pi a^2 \rho_0 c$	$r_{S2} = 1/\rho_0 c$	$r_{A2} = \pi a^2/\rho_0 c$	

† This table gives element sizes for analogous circuits in the region where $ka < 0.5$ and $ka > 5.0$. All constants are dimensionless. In the region between 0.5 and 5.0, the approximate circuits of Fig. 5.5 or the exact charts of Figs. 5.3 and 5.4 should be used.

TABLE 5.2. Radiation Impedance and Mobility for the Outer Side of a Plane Piston in End of Long Tube†

Impedance	Mechanical	Specific acoustic	Acoustic	Analogous circuits
	f = drop u = flow	p = drop u = flow	p = drop U = flow	
$ka < 0.5$: Series resistance, $\Re$ Shunt resistance, R Mass, M_1	$\Re_M = 0.7854\omega^3 a^4\rho_0/c$ $R_M = 4.73a^2\rho_0 c$ $M_{M1} = 1.927a^3\rho_0$	$\Re_S = 0.247\omega^2 a^2\rho_0/c$ $R_S = 1.505\rho_0 c$ $M_{S1} = 0.6133a\rho_0$	$\Re_A = 0.0796\omega^2\rho_0/c$ $R_A = 0.479\rho_0 c/a^2$ $M_{A1} = 0.1952\rho_0/a$	
$ka > 5$: Resistance, R_2	$R_{M2} = \pi a^2\rho_0 c$	$R_{S2} = \rho_0 c$	$R_{A2} = \rho_0 c/\pi a^2$	
Mobility	u = drop f = flow	u = drop p = flow	U = drop p = flow	
$ka < 0.5$: Series responsiveness, r Mass, M_1	$r_M = 0.2116/a^2\rho_0 c$ $M_{M1} = 1.927a^3\rho_0$	$r_S = 0.665/\rho_0 c$ $M_{S1} = 0.6133a\rho_0$	$r_A = 2.09a^2/\rho_0 c$ $M_{A1} = 0.1952\rho_0/a$	
$ka > 5$: Responsiveness, r_2	$r_{M2} = 1/\pi a^2\rho_0 c$	$r_{S2} = 1/\rho_0 c$	$r_{A2} = \pi a^2/\rho_0 c$	

† This table gives element sizes for analogous circuits in the regions where $ka < 0.5$ and $ka > 5$. All constants are dimensionless. For the region between 0.5 and 5.0, the approximate circuits of Fig. 5.5 or the exact charts of Figs. 5.7 and 5.8 should be used.

TABLE 5.3. Radiation Impedance for Both Sides of a Plane Circular Disk in Free Space†

Impedance	Mechanical	Specific acoustic	Acoustic	Analogous circuits
	f = drop u = flow	p = drop u = flow	p = drop U = flow	
ka < 0.5: Series resistance, $\Re$ Mass (both sides of piston), M_1	$\Re_M = 0.1886 a^6 \rho_0 \omega^4/c^3$ $M_{M1} = 2.67 a^3 \rho_0$	$\Re_S = 0.0600 a^4 \rho_0 \omega^4/c^3$ $M_{S1} = 0.850 a \rho_0$	$\Re_A = 0.01901 a^2 \rho_0 \omega^4/c^3$ $M_{A1} = 0.2705 \rho_0/a$	
ka > 5: Resistance (both sides of piston), R	$R_M = 2\pi a^2 \rho_0 c$	$R_S = 2\rho_0 c$	$R_A = 2\rho_0 c/\pi a^2$	
Mobility	u = drop f = flow	u = drop p = flow	U = drop p = flow	
ka < 0.5: Series responsiveness, r Mass (both sides of piston), M_1	$r_M = 0.0265 \omega^2/\rho_0 c^3$ $M_{M1} = 2.67 a^3 \rho_0$	$r_S = 0.0832 a^2 \omega^2/\rho_0 c^3$ $M_{S1} = 0.850 a \rho_0$	$r_A = 0.261 \omega^2 a^4/\rho_0 c^3$ $M_{A1} = 0.2705 \rho_0/a$	
ka > 5: Responsiveness (both sides of piston), r	$r_M = 1/2\pi a^2 \rho_0 c$	$r_S = 1/2\rho_0 c$	$r_A = \pi a^2/2\rho_0 c$	

† This table gives element sizes for analogous circuits in the region where $ka < 0.5$ and $ka > 5$. All constants are dimensionless. For the region between 0.5 and 5.0, the chart of Fig. 5.9 should be used.

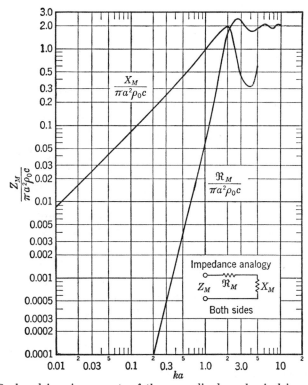

FIG. 5.9. Real and imaginary parts of the normalized mechanical impedance $(Z_M/\pi a^2\rho_0 c)$ of the air load upon *both* sides of a plane circular disk of radius a in free space. Frequency is plotted on a normalized scale, where $ka = 2\pi f a/c = 2\pi a/\lambda$. Note also the ordinate is equal to $Z_A \pi a^2/\rho_0 c$, where Z_A is the normalized acoustic impedance.

the same analogous circuits used for the piston in an infinite baffle and shown in Fig. 5.5, where the elements now are

$$R_{M2} = \pi a^2 \rho_0 c \text{ mks mechanical ohms} \tag{5.20}$$

$$R_M = R_{M2} + R_{M1} = 4\pi(0.6133)^2 a^2 \rho_0 c$$
$$= 4.73 a^2 \rho_0 c \text{ mks mechanical ohms} \tag{5.21}$$

$$R_{M1} = 0.504\pi a^2 \rho_0 c = 1.58 a^2 \rho_0 c \text{ mks mechanical ohms} \tag{5.22}$$

$$C_{M1} = 0.55/a\rho_0 c^2 \text{ m/newton} \tag{5.23}$$

$$M_{M1} = 0.6133\pi a^3 \rho_0 = 1.927 a^3 \rho_0 \text{ kg} \tag{5.24}$$

$$r_{M2} = 1/\pi a^2 \rho_0 c = 0.318/a^2 \rho_0 c \text{ mks mechanical mohms} \tag{5.25}$$

$$r_{M1} = 0.633/a^2 \rho_0 c \text{ mks mechanical mohms} \tag{5.26}$$

$$R_{A2} = \rho_0 c/\pi a^2 = 0.318\rho_0 c/a^2 \text{ mks acoustic ohms} \tag{5.27}$$

$$R_A = R_{A2} + R_{A1} = (4)(0.6133)^2 \rho_0 c/\pi a^2$$
$$= 0.479\rho_0 c/a^2 \text{ mks acoustic ohms} \tag{5.28}$$

$$R_{A1} = 0.504\rho_0 c/\pi a^2 = 0.1604\rho_0 c/a^2 \text{ mks acoustic ohms} \tag{5.29}$$

$$C_{A1} = 5.44 a^3/\rho_0 c^2 \text{ m}^5/\text{newton} \tag{5.30}$$

$$M_{A1} = 0.1952\rho_0/a \text{ kg/m}^4 \tag{5.31}$$

$r_{A2} = \pi a^2/\rho_0 c$ mks acoustic mohms   (5.32)

$r_{A1} = 6.23a^2/\rho_0 c$ mks acoustic mohms   (5.33)

In the frequency ranges where $ka < 0.5$ and $ka > 5$, analogous circuits of the type shown in Table 5.2 may be used.

5.4. Plane Circular Free Disk.[6] A disk in free space without surrounding structure is a suitable model, at low frequencies, for a direct-radiator loudspeaker without a baffle of any sort. A simple equivalent circuit, approximately valid for all frequencies like those shown in Fig. 5.5, cannot be drawn for this case. At very low frequencies, however, it is possible to represent the impedance by a series combination of a mass and a frequency-dependent resistance just as was done for the pistons in baffles.

Graphs of the real and imaginary parts of the normalized mechanical-impedance load *on both sides of the diaphragm*, $Z_M/\pi a^2\rho_0 c$, as a function of ka for the free disk, are shown in Fig. 5.9. The data of Fig. 5.9 are used in dealing with impedance analogies. The complex mobility can be obtained by taking the reciprocal of the complex impedance.

In the frequency ranges where $ka < 0.5$ and $ka > 5$, analogous circuits of the type shown in Table 5.3 may be used.

PART **XIII** *Acoustic Elements*

5.5. Acoustic Compliances. *Closed Tube.* In Eq. (2.50) we showed that a length of tube, rigidly closed on one end, with a radius in meters greater than $0.05/\sqrt{f}$ and less than $10/f$ had an input acoustic impedance (at the open end) of

$$Z_A = \frac{Z_s}{\pi a^2} = \frac{-j\rho_0 c}{\pi a^2} \cot kl' \tag{5.34}$$

where Z_A = acoustic impedance in mks acoustic ohms

Z_s = specific acoustic impedance in mks rayls

a = radius of tube in meters

l' = length of tube in meters

For values of k that are not too large, the cotangent may be replaced by the first two terms of its equivalent-series form

$$\cot kl' = \frac{1}{kl'} - \frac{kl'}{3} - \frac{(kl')^3}{45} - \cdots \tag{5.35}$$

[6] F. M. Wiener, On the Relation between the Sound Fields Radiated and Diffracted by Plane Obstacles, *J. Acoust. Soc. Amer.*, **23**: 697–700 (1951).

Equation (5.34) becomes

$$Z_A \doteq -j\frac{1}{\omega(V/\rho_0 c^2)} + j\omega\frac{l'\rho_0}{3\pi a^2} \cdots \qquad (5.36)$$

where $V = l'\pi a^2$ = volume of air in the tube. The acoustic impedance Z_A is a series combination of an acoustic mass ($l'\rho_0/3\pi a^2$) and an acoustic compliance ($V/\rho_0 c^2$). Equation (5.36) is valid within 5 per cent for l's up to about $\lambda/7$.†

If the impedance of the cavity is represented by the series combination of an acoustic mass and an acoustic compliance as is shown by Eq. (5.36), a lumped-circuit representation is permissible out to dimensions of $l = \lambda/8$. In this case the shape of the cavity is important because the magnitude of the inductance involves the ratio $l/\pi a^2$, that is, length divided by cross-sectional area. Such a series combination of mass and compliance is shown in Fig. 5.11a.

Closed Volume. If the second term of Eq. (5.36) has a magnitude that is small compared with that of the first term, 5 per cent perhaps that of the first term, we may neglect it and

$$Z_A = -j\frac{1}{\omega(V/\rho_0 c^2)} = -j\frac{1}{\omega C_A} \qquad (5.37a)$$

where

$$C_A = \frac{V}{\rho_0 c^2} = \frac{V}{\gamma P_0} \qquad (5.38)$$

is an acoustic compliance with the units of m^5/newton. The 5 per cent restriction given above means that l' should be less than $(\lambda/16)$, where λ is the wavelength in meters. Such a compliance is shown in Fig. 5.11b. For quite small pistons, provided the largest dimension of the cavity does not exceed about one-eighth wavelength, the acoustic impedance presented to the piston is

$$Z_A = -j\frac{1}{\omega(V/\gamma P_0)} + j\omega\frac{0.85\rho_0}{\pi a} \qquad (5.37b)$$

For the second term to be negligible within 5 per cent ($a\lambda^2/V$) should be greater than about 200. For intermediate sized pistons, the discussion on page 217 should be studied.

An acoustic compliance obtained by compressing air in a closed volume is a two-terminal device as shown in Fig. 5.11b, but *one terminal must always be at "ground potential."* That is to say, one terminal is the outside of the enclosure housing the cavity, and it is usually at atmospheric pressure. With such an arrangement for an acoustic compliance (see Figs. 3.25 and 3.31 for examples) it is *never* possible to insert a compliance between acoustic resistors or acoustic masses as, for example, in the upper branches of Fig. 3.32.

† In Eq. (5.42) we define the acoustic mass of a short open-ended tube as equaling $M_A \equiv \rho_0 l'/\pi a^2$. Hence, to this approximation, the mass for a closed tube is one-third that for an open tube.

Stiffness-controlled Diaphragm.   To obtain a "series" type of acoustic compliance a diaphragm or stretched membrane must be used.   Diaphragms and membranes resonate at various frequencies so that the frequency range where they act as compliances is restricted to that region well below the lowest frequency of resonance.   A combination of series acoustic compliance, resistance, and mass is shown in Fig. 5.10.

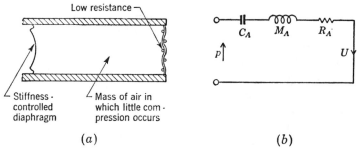

FIG. 5.10. Example of a series acoustic compliance obtained with a stiffness-controlled diaphragm.

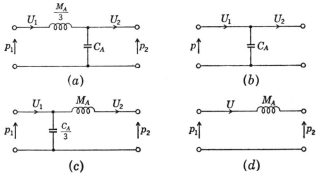

FIG. 5.11. Approximate analogous circuits for a short tube of medium diameter. (a) and (b) Circuits used when p_2/U_2 is very large (closed end).   (c) and (d) Circuits used when p_2/U_2 is very small (open end).   Circuits (a) and (c) yield the impedance within about 5 per cent for a tube length l' that is less than $\lambda/8$.   Circuits (b) and (d) yield the impedance within about 5 per cent for $l' < \lambda/16$.

Example 5.1.   The old-fashioned jug of Fig. 5.12 is used in a country dance band as a musical instrument.   You are asked to analyze its performance acoustically.   If the inside dimensions of the jug are diameter = 8 in. and air-cavity height = 10 in., give the analogous circuit, the element sizes and the acoustic impedance for the air-cavity portion of the jug at 50, 100, and 300 cps.   Assume $T = 23°C$ and $P_0 = 10^5$ newton/m^2.   (NOTE: The neck portion will be discussed later in this part.)

Solution.   The speed of sound at 23°C is about 1133 ft/sec.   Hence,

$$\lambda_{50} = 22.66 \text{ ft}$$
$$\lambda_{100} = 11.33 \text{ ft}$$
$$\lambda_{300} = 3.78 \text{ ft}$$

The length l of the jug is 0.833 ft. Hence,

$$l = \frac{\lambda_{50}}{27} = \frac{\lambda_{100}}{13.6} = \frac{\lambda_{300}}{4.5} = 0.833 \text{ ft}$$

At 50 cps, where $l/\lambda = \frac{1}{27}$, the cavity portion of the jug may be represented by an acoustic compliance

$$C_A = \frac{V}{\gamma P_0} = \frac{8.24 \times 10^{-3}}{1.4 \times 10^5} = 5.89 \times 10^{-8} \text{ m}^5/\text{newton}$$

$$Z_A = -j \frac{10^8}{314 \times 5.89} = -j5.4 \times 10^4 \text{ mks acoustic ohms}$$

At 100 cps, where $l/\lambda = \frac{1}{14}$, the cavity portion of the jug may be represented by a series acoustic mass and acoustic compliance.

$$M_A = \frac{l\rho_0}{3\pi a^2} = \frac{0.254 \times 1.19}{3\pi (0.1016)^2} = 3 \text{ kg/m}^4$$

$$C_A = 5.89 \times 10^{-8} \text{ m}^5/\text{newton}$$

$$Z_A = j \left(628 \times 3 - \frac{10^8}{628 \times 5.89} \right) = -j2.51 \times 10^4$$

At 300 cps, where $l/\lambda = 1/4.5$, the acoustic impedance of the cavity portion of the jug must be solved for directly from Eq. (5.34).

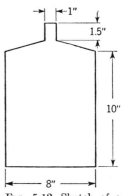

$$Z_A = \frac{-j\rho_0 c}{\pi a^2} \cot kl$$

$$= \frac{-j(1.19 \times 345.4)}{\pi (0.1016)^2} \cot \frac{2\pi \times 300 \times 0.254}{345.4}$$

$$Z_A = -j4.5 \times 10^3 \text{ mks acoustic ohms}$$

FIG. 5.12. Sketch of a musical jug.

5.6. Acoustic Mass (Inertance). A tube open at both ends and with rigid walls behaves as an acoustic mass if it is short enough so that the air in it moves as a whole without appreciable compression. In setting up the boundary condition, the assumption is made that the sound pressure at the open end opposite the source is nearly zero. This assumption would be true if it were not for the radiation impedance of the open end, which acts very much like a piston radiating into open air. However, this radiation impedance is small for a tube of small diameter and acts only to increase the apparent length of the tube slightly. Therefore, the radiation impedance will be added as a correction factor later.

Tube of Medium Diameter. In order to be able to neglect viscous losses inside the tube, the radius of the tube a in meters must not be too small. Also, in order to be able to neglect transverse resonances in the tube, the radius must not be too large. The equations which follow are valid for a radius in meters greater than about $0.05/\sqrt{f}$ and less than about $10/f$. Solution of the one-dimensional wave equation of Part III, with the boundary condition $p = 0$ at $x = l$, that is, at the open end, yields

$$Z_A = j \frac{\rho_0 c}{\pi a^2} \tan kl' \tag{5.39}$$

where Z_A = acoustic impedance in mks acoustic ohms of the open tube
of length l'

ρ_0 = density of the gas in kilograms per cubic meter

c = speed of sound in meters per second

a = radius of tube in meters

k = wave number = ω/c in reciprocal meters

l' = length of tube in meters measured from the open end to the
plane where Z_A is being determined

For small values of kl', the tangent may be replaced by the first terms
of its equivalent series form.

$$\tan kl' = kl' + \frac{(kl')^3}{3} + \frac{2(kl')^5}{15} + \cdots \tag{5.40}$$

Equation (5.39) becomes

$$Z_A(l') \doteq j\omega \frac{\rho_0 l'}{\pi a^2} + j \frac{\omega^3 \rho_0 l'^3}{3\pi a^2 c^2} + \cdots \tag{5.41}$$

An equivalent circuit for Eq. (5.41) valid within about 5 per cent for
the frequency range where $l' < \lambda/8$ is given in Fig. 5.11c. In that
analogous circuit,

$$M_A = \frac{\rho_0 l'}{\pi a^2} \tag{5.42}$$

and C_A is given by Eq. (5.38). The quantity M_A is an acoustic mass with
the units of kilograms per meter[4].

For values of $l' < \lambda/16$, the second term of Eq. (5.41) will be less than
5 per cent as large as the first, so that

$$Z_A = j\omega \frac{\rho_0 l'}{\pi a^2} = j\omega M_A \tag{5.43}$$

The equivalent circuit for this impedance is given in Fig. 5.11d.

End Corrections. END CORRECTIONS IF THE END OF TUBE TERMINATES
IN INFINITE BAFFLE—FLANGED TUBE. Most acoustic masses terminate at
either one end or the other in open air or in a larger cavity. This means
that corrections must be added to the length l' above if the value of M_A is
to be correct when l' is not large. The correction is the impedance given
in Table 5.1 by the circuits for $ka < 0.5$. For the case that a is less than
$\lambda/25$, the circuits of Table 5.1 reduce simply to one element of magnitude
M_{A1} [see Eq. (5.17)]. Hence the end correction l'' necessary for each
baffled end of a vibrating "plug" of air in a tube is equal to

$$l'' = \frac{M_{A1}\pi a^2}{\rho_0} = \frac{0.27\rho_0}{a}\frac{\pi a^2}{\rho_0} \doteq 0.85a \qquad \text{meters} \tag{5.44}$$

Hence, (5.42) becomes

$$M_A = \frac{\rho_0[l' + (2)l'']}{\pi a^2} \equiv \frac{\rho_0 l}{\pi a^2} \quad \text{kg/m}^4 \qquad (5.45)$$

where the quantity l is by definition equal to the effective length of the tube in meters and is the sum of the actual length l' plus the end correction(s) l'' (or $2l''$). The numeral 2 in the parentheses will be used only if there are two free ends to the vibrating plug of air.

If the tube is not round, we may replace a by $\sqrt{S/\pi}$, where S is the cross sectional area of the tube.

END CORRECTIONS IF END OF TUBE IS FREE (UNFLANGED TUBE)[5] (see Fig. 5.6). In this case, the correction at each free end of a tube is the impedance given in Table 5.2 by the circuits for $ka < 0.5$. If a is less than $\lambda/25$, the circuit of Table 5.2 reduces simply to one element of magnitude M_{A1} as given by Eq. (5.31). Hence, the end correction l'' necessary for each free end of a vibrating plug of air in a tube is equal to

$$l'' = \frac{M_{A1}\pi a^2}{\rho_0} = \frac{0.1952\rho_0}{a}\frac{\pi a^2}{\rho_0} = 0.613a \qquad \text{m} \qquad (5.46)$$

and Eq. (5.42) becomes

$$M_A = \frac{\rho_0[l' + (2)l'']}{\pi a^2} \equiv \frac{\rho_0 l}{\pi a^2} \quad \text{kg/m}^4 \qquad (5.47)$$

If the tube is not round we may replace a by $\sqrt{S/\pi}$, where S is the cross-sectional area of the tube. Cases of elements combining both acoustic mass and acoustic resistance will be discussed later in this part.

Example 5.2. The jug of Example 5.1 has a neck with a diameter of 1 in. and a length of 1.5 in. (see Fig. 5.12). At what frequency will the jug resonate?

Solution. First, let us assume that the frequency of resonance is so low that the length of the neck l' is small compared with $\lambda/16$. Then, because the air in it is not constrained, it will be an acoustic mass.

$$M_A = \frac{\rho_0(l' + 0.85a + 0.61a)}{\pi a^2}$$

$$= 1.19\left[\frac{0.0381 + 1.46 \times 0.0127}{\pi(0.0127)^2}\right]$$

$$= 133 \text{ kg/m}^4$$

The volume velocity through the neck of the jug is the same as that entering the air cavity inside. Hence, the two elements are in series and will resonate at

$$f = \frac{1}{2\pi \sqrt{M_A C_A}} = \frac{10^4}{2\pi \sqrt{133 \times 5.89}}$$

$$= 57 \text{ cps}$$

5.7. Acoustic Resistances. Any device in which the flow of gas occurs in phase with and directly proportional to the applied pressure may be represented as a pure acoustic resistance. In other words, there is no stored (reactive) energy associated with the flow. Four principal forms of acoustic resistance are commonly employed in acoustic devices: fine-meshed screens made of metal or cloth, small-bore tubes, narrow slits, and porous acoustical materials.

Screens are often used in acoustic transducers because of their low cost, ease of selection and control in manufacture, satisfactory stability, and relative freedom from inductive reactance. Slits are often used where an adjustable resistance is desired. This is accomplished by changing the width of the slit. Tubes have the disadvantage that unless their diameter is very small, which in turn results in a high resistance, there is usually appreciable inductive reactance associated with them. However, if a combination of resistance and inductance is desired, they are useful. Such combinations will be treated later in this part. Fibrous or porous

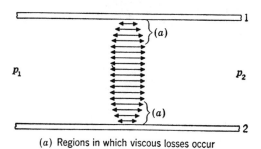

(a) Regions in which viscous losses occur

Fig. 5.13. Sketch showing the diminution of the amplitude of vibration of air particles in a sound wave near a surface. The letters (a) show the regions in which viscous losses occur.

acoustic materials, porous ceramics, and sintered metals are often used in industrial applications and are mixtures of mass and resistance. In all four forms of acoustic resistance, the frictional effects producing the resistance occur in the same manner.

In Fig. 5.13, we see the opposite sides 1 and 2 of a slit, or tube, or of one mesh of screen. An alternating pressure difference $(p_2 - p_1)$ causes a motion of the air molecules in the space between the sides 1 and 2. At 1 and 2 the air particles in contact with the sides must remain at rest. Halfway between the sides, the maximum amplitude of motion will occur. Frictional losses occur in a gas whenever adjacent layers of molecules move over each other with different velocities. Hence frictional losses occur in the example of Fig. 5.13 near each of the walls as marked by the letter (a). In any tube, slit, mesh, or interstice the losses become appreciable when the regions in which adjacent layers differ in velocity extend over the entire space.

Screens. The specific acoustic resistances of a variety of screen sizes commonly manufactured in the United States are shown in Table 5.4.

The acoustic resistance is obtained by dividing the values of Z_s in this table by the area of screen being considered.

The acoustic resistances of screens are generally determined by test and not by calculations.

TABLE 5.4.   Specific Acoustic Resistance of Single Layers of Screens

No. of wires per linear in.	Approx. diam of wire, cm	Rs, rayls, dyne-sec/cm³	Rs, mks rayls, newton-sec/m³
30	0.033	0.567	5.67
50	0.022	0.588	5.88
100	0.0115	0.910	9.10
120	0.0092	1.35	13.5
200	0.0057	2.46	24.6

Tube of Very Small Diameter[7] [*Radius a(meters)* $< 0.002/\sqrt{f}$]. The acoustic impedance of a tube of *very small* diameter, neglecting the end corrections, is

$$Z_A = \frac{8\eta l}{\pi a^4} + j\frac{4}{3} M_A \omega \qquad \text{newton-sec/m}^5 \qquad (5.48)$$

$$M_A = \frac{\rho_0 l}{\pi a^2} \qquad \text{kg/m}^4 \qquad (5.49)$$

where η = viscosity coefficient.   For air $\eta = 1.86 \times 10^{-5}$ newton-sec/m² at 20°C and 0.76 m Hg.   This quantity varies with temperature, that is, $\eta \propto T^{0.7}$, where T is in degrees Kelvin.

l = length of tube in meters.

a = radius of tube in meters.

M_A = acoustic mass of air in tube in kilograms per meter⁴.

ρ_0 = density of gas in kilograms per cubic meter.

The mechanical impedance of a very small tube is found by multiplying Eq. (5.48) by $(\pi a^2)^2$, which yields

$$Z_M = 8\pi\eta l + j\tfrac{4}{3} M_M \omega \qquad (5.50)$$

where $M_M = \rho_0 \pi a^2 l$ = mass of air in the tube in kilograms.

Narrow Slit[8] [*t(meters)* $< 0.003/\sqrt{f}$].   The acoustic impedance of a very narrow slit, neglecting end corrections, is

$$Z_A = \frac{12\eta l}{t^3 w} + j\frac{6\rho_0 l \omega}{5wt} \qquad \text{newton-sec/m}^5 \qquad (5.51)$$

[7] Lord Rayleigh, "Theory of Sound," 2d ed., Vol. II, pp. 323–328, Macmillan & Co., Ltd., London, 1929.   Also see L. L. Beranek, "Acoustic Measurements," pp. 187–186, John Wiley & Sons, Inc., New York, 1949.

[8] I. B. Crandall, "Vibrating Systems and Sound," D. Van Nostrand Company, Inc., New York, 1926.

where l = length of slit in meters in direction in which the sound wave is traveling (see Fig. 5.14)

  w = width of slit in meters as viewed from the direction from which the wave is coming (see Fig. 5.14)

  t = thickness of slit in meters (see Fig. 5.14)

The mechanical impedance of a very narrow slit is given by multiplying Eq. (5.51) by t^2w^2.

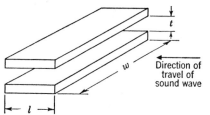

$$Z_M = \frac{12\eta lw}{t} + j\frac{6}{5}M_M\omega \quad (5.52)$$

where $M_M = \rho_0 lwt$ = mass of air in the slit in kilograms.

Fig. 5.14. Dimension of a narrow slit.

Example 5.3. An acoustic resistance of 1 acoustic ohm is desired as the damping element in an earphone. Select a screen and the diameter of hole necessary to achieve this resistance.

Solution. As the resistance is needed for an earphone, it should be quite small. If we select a 200-mesh screen (see Table 5.4), the acoustic resistance for unit area is 2.46 rayls. For 1 acoustic ohm, an area S is needed of

$$S = \frac{1}{2.46} = 0.406 \text{ cm}^2$$

The diameter d of the hole required for this area is

$$d = 2a = 2\sqrt{\frac{S}{\pi}} = 0.72 \text{ cm}$$

5.8. Cavity with Holes on Opposite Sides—Mixed Mass-Compliance Element. A special case of an element that is frequently encountered in

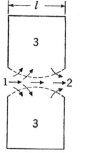

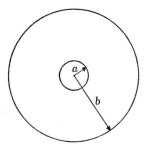

Fig. 5.15. Example of a mixed mass-compliance element made from a cavity with holes on opposite sides.

acoustical devices and that has often led to confusion in analysis is that shown in Fig. 5.15. Imagine this to be a doughnut-shaped element, each side of which has a hole of radius a bored in it. When a flow of air with a volume velocity U_1 enters opening 1, all the air particles in the vicinity of the opening will move with a volume velocity U_1. Part of this velocity goes to compress the air in the cylindrical space 3, and part of it appears as a movement of air that is not appreciably compressed. It was pointed out earlier that a portion of a gas that compresses without appreciable motion of the particles is to be treated as an acoustic compliance.

By inspection of Fig. 5.15 we see that the portion enclosed approximately by the dotted lines moves without appreciable compression and, hence, is an acoustic mass. That lying outside the dotted lines is an acoustic compliance. The analogous circuit for this acoustic device is given in Fig. 5.16. The volume velocity U_1 entering opening 1 divides into two parts, one to compress the air (U_3) and the other (U_2) to leave opening 2. By judicious estimation we arrive at values for M_A. If the length l of the cylinder is fairly long and the volume 3 is large, M_A is merely the end correction l'' of Eq. (5.44). If the volume 3 is small, then M_A becomes nearly the

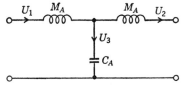

FIG. 5.16. Analogous circuit for the device of Fig. 5.15.

acoustic mass of a tube of radius a and length $l/2$. The acoustic compliance is determined by the volume of air lying outside of the estimated dotted lines of Fig. 5.15.

5.9. Intermediate-sized Tube—Mixed Mass-Resistance Element[9] [a(meters) $> 0.01\sqrt{f}$ and $a < 10/f$]. The acoustic impedance for a tube with a radius a (in meters) that is less than $0.002/\sqrt{f}$ was given by Eqs. (5.48) and (5.49). Here we shall give the acoustic impedance for a tube whose radius (in meters) is greater than $0.01/\sqrt{f}$ but still less than $10/f$. For a tube whose radius lies between $0.002/\sqrt{f}$ and $0.01/\sqrt{f}$ interpolation must be used. The acoustic impedance of the intermediate-sized tube is equal to

$$Z_A = R_A + j\omega M_A \qquad (5.53)$$

where

$$R_A = \frac{1}{\pi a^2}\rho_0\sqrt{2\omega\mu}\left[\frac{l'}{a} + (2)\right] \qquad \text{mks acoustic ohms} \qquad (5.54)$$

$$M_A = \frac{\rho_0[l' + (2)l'']}{\pi a^2} \qquad \text{kg/m}^4 \qquad (5.55)$$

a = radius of tube in meters.

ρ_0 = density of air in kilograms per cubic meter.

μ = kinematic coefficient of viscosity. For air at 20°C and 0.76 m Hg, $\mu \doteq 1.56 \times 10^{-5}$ m²/sec. This quantity varies about as $T^{1.7}/P_0$, where T is in degrees Kelvin and P_0 is atmospheric pressure.

l' = actual length of the tube.

[9] U. Ingard, "Scattering and Absorption by Acoustic Resonators," doctoral dissertation, Massachusetts Institute of Technology, 1950, and J. Acoust. Soc. Amer., **25**: 1044–1045 (1953).

l'' = end correction for the tube.   It is given by Eq. (5.44) if the tube is flanged or Eq. (5.46) if the tube is unflanged.   The numbers (2) in parentheses in Eqs. (5.54) and (5.55) must be used if both ends of the tube are being considered.   If only one end is being considered, replace the number (2) with the number 1.

ω = angular frequency in radians per second.

5.10. Perforated Sheet—Mixed Mass-Resistance Element[9] **[a(meters) > $0.01/\sqrt{f}$ and $a < 10/f$].**   Many times, in acoustics, perforated sheets are used as mixed acoustical elements.   We shall assume a perforated sheet with the dimensions shown in Fig. 5.17 and holes whose centers are spaced more than one diameter apart.   For this case the acoustic impedance for each area b^2, that is, each hole, is given by

$$Z_A = R_A + j\omega M_A$$

where

$$R_A = \frac{1}{\pi a^2} \rho_0 \sqrt{2\omega\mu} \left[\frac{t}{a} + 2 \left(1 - \frac{A_h}{A_b} \right) \right] \qquad \text{mks acoustic ohms} \qquad (5.56)$$

$$M_A = \frac{\rho_0}{\pi a^2} \left[t + 1.7a \left(1 - \frac{a}{b} \right) \right] \qquad (5.57)$$

$A_h = \pi a^2$ = area of hole in square meters

$A_b = b^2$ = area of a square around each hole in square meters

t = thickness of the sheet in meters

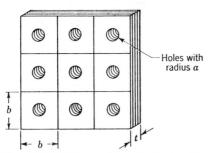

FIG. 5.17. Thin perforated sheet with holes of radius a, and length t, spaced a distance b on centers.

If there are n holes, the acoustic impedance is approximately equal to $1/n$ times that for one hole.

If this mass-resistance element is used with a compliance to form a resonant circuit, we are often interested in the ratio of the angular frequency of resonance ω_0 to the angular bandwidth w (radians per second) measured at the half-power points.   This ratio is called the "Q" of the circuit and is a measure of the sharpness of the resonance curve.

The "Q_A" of a perforated sheet when used with a compliance of such size as to produce resonance at angular frequency ω_0 is

$$Q_A = \frac{\omega_0 M_A}{R_A} = \sqrt{\frac{\omega_0}{2\mu}} \, a \, \frac{t + 1.7a[1 - (a/b)]}{t + 2a[1 - (\pi a^2/b^2)]} \qquad (5.58)$$

The Q_A is independent of the number of holes in the perforated sheet.

We repeat that these formulas are limited to cases where the centers of the holes are spaced more than a diameter apart.

5.11. Acoustic Transformers. As for the other acoustical elements, there is no configuration of materials that will act as a "lumped" transformer over a wide frequency range. Also, what may appear to be an acoustic transformer when impedances are written as mechanical impedances may not appear to be one when written as acoustic impedances, and vice versa. As an example of this situation, let us investigate the case of a simple discontinuity in a pipe carrying an acoustic wave.

Junction of Two Pipes of Different Areas. A junction of two pipes of different areas is equivalent to a discontinuity in the area of a single pipe (see Fig. 5.18a). If we assume that the diameter of the larger pipe is less

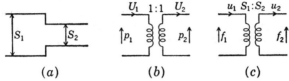

$$(a) \qquad\qquad (b) \qquad\qquad (c)$$

Fig. 5.18. (a) Simple discontinuity between two pipes. (b) Acoustic-impedance transducer representation of (a); because the transformation ratio is unity, no transformer is required. (c) Mechanical-impedance transducer representation of (a).

than $\lambda/16$, then we may write the following two equations relating the pressure and volume velocities at the junction:

$$p_1 = p_2 \qquad\qquad (5.59)$$
$$U_1 = U_2 \qquad\qquad (5.60)$$

Equation (5.59) says that the sound pressure on both sides of the junction is the same. Equation (5.60) says that the volume of air leaving one pipe in an interval of time equals that entering the other pipe in the same interval of time. The transformation ratio for acoustic impedances is unity so that no transformer is needed.

For the case of a circuit using lumped *mechanical* elements the discontinuity appears to be a transformer with a turns ratio of $S_1:S_2$ because, from Eq. (5.59),

$$\frac{f_1}{S_1} = \frac{f_2}{S_2} \qquad\qquad (5.61)$$

and, from Eq. (5.60),

$$u_1 S_1 = u_2 S_2 \qquad\qquad (5.62)$$

where f_1 and f_2 are the forces on the two sides of the junction and u_1 and u_2 are the average particle velocities over the areas S_1 and S_2. We have

$$f_1 = \frac{S_1}{S_2} f_2 \qquad\qquad (5.63)$$

and

$$u_2 = \frac{S_1}{S_2} u_1 \qquad (5.64)$$

so that

$$Z_{M1} = \frac{f_1}{u_1} = \left(\frac{S_1}{S_2}\right)^2 \frac{f_2}{u_2} = \left(\frac{S_1}{S_2}\right)^2 Z_{M2} \qquad (5.65)$$

A transformer is needed in this case and is drawn as shown in Fig. 5.18c.

It must be noted that a reflected wave will be sent back toward the source by the simple discontinuity. We saw in Part IV that, in order that there be no reflected wave, the specific acoustic impedance in the second tube (p_2/u_2) must equal that in the first tube (p_1/u_1). This is possible only if $S_1 = S_2$, that is, if there is no discontinuity.

To find the magnitude and phase of the reflected wave in the first tube resulting from the discontinuity, we shall use material from Part IV. Assume that the discontinuity exists at $x = 0$. The specific acoustic impedance in the first tube is

$$Z_{S1} = \frac{p_1}{u_1} \qquad (5.66)$$

If the second tube is infinitely long, the specific acoustic impedance for it at the junction will be

$$Z_{S2} = \frac{p_2}{u_2} = \rho_0 c \qquad (5.67)$$

[see Eq. (2.60)]. The impedance Z_{S1} at the junction is, from Eqs. (5.59) and (5.64),

$$Z_{S1} = \frac{p_1}{\frac{S_2 u_2}{S_1}} = \frac{p_2}{\frac{S_2 u_2}{S_1}} \qquad (5.68)$$

From Eqs. (5.67) and (5.68),

$$Z_{S1} = \frac{S_1}{S_2} \rho_0 c \qquad (5.69)$$

Using Eqs. (2.37) and (2.58), setting $x = 0$, we may solve for the rms reflected wave p_- in terms of the incident wave p_+.

$$p_1 = p_+ + p_- \qquad (5.70)$$

$$u_1 = \frac{1}{\rho_0 c} (p_+ - p_-) \qquad (5.71)$$

$$\frac{p_1}{u_1} = \frac{S_1}{S_2} \rho_0 c = \rho_0 c \frac{p_+ + p_-}{p_+ - p_-} \qquad (5.72)$$

$$p_- = \frac{S_1 - S_2}{S_1 + S_2} p_+ \qquad (5.73)$$

The sound pressure p_T of the transmitted wave in the second tube at the junction point must equal the sound pressure in the first tube at that point,

$$p_T = p_+ + p_-$$ (5.74)

so that

$$p_T = \frac{2S_1}{S_1 + S_2} p_+$$ (5.75)

If S_1 equals S_2, there is no reflected wave p_- and then $p_+ = p_T$.

Note also that if S_2 becomes vanishingly small, this case corresponds to the case of a rigid termination at the junction. For this case,

$$p_- = p_+$$ (5.76)

and

$$p_+ + p_- = 2p_+$$ (5.77)

This equation illustrates the often-mentioned case of *pressure doubling*. That is to say, when a plane sound wave reflects from a plane rigid surface, the sound pressure at the surface is double that of the incident wave.

Two Pipes of Different Areas Joined by an Exponential Connector.[10] An exponential connector may be used to join two pipes of different areas. Such a connector (see Fig. 5.19) acts as a simple discontinuity when its

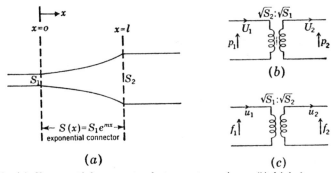

FIG. 5.19. (a) Exponential connector between two pipes; (b) high-frequency representation of (a) using acoustic-impedance transducer; (c) high-frequency representation of (a) using mechanical-impedance transducer.

length is short compared with a wavelength and as a transformer for *acoustic* impedances when its length is greater than a half wavelength. If the second tube is infinitely long, then at $x = l$ (see Fig. 5.19),

$$\frac{p_2}{u_2} = \rho_0 c$$ (5.78)

If the cross-sectional area of the exponential connector is given by

$$S(x) = S_1 e^{mx}$$ (5.79)

[10] H. F. Olson, "Elements of Acoustical Engineering," 2d ed., pp. 109–111, D. Van Nostrand, Company, Inc., New York, 1947.

and the length of the connector is l, then the specific acoustic impedance at $x = 0$ is

$$Z_{S1} = \frac{p_1}{u_1} = \rho_0 c\, \frac{\cos{(bl + \theta)} + j\sin{(bl)}}{\cos{(bl - \theta)} + j\sin{(bl)}} \tag{5.80}$$

where $b = \frac{1}{2}\sqrt{(4\omega^2/c^2) - m^2}$ in meters^{-1}
  m = flare constant in meters^{-1} [see Eq. (5.79)]
  $\theta = \tan^{-1}{(m/2b)}$
  c = speed of sound in meters per second
  l = length of the exponential connector in meters
  ρ_0 = density of air in kilograms per cubic meter
At low frequencies (b imaginary and λ/l large)

$$\frac{p_1}{U_1} = \frac{p_2}{U_2} \quad \text{or} \quad Z_{A1} = Z_{A2} \tag{5.81}$$

at high frequencies (b real and $l/\lambda > 1$)

$$\frac{p_1}{u_1} = \frac{p_2}{u_2} \tag{5.82}$$

or

$$Z_{A1} = \frac{S_2}{S_1} Z_{A2} \tag{5.83}$$

At intermediate frequencies the transformer introduces a phase shift, and the transformation ratio varies between the limits set by the two equations above.

The transformation ratio for acoustic impedance at high frequencies is seen from Eq. (5.83) to be $\sqrt{S_2/S_1}$ (see Fig. 5.19b). That is to say,

$$Z_{A1} = \left(\sqrt{\frac{S_2}{S_1}}\right)^2 Z_{A2} \tag{5.84}$$

For mechanical impedance at high frequencies the transformation ratio is seen from Eq. (5.83) to be $\sqrt{S_1/S_2}$ (see Fig. 5.19c). That is to say,

$$Z_{M1} = \left(\sqrt{\frac{S_1}{S_2}}\right)^2 Z_{M2}$$

Example 5.4. It is desired to resonate a bass-reflex loudspeaker box to 100 cps by drilling a number of holes in it. The box has a volume of 1 ft^3 and a wall thickness of $\frac{1}{2}$ in. Determine the size and number of holes needed, assuming a $Q_A = 4$ and a ratio of hole diameter to on-center spacing of 0.5.

Solution. From Eq. (5.58) we see that, approximately,

$$Q_A \doteq \frac{a\sqrt{\omega_0}}{(0.00395)\sqrt{2}}$$

$$a = \frac{(4)(0.00395)\sqrt{2}}{\sqrt{628}} = 0.000893 \text{ m}$$

The diameter of the hole in inches is $2a/0.0254 = 0.0703$ in.

The reactance of the box at resonance equals

$$X_B = \frac{1}{\omega_0 C_A} = \frac{\gamma P_0}{628 V} = \frac{1.4 \times 10^5}{628 \times 0.02832} = 7880 \text{ mks acoustic ohms}$$

The desired acoustic mass of the holes is

$$M_A = \frac{X_B}{\omega_0} = \frac{7880}{628} = 12.55 \text{ kg/m}^4$$

If there are n holes, the acoustic mass for each hole equals

$$n M_A = n(12.55) \text{ kg/m}^4$$

From Eq. (5.57),

$$n(12.55) = \frac{1.18}{\pi (0.000893)^2} [0.0125 + (0.00152)(0.783)]$$

$$n = \frac{(1.18)(0.0137)}{\pi (12.55)(8 \times 10^{-7})} \doteq 510 \text{ holes}$$

Example 5.5. Design a single-section T constant-k low-pass wave filter with a cutoff frequency of 100 cps and a design impedance of 10^3 mks acoustic ohms.

Solution. A single T section of this type is shown in Figs. 5.15 and 5.16 except that the acoustic masses for one section have the values $M_A/2$, because each M_A has to serve as the acoustic mass both for the section in question and the adjacent section. By definition, the cutoff frequency is

$$f_0 = \frac{1}{\pi \sqrt{M_A C_A}}$$

The design impedance is equal to

$$R_0 = \sqrt{\frac{M_A}{C_A}}$$

From these two equations we can solve for C_A and M_{A1}.

$$M_A = \frac{1}{\pi^2 f_0^2 C_A}$$
$$= C_A R_0^2$$

So

$$C_A = \frac{1}{\pi f_0 R_0}$$
$$= \frac{1}{\pi \times 100 \times 10^3} = 3.18 \times 10^{-6} \text{ m}^5/\text{newton}$$
$$M_A = 3.18 \times 10^{-6} \times 10^6 = 3.18 \text{ kg/m}^4$$

From Par. 5.8 and Eq. (5.55), with l' equal to zero, we get the size of the hole in the device of Fig. 5.15.

$$M_A = \frac{\rho_0(0.85a)}{\pi a^2}$$

$$\frac{(1.7)(1.18)}{\pi a} = 3.18$$

$$a = 0.2 \text{ m}$$

The diameter of the hole is 0.4 m. The volume of the cavity is

$$V = C\gamma P_0 = 1.4 \times 3.18 \times 10^{-6} \times 10^5$$
$$= 0.445 \text{ m}^3$$

The elements for the T section are thereby determined.

CHAPTER 6

MICROPHONES

PART **XIV** *General Characteristics of Microphones*

Microphones are electroacoustic transducers for converting acoustic energy into electric energy. They serve two principal purposes. First, they are used for converting music or speech into electric signals which are transmitted or processed in some manner and then reproduced. Second, they serve as measuring instruments, converting acoustic signals into electric currents which actuate indicating meters. In some applications like the telephone, high electrical output, low cost, and durability are greater considerations than fidelity of reproduction. In other applications, small size and high fidelity are of greater importance than high sensitivity and low cost. In measurement applications we may be interested in determining the sound pressure or the particle velocity. In some applications the microphone must operate without appreciable change in characteristics regardless of major changes in temperature and barometric pressure.

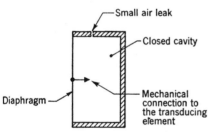

FIG. 6.1. Sketch of a pressure-actuated microphone consisting of a rigid enclosure, in one side of which there is a flexible diaphragm connected to a transducing element.

For these different applications, a variety of microphones have been developed. For purposes of discussion in this part they are divided into three broad classes in each of which there are a number of alternative constructions. The classes are:

1. Pressure microphones.
2. Pressure-gradient microphones.
3. Combinations of (1) and (2).

In this part we shall describe the distinguishing characteristics of these

144

three types. In the next two parts we shall discuss in detail several examples of each type involving electromagnetic, electrostatic, and piezoelectric types of transduction.

6.1. Pressure Microphones. A pressure microphone is one that responds to changes in sound pressure. A common example of a pressure microphone is one with a diaphragm, the back side of which is terminated in a closed cavity (see Fig. 6.1). A tiny hole through the wall of the cavity keeps the average pressure inside of the cavity at atmospheric pressure. However, rapid changes in pressure, such as those produced by a sound wave, cause the diaphragm to move backward and forward.

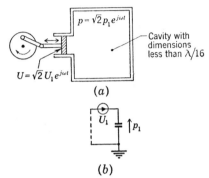

(a)

(b)

FIG. 6.2. Sketch of a pressure chamber.

If a pressure microphone is placed in a small cavity in which the pressure is varied, as shown in Fig. 6.2, the output voltage will be the same regardless of what position the microphone occupies in the cavity. On the other hand, if a pressure microphone is placed at successive points 1, 2, 3, and 4 of Fig. 6.3a, it will respond differently at each of these points for reasons that can be seen from

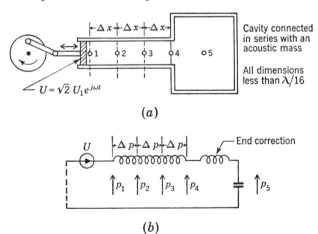

(a)

(b)

FIG. 6.3. Sketch of an arrangement in which a pressure gradient is produced.

Fig. 6.3b. The pressure drops p_1, p_2, p_3, and p_4 are different from each other by an amount Δp, if the spacings Δx are alike.

If a pressure microphone is placed in a plane sound wave of constant intensity I (watts flowing through a unit area in the plane of the wave front), the force acting to move the diaphragm will be independent of frequency because $p = \sqrt{I\rho_0 c}$ [see Eq. (2.88)].

6.2. Pressure-gradient Microphones. A pressure-gradient micro-phone is one that responds to a difference in pressure at two closely spaced points. A common example of this type of microphone has a diaphragm, both sides of which are exposed to the sound wave. Such a construction is shown in Fig. 6.4.

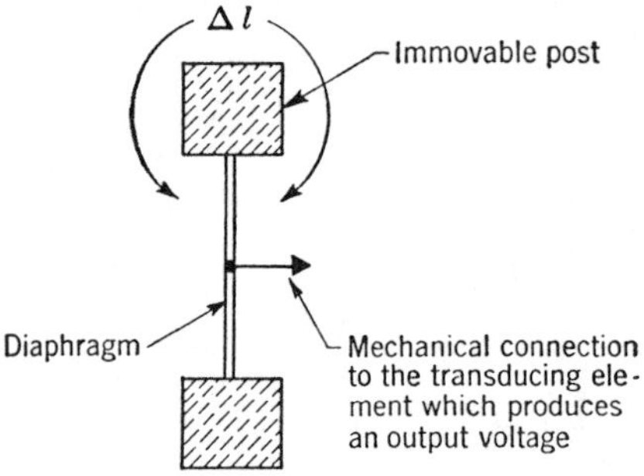

If a pressure-gradient micro-phone is placed in the cavity of Fig. 6.2a, there will be no net force acting on the diaphragm and its output will be zero. This happens because there is no pressure gradient in the cavity. In contrast, if a pressure-gradient micro-phone is placed at the successive positions 1 to 4 of Fig. 6.3a, it will produce an output voltage proportional to the pressure gradient $\Delta p/\Delta x$. In other words, if Δx is the same between successive points, the microphone output will be independent of whichever of the four positions it occupies in Fig. 6.3a.

Fig. 6.4. Sketch of a pressure-gradient microphone consisting of a movable diaphragm, both sides exposed, connected to a transducing element.

If a very small pressure-gradient microphone is placed in a plane sound wave traveling in the x direction, the complex rms force f_D acting to move

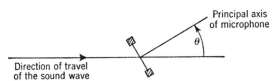

Fig. 6.5. Pressure-gradient microphone with principal axis located at an angle θ with respect to the direction of travel of the sound wave.

the diaphragm will be

$$f_D = -S \frac{\partial p}{\partial x} \Delta l \cos \theta \qquad (6.1)$$

where p = rms sound pressure

$\frac{\partial p}{\partial x} \cos \theta$ = component of the x gradient of pressure acting across the faces of the diaphragm

θ = angle the normal to the diaphragm makes with the direction of travel of the wave (see Fig. 6.5)

Δl = effective distance between the two sides of the diaphragm (see Fig. 6.4)

S = area of diaphragm

The equation for a plane traveling sound wave has already been given [Eq. (2.57)]; it is

$$p = p_0 e^{-jkx} \qquad (6.2)$$

where $k = \omega/c$

p_0 = rms pressure at $x = 0$

If we assume that the introduction of the microphone into the sound field does not affect the pressure gradient, we may substitute Eq. (6.2) into Eq. (6.1) and get

$$f_D = \frac{jp_0\omega S \, \Delta l \, \cos\theta}{c} e^{-jkx} \tag{6.3}$$

The magnitude of the force at any point x is

$$|f_D| = \frac{|p|\omega S \, \Delta l \, \cos\theta}{c} \tag{6.4}$$

It should be remembered [see Eq. (2.4)] that in the steady state the pressure gradient is proportional to $j\omega\rho_0$ times the component of particle velocity in the direction the gradient is being taken. The force f_D is, therefore, proportional to the particle velocity at any given frequency. Reference to Fig. 6.5 is sufficient to convince one that when $\theta = 90°$, the force acting on the diaphragm will be zero, because conditions of symmetry require that the pressure be the same on both sides of the diaphragm. From Eq. (6.4) we also see that the effective force acting on the diaphragm is proportional to frequency and to the rms sound pressure.

In spherical coordinates, for a microphone whose dimensions are small compared with r, Eq. (6.1) becomes

$$f_D = -S \frac{\partial p}{\partial r} \, \Delta l \, \cos\theta \tag{6.5}$$

The equation for a spherical wave is found from Eq. (2.62).

$$p = \frac{A_0}{r} e^{-jkr} \tag{6.6}$$

Substituting (6.6) in (6.5) gives

$$f_D = \frac{A_0(1 + jkr)}{r^2} e^{-jkr} S(\Delta l \, \cos\theta) \tag{6.7}$$

This yields

$$|f_D|_{\text{rms}} = \frac{|p|\omega S \, \Delta l \, \cos\theta}{c} \frac{\sqrt{1 + k^2 r^2}}{kr} \tag{6.8}$$

However, we see from Eq. (2.60) that in a plane wave the rms velocity is related to the rms pressure by

$$|u| = \frac{|p|}{\rho_0 c} \tag{6.9}$$

and in a spherical wave [Eq. (2.64)]

$$|u| = \frac{|p|}{\rho_0 c} \frac{\sqrt{1 + k^2 r^2}}{kr} \tag{6.10}$$

where $|u|$ is the rms particle velocity in the direction of travel of the sound wave. Hence, Eqs. (6.4) and (6.8) become

$$|f_D| = |u|\omega\rho_0 S \, \Delta l \cos \theta \qquad (6.11)$$

In other words, the effective (rms) force f_D acting on the diaphragm of a pressure-gradient microphone is directly proportional to the effective particle velocity in the direction of propagation of the wave, to the frequency, to the density of the air, to the size and area of the diaphragm, and to the angle it makes with the direction of propagation of the sound wave. This statement is true for any type of wave front, plane, spherical, cylindrical, or other, provided the microphone is so small that its presence does not disturb the sound wave.

At any given frequency, the response of the microphone is proportional to the cos θ, which yields the *directivity pattern* shown in Fig. 6.6a. This

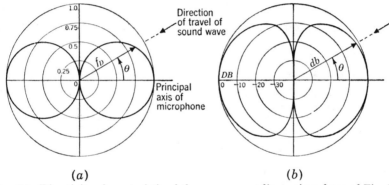

(a)   (b)

Fig. 6.6. Directivity characteristic of the pressure-gradient microphone of Fig. 6.4.

shape of plot is commonly referred to as a "figure 8" pattern. The same pattern, plotted in decibels relative to the force at $\theta = 0$, is given in (b). It is interesting to observe that the pattern is the same as that for an acoustic doublet or for an unflanged diaphragm at low frequencies (see Figs. 4.8 and 4.13).

The frequency response of a pressure-gradient (particle-velocity) microphone, when placed in a spherical wave, is a function of the curvature of the wave front. That is to say, from Eq. (6.10) we see that for values of $k^2 r^2$ (kr equals $\omega r/c$) large compared with 1 the particle velocity is linearly related to the sound pressure. A large value of kr means that either the frequency is high or the radius of curvature of the wave front is large. However, for values of $k^2 r^2$ small compared with 1, which means that the radius of curvature is small or the frequency is low, or both, the particle velocity is proportional to $|p|/(\omega r)$. As a result, when a person talking or singing moves near to a pressure-gradient microphone so that r is small his voice seems to become more "boomy" or "bassy" because the output of the microphone increases with decreasing frequency.

6.3. Combination Pressure and Pressure-gradient Microphones. A combination pressure and pressure-gradient microphone is one that responds to both the pressure and the pressure gradient in a wave. A common example of such a microphone is one having a cavity at the back side of the diaphragm that has an opening to the outside air containing an acoustic resistance (see Fig. 6.7a).

The analogous circuit for this device is shown in Fig. 6.7b. If we let

$$p_1 = p_0 e^{-jkx} \qquad (6.12)$$

$$p_2 = p_1 + \frac{\partial(p_0 e^{-jkx})}{\partial x} \Delta l \cos \theta$$

$$= p_1 \left(1 - j \frac{\omega}{c} \Delta l \cos \theta\right) \quad (6.13)$$

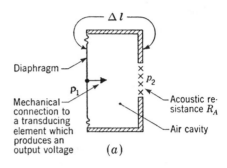

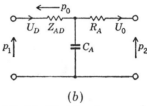

Let us say that U_D is the rms volume velocity of the diaphragm; U_0 is the rms volume velocity of the air passing through the resistance; p_D is the net rms pressure acting to move the diaphragm; and Z_{AD} is the diaphragm impedance.   Then we can write the following equations from Fig. 6.7b:

FIG. 6.7. (a) Sketch of a combination pressure and pressure-gradient microphone consisting of a right enclosure in one side of which is a movable diaphragm connected to a transducing element and in another side of which is an opening with an acoustic resistance R_A. (b) Acoustic-impedance circuit for (a).

$$U_D \left(Z_{AD} + \frac{1}{j\omega C_A}\right) - \frac{U_0}{j\omega C_A} = p_1$$

$$- \frac{U_D}{j\omega C_A} + U_0 \left(R_A + \frac{1}{j\omega C_A}\right) = -p_2$$

$$(6.14)$$

The pressure difference across the diaphragm is

$$p_D = U_D Z_{AD} = \frac{Z_{AD}\left(p_1 R_A + \dfrac{p_1 - p_2}{j\omega C_A}\right)}{Z_{AD}R_A - j[(R_A + Z_{AD})/\omega C_A]} \qquad (6.15)$$

Substitution of (6.13) in (6.15) yields

$$p_D = p_1 \frac{Z_{AD}\left(R_A + \dfrac{\Delta l \cos \theta}{c C_A}\right)}{Z_{AD}R_A - j[(R_A + Z_{AD})/\omega C_A]} \qquad (6.16)$$

Let

$$\frac{\Delta l}{cC_A R_A} \equiv B \tag{6.17}$$

where B is an arbitrarily chosen dimensionless constant. Since $f_D = p_D S$, where S is the effective area of the diaphragm, we have

$$|f_D| = p_1|\alpha|S(1 + B \cos \theta) \tag{6.18}$$

where α is the ratio

$$\alpha = \frac{Z_{AD} R_A}{Z_{AD} R_A - j[(R_A + Z_{AD})/\omega C_A]} \tag{6.19}$$

A plot of the force $|f_D|$ acting on the diaphragm as a function of θ for $B = 1$ is shown in Fig. 6.8a. The same pattern plotted in decibels is

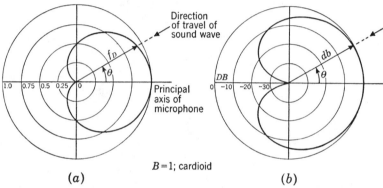

FIG. 6.8. Directivity characteristic of the combination pressure and pressure-gradient microphone of Fig. 6.7.

given in (b). The *directivity pattern* for $B = 1$ is commonly called a *cardioid pattern*. Other directivity patterns are shown in Fig. 6.46 for $B = 0, 0.5, 1, 5,$ and ∞.

PART XV *Pressure Microphones*

Pressure microphones are the most widely used of the three basic types discussed in the preceding part. They are applicable to acoustic measuring systems and to the pickup of music and speech in broadcast studios, in public-address installations, and in hearing aids. Many engineers and artists believe that music reproduced from the output of a well-designed pressure microphone is superior to that from the more directional types of microphone because the quality of the reverberation in the auditorium or studio is fully preserved, because undesirable wave-form distortion is minimized, and because the quality of the reproduced sound is not as strongly dependent as for other types upon how close the talker or the musical instrument is to the microphone.

Three principal types of pressure microphones are commonly found in broadcast and public-address work and in hearing aids. They are the electromagnetic, electrostatic, and piezoelectric types. We shall analyze one commercially available microphone of each of these three types in the next few sections of this part. Various other types of microphones are used in other applications, such as the carbon microphone in telephone systems, the hot-wire microphone in aerodynamic measurements, and the Rayleigh disk in absolute particle-velocity measurements. Lack of space precludes their inclusion here.

6.4. Electromagnetic Moving-coil Microphone (Dynamic Microphone).

General Features. The moving-coil electromagnetic microphone is a medium-priced instrument of high sensitivity. It is principally used in broadcast work and in applications where long cables are required or where rapid fluctuations or extremes in temperature and humidity are expected.

The best designed moving-coil microphones have open-circuit voltage responses to sounds of random incidence that are within 5 db of the average response over the frequency range between 40 and 8000 cps. Sound pressures as low as 20 db and as high as 140 db *re* 0.0002 microbar can be measured. Changes of response with temperature, pressure, and humidity are believed to be, in the better instruments, of the order of 3 to 5 db maximum below 1000 cps for the temperature range of 10 to 100°F, pressure range of 0.65 to 0.78 m Hg, and humidity range of 0 to 90 per cent relative humidity.

The electrical impedance is that of a coil of wire. Below 1000 cps, the resistive component predominates over the reactive component. Most moving-coil microphones have a nominal electrical impedance of about 20 ohms. The mechanical impedance is not high enough to permit use in a closed cavity without seriously changing the sound pressure therein.

To connect a dynamic microphone to the grid of a vacuum tube, a transformer with a turns ratio of about 30:1 is required.

Construction. The electromagnetic moving-coil microphone consists of a diaphragm that has fastened to it a coil of wire situated in a magnetic field (see Fig. 6.9a). In addition, there are acoustical circuits behind and in front of the diaphragm to extend the response of the microphone over a greater frequency range. A cross-sectional sketch of a widely used type of moving-coil microphone is shown in Fig. 6.9b.

Electro-mechano-acoustical Relations. The sound passes through the protective screens and arrives at a grid in front of the diaphragm. This grid has a number of small holes drilled in it which form a small acoustic mass and a small acoustic resistance. This grid is so small that its radia-

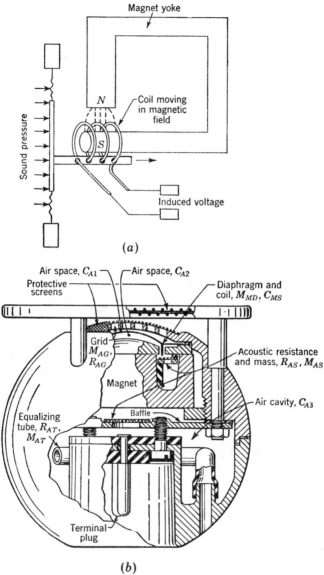

FIG. 6.9. (a) Diagrammatic representation of the essential elements of a moving-coil (dynamic) microphone. (*From Beranek, "Acoustic Measurements," John Wiley & Sons, Inc., New York*, 1949.)   (b) Cross-sectional sketch of a commercially available moving-coil microphone. [*From Marshall and Romanow, A Non-directional Microphone, Bell Syst. Tech. J.*, **15**: 405–423 (1936).]

tion impedance, operating as a loudspeaker, is essentially reactive over the whole frequency range (see Fig. 5.3). The space between the grid and diaphragm is a small acoustic compliance. Hence, the total acoustical circuit *in front of the diaphragm* is that of Fig. 6.10. The pressure p_{B1} is that which the sound wave would produce at the face of the grid if the holes of the grid were closed off. U_G is the volume velocity of the air that moves through the grid. U_D is the volume velocity of the diaphragm and is equal to the effective linear velocity u_D of the diaphragm times its

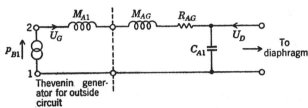

FIG. 6.10. Acoustical circuit for the elements in front of the diaphragm of the microphone of Fig. 6.9 (acoustic-impedance analogy).

effective area S_D. The radiation mass looking outward from the grid openings is M_{A1}. The acoustic mass and resistance of the grid openings are M_{AG} and R_{AG}. The compliance of the air space in front of the diaphragm is C_{A1}. At all frequencies, except the very highest, the effect of the protective screen can be neglected.

Behind the diaphragm the acoustical circuit is more complicated. First there is an air space between the diaphragm and the magnet that forms an acoustic compliance (see Fig. 6.9b). This air space connects

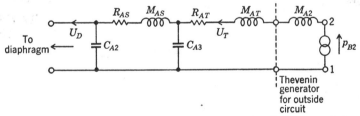

FIG. 6.11. Acoustical circuit for the elements behind the diaphragm of the microphone of Fig. 6.9 (acoustic-impedance analogy).

with a large air cavity that is also an acoustic compliance. In the connecting passages there are screens that serve as acoustic resistances. Also, the interconnecting passages form an acoustic mass. The large air cavity connects to the outside of the microphone through an "equalizing" tube.

The complete acoustical circuit behind the diaphragm is given in Fig. 6.11. The acoustic compliance directly behind the diaphragm is C_{A2}; the acoustic resistances of the screens are R_{AS}; the acoustic mass of the interconnecting passage is M_{AS}; the acoustic compliance of the large air space is C_{A3}; the acoustic mass of the tube is M_{AT}, and its resistance is R_{AT}; the radiation mass looking out from the tube is M_{A2}. The pressure p_{B2} is the sound pressure that would be produced at the outer end of the tube if.it were blocked so that no air can move through it, that is, $U_T = 0$.

At low frequencies, where the separation between the outer end of the equalizing tube and the grid is small compared with a wavelength, p_{B1} equals p_{B2}. At high frequencies where p_{B1} would be expected to differ from p_{B2}, ωM_{AT} becomes so large and $1/\omega C_{A3}$ so small that the movement

MECHANICAL MOBILITY ANALOGY

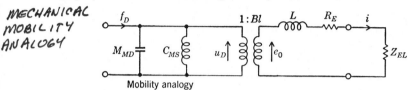

Mobility analogy

Fig. 6.12. Mechano-electrical circuit of diaphragm, voice coil, and magnetic field of the microphone of Fig. 6.9 (mechanical-mobility analogy).

of the diaphragm U_D is independent of p_{B2}. Hence, we can draw our circuit as though $p_{B1} = p_{B2}$ over the entire frequency range with negligible error. This will be done in Fig. 6.14.

The electromechanical circuit (mechanical-mobility analogy) for the diaphragm and voice coil is given in Fig. 6.12. The force exerted on the diaphragm is f_D, and its resulting velocity is u_D. Here, M_{MD} = mass of diaphragm and voice coil; C_{MS} = compliance of the suspension; L = inductance of voice coil; and R_E = electric resistance of the voice coil.

MECHANICAL IMPEDANCE ANALOGY

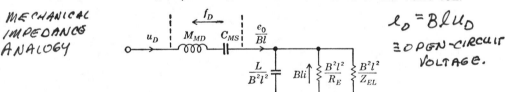

$e_0 = Bl u_D$ ≡ OPEN-CIRCUIT VOLTAGE.

Fig. 6.13. Mechano-electrical circuit of the diaphragm, voice coil, and magnetic field of the microphone of Fig. 6.9 (mechanical-impedance analogy). Note that u_D is also equal to e_0/Bl.

Z_{EL} is the electric impedance of the electric load to which the microphone is connected. The quantity $e_0 = Blu_D$ is the open-circuit voltage produced by the microphone.

In order to combine Figs. 6.10, 6.11, and 6.12, the dual of Fig. 6.12 must first be taken; it is shown in Fig. 6.13. Now, in order to join Figs. 6.10, 6.11, and 6.13, all forces in Fig. 6.13 must be divided by the area of the diaphragm S_D and all velocities multiplied by S_D. This can be done by inserting an area transformer into the circuit. Allowing by definition $p_{B1} = p_{B2} \equiv p_B$, and recognizing that U_D must be the same for all three component circuits, we get the circuit of Fig. 6.14 for the moving-coil microphone.

Performance. The performance of the circuit of Fig. 6.14 can best be understood by reference to Fig. 6.15, which is derived from 6.14. Let us assume from now on that $Z_{EL} \rightarrow \infty$. This means that the electrical terminals are open-circuited so that the voltage appearing across them is the open-circuit voltage e_o (see Fig. 6.12). In the circuit of Fig. 6.14, the "short-circuit" velocity is equal to e_o/Bl.

At very low frequencies Fig. 6.14 reduces to Fig. 6.15a. The generator p_B is effectively open-circuited by the three acoustic compliances C_{A1}, C_{A2}, and C_{A3}, of which only C_{A3} has appreciable size. Also, the three resistances R_{AT}, R_{AG}, and R_{AS} are small compared with the reactances of C_{A3} and $C_{MS}S_D{}^2$. Physically, the fact that u_D, the diaphragm velocity, is so small means that the pressure on the two sides of the diaphragm is the same, which obviously is true for slow changes in the atmospheric pressure. Hence, e_o is very small. This region is marked (a) in Fig. 6.16, where we

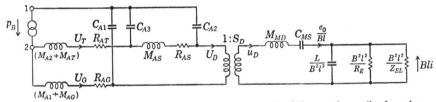

Fig. 6.14. Complete electro-mechano-acoustical circuit of the moving-coil microphone of Fig. 6.9 (impedance analogy). The electromechanical transformer has been cleared from the circuit.

see the voltage response in decibels as a function of frequency. In region (a) the response increases at the rate of 12 db per octave increase in frequency.

At a higher frequency (see Fig. 6.15b) a resonance condition develops involving the large cavity behind the diaphragm with compliance C_{A3} and the equalizing tube with mass and resistance M_{AT} and R_{AT}. Here, the forces on the rear of the diaphragm exceed those on the front. This region is marked (b) on Fig. 6.16.

As the frequency increases (see Fig. 6.15c), a highly damped resonance condition occurs involving the resistance and mass of the screens behind the diaphragm, R_{AS} and M_{AS}, and the diaphragm constants themselves, M_{MD} and C_{MS}. This is region (c) of Fig. 6.16. A highly important design feature, therefore, is a resistance of the screens R_{AS} large enough so that the response curve in region (c) is as flat as possible.

Above region (c) (see Fig. 6.15d), a resonance condition results that involves primarily the mass of the diaphragm M_{MD} and the stiffness of the air immediately behind it, C_{A2}. This yields the response shown in region (d) of Fig. 6.16.

Finally, a third resonance occurs involving primarily the acoustical elements in front of the diaphragm [see Fig. 6.15e and region (e) of Fig.

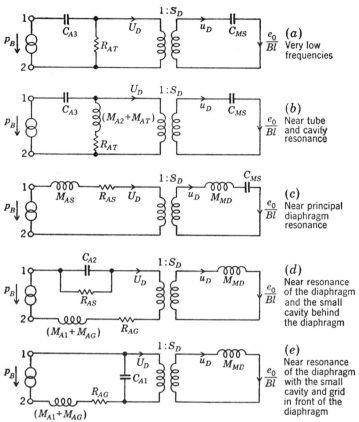

FIG. 6.15. Moving-coil microphones. Simplified circuits for five frequency regions (impedance analogy). The excess pressure produced by the sound wave at the grid of the microphone with the holes in the grid blocked off is p_B, and the open-circuit voltage is e_o.

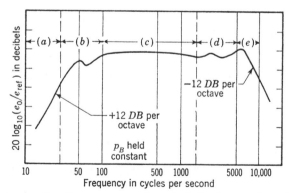

FIG. 6.16. Open-circuit-voltage response characteristic of a moving-coil microphone of the type shown in Fig. 6.9. Normally the reference voltage e_{ref} in the argument of the logarithm is taken to be 1 volt.

6.16]. The response then drops off at the rate of -12 db per doubling of frequency.

These various resonance conditions result in a microphone whose response is substantially flat from 50 to 8000 cps except for diffraction effects around the microphone. These diffraction effects will influence the response in different ways, depending on the direction of travel of the sound wave relative to the position of the microphone. The usual effect is that the response is enhanced in regions (d) and (e) if the sound wave impinges on the microphone grid at normal (perpendicular) incidence compared with grazing incidence. One purpose of the outer protective screen is to minimize this enhancement.

6.5. Electrostatic Microphone (Capacitor Microphone). *General Features.* The electrostatic type of microphone is used extensively as a standard microphone for the measurement of sound pressure and as a studio microphone for the high-fidelity pickup of music. It can be made

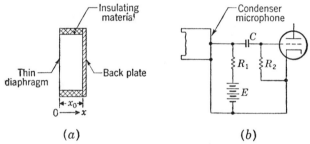

(a) (b)

Fig. 6.17. (a) Schematic representation of an electrostatic microphone. (b) Simple vacuum-tube circuit for use with capacitor microphone.

small in size so it does not disturb the sound field appreciably in the frequency region below 1000 cps.

Sound-pressure levels as low as 35 db and as high as 140 db *re* 0.0002 microbar can be measured with standard instruments. The mechanical impedance of the diaphragm is that of a stiffness and is high enough so that measurement of sound pressures in cavities is possible. The electrical impedance is that of a pure capacitance.

The temperature coefficient for a well-designed capacitor microphone is less than 0.025 db for each degree Fahrenheit rise in temperature.

Continued operation at high relative humidities may give rise to noisy operation because of leakage across the insulators inside. Quiet operation can be restored by desiccation.

Construction. In principle, the electrostatic microphone consists of a thin diaphragm, a very small distance behind which there is a back plate (see Fig. 6.17). The diaphragm and back plate are electrically insulated from each other and form an electric capacitor. Two commercial forms of this type of microphone are shown in Fig. 6.18. The principal devia-

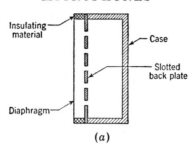

(a)

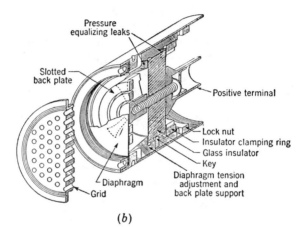

(b)

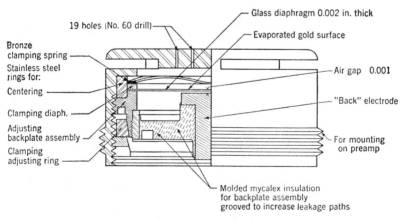

(c)

Fig. 6.18. (a) Schematic representation of a microphone with a slotted back plate. (b) Cross-sectional drawing of the W.E. 640-AA capacitor microphone. The slotted back plate serves both as the second terminal of the condenser and as a means for damping the principal resonant mode of the diaphragm. (c) Cross-sectional drawing of the Altec 21-C capacitor microphone. The cap with holes in it serves both for protection and as an acoustic network at high frequencies. (*From Beranek, "Acoustic Measurements," John Wiley & Sons, Inc., New York, 1949, and courtesy of Altec Lansing Corporation.*)

tion in construction of the 640-AA from the 21-C is that the back plate in the former has several annular slots cut in it. These slots form an acoustic resistance that serves to damp the diaphragm at resonance. One manner in which the microphones are operated is shown in Fig. 6.17b. The resistance R_1 is very large. The direct voltage E is several hundred volts and acts to polarize the microphone.

Electromechanical Relations. Electrically, the electrostatic microphone is a capacitor with a capacitance that varies with time so that the total charge $Q(t)$ is

$$Q(t) = q_0 + q(t) = C_E(t)[E + e(t)] \tag{6.20}$$

where q_0 is the quiescent charge in coulombs, $q(t)$ is the incremental charge in coulombs, $C_E(t)$ is the capacitance in farads, E is the quiescent polarizing voltage in volts, and $e(t)$ is the incremental voltage in volts.

The capacitance $C_E(t)$ in farads is equal to (see Fig. 6.17a)

$$C_E(t) = C'_{E0} + C_{E1}(t) = \frac{\epsilon_0 S}{x_0 - x(t)} \doteq \frac{\epsilon_0 S}{x_0}\left[1 + \frac{x(t)}{x_0}\right]$$
$$\doteq C'_{E0}\left[1 + \frac{x(t)}{x_0}\right] \tag{6.21}$$

where C'_{E0} is the capacitance in farads for $x(t) = 0$ and $C_{E1}(t)$ is the incremental capacitance in farads, ϵ_0 is a factor of proportionality that for air equals 8.85×10^{-12}, S is the effective area of one of the plates in square meters, x_0 is the quiescent separation in meters, and $x(t)$ is the incremental separation in meters. It is assumed in writing the right-hand term of (6.21) that the square of the maximum value of $x(t)$ is small compared with x_0^2.

If we similarly assumed that $[e(t)]^2_{max} \ll E^2$, then (6.20) and (6.21) yield

$$q_0 + q(t) = C'_{E0}E + C'_{E0}E\left[\frac{e(t)}{E} + \frac{x(t)}{x_0}\right] \tag{6.22}$$

so that

$$q(t) = C'_{E0}\left[e(t) + \frac{E}{x_0}x(t)\right] \tag{6.23}$$

The total stored potential energy $W(t)$ at any instant is equal to the sum of the stored electrical and mechanical energies, $\frac{1}{2}Q(t)^2/C_E(t)$ plus $\frac{1}{2}x(t)^2/C_{MS}$, where C_{MS} is the mechanical compliance of the moving plate in meters per newton. That is,

$$W(t) = \frac{1}{2}\frac{[q_0 + q(t)]^2}{C'_{E0} + C_{E1}(t)} + \frac{1}{2}\frac{x(t)^2}{C_{MS}} \doteq \frac{1}{2}\frac{q_0^2 + 2q_0q(t)}{C'_{E0}\left[1 + \dfrac{x(t)}{x_0}\right]}$$
$$+ \frac{1}{2}\frac{x(t)^2}{C_{MS}} = \frac{1}{2}\frac{q_0}{C'_{E0}}[q_0 + 2q(t)]\left[1 - \frac{x(t)}{x_0}\right] + \frac{1}{2}\frac{x(t)^2}{C_{MS}} \tag{6.24}$$

The force at any instant in newtons acting to move the plate is, from the equation for work, $dW = f\,dx$,

$$f_0 + f(t) = \frac{dW(t)}{dx} \qquad (6.25)$$

so that, by differentiation of Eq. (6.24),

$$f_0 + f(t) \doteq -\frac{q_0}{2x_0 C'_{E0}}\,[q_0 + 2q(t)] + \frac{x(t)}{C_{MS}}$$

$$= -\frac{q_0^2}{2x_0 C'_{E0}} + \left[\frac{x(t)}{C_{MS}} - \frac{q(t)q_0}{x_0 C'_{E0}}\right] \qquad (6.26)$$

Hence, because $E = q_0/C'_{E0}$,

$$f(t) = \frac{x(t)}{C_{MS}} - \frac{Eq(t)}{x_0} \qquad (6.27)$$

Rearranging Eq. (6.23) gives

$$e(t) \doteq -\frac{Ex(t)}{x_0} + \frac{q(t)}{C'_{E0}} \qquad (6.28)$$

In the steady state,

$$j\omega q = i$$
$$j\omega x = u \qquad (6.29)$$

where q, i, x, and u are now taken to be complex rms quantities; so Eqs. (6.27) and (6.28) become

$$f = \frac{1}{j\omega C_{MS}}\,u - \frac{E}{j\omega x_0}\,i \qquad (6.30)$$

$$e = -\frac{E}{j\omega x_0}\,u + \frac{1}{j\omega C'_{E0}}\,i \qquad (6.31)$$

with e and f also being complex rms quantities.

Analogous Circuits. Equations (6.30) and (6.31) may be represented by either of the networks shown in Fig. 6.19, where

$$C'_{E0} = \frac{C_{E0}C_{MS}(x_0^2/E^2C_{MS}^2)}{C_{E0} + C_{MS}(x_0^2/E^2C_{MS}^2)} = \frac{C_{E0}x_0^2}{E^2C_{MS}C_{E0} + x_0^2}$$

$$= \frac{C_{E0}}{(E^2/x_0^2)C_{MS}C_{E0} + 1} \qquad (6.32)$$

$$C_{MS} = \frac{C'_{MS}}{(E^2/x_0^2)C'_{MS}C_{E0} + 1} \qquad (6.33)$$

Note in particular that C_{MS} is the mechanical compliance measured with the electric current $i = 0$; C_{E0} is the electric capacitance measured with the force f equal to zero; C'_{E0} is the electric capacitance measured with the velocity u equal to zero; and C'_{MS} is the mechanical compliance measured with the voltage $e = 0$. These circuits were first shown as Fig. 3.37, and the element sizes were given in Eqs. (3.34) to (3.37).   In practice, the circuit of Fig. 6.19b is ordinarily used for electrostatic microphones.

When one of the microphones shown in Fig. 6.18 is radiating sound into air, the force built up at the face of the microphone when a voltage is applied to electrical terminals (3-4 of Fig. 6.19) is very small.   Hence,

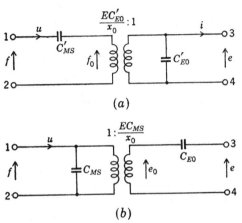

(a)

(b)

FIG. 6.19. Alternate electromechanical analogous circuits for electrostatic microphones (impedance analogy).

when an electric-impedance bridge is used to measure the capacitance of the microphone, the capacitance obtained is approximately equal to C_{EO}.

By Thévenin's theorem, the capacitor microphone in a free field can be represented by Fig. 6.20.   The quantity e_0 is the rms open-circuit voltage produced at the terminals of the microphone by the sound wave and equals [from Eq. (6.31) and Fig. 6.19]

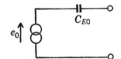

$$e_0 = -\frac{uE}{j\omega x_0} = -\frac{C_{MS}S p_B E}{x_0} \qquad (6.34)$$

FIG. 6.20. Thévenin's circuit of a capacitor microphone of the type shown in Fig. 6.17 situated in free space.

where the force f_B, acting on the microphone with the diaphragm blocked so that $u = 0$, is equal to the blocked pressure p_B times the area of the diaphragm S.

Acoustical Relations.   The microphones of Fig. 6.18 have diaphragms with the property of mass M_{MD} in addition to the mechanical compliance C_{MS} assumed so far.   For the 640-AA microphone, the internal acoustical circuit consists of an air space directly behind the diaphragm with an acoustic compliance, C_{A1}; a back plate and slots with acoustic resistance and mass, R_{AS} and M_{AS}; and an air cavity around and behind the plate with an acoustic stiffness, C_{A2}.   The radiation impedance looking outward from the front side of the diaphragm is $j\omega M_{AA}$, where M_{AA} is found from Eq. (5.31).   The complete acoustical and mechanical circuit in **the** impedance-type analogy is seen in Fig. 6.21. In this circuit p_B equals the rms pressure at the diaphragm when it is restrained from moving, $M_{AD} = M_{MD}/S^2$ = acoustic mass of the diaphragm, S = effective area of the diaphragm, and $U_D = Su_D$ = rms volume velocity of the diaphragm. All units are of the mks system.

When Fig. 6.19 is combined with Fig. 6.21, the complete circuit for the electrostatic microphone shown in Fig. 6.22 is obtained.

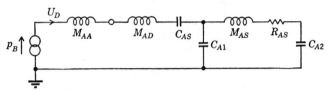

Fig. 6.21. Acoustical circuit of a capacitor microphone including the radiation mass and the acoustical elements behind the diaphragm (impedance analogy).

Performance. The performance of the capacitor microphone shown in Fig. 6.18b, viz., the Western Electric type 640-AA, can best be understood by reference to Figs. 6.23 and 6.24, which are derived from Fig. 6.22. At low frequencies the circuit is essentially that of Fig. 6.23a. From this circuit, the open-circuit voltage e_o is equal to

$$e_o = p_B \frac{E}{x_0 S} \frac{C_{A2}C_{AS}}{C_{A2} + C_{AS}} \tag{6.35}$$

At low frequencies, therefore, e_o is independent of frequency. This is the frequency region shown as (a) in Fig. 6.24.

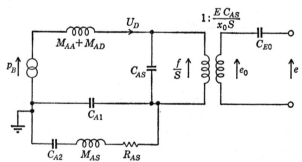

Fig. 6.22. Complete electroacoustical circuit of a capacitor microphone (impedance analogy).

In the vicinity of the first major resonance, the circuit becomes that of Fig. 6.23b. At resonance, the volume velocity through the compliance C_{AS} is limited only by the magnitude of the acoustic resistance R_{AS}. In general, this resistance is chosen to be large enough so that the resonance peak is only about 2 db (26 per cent) higher than the response at lower frequencies. The response near resonance is shown at (b) in Fig. 6.24.

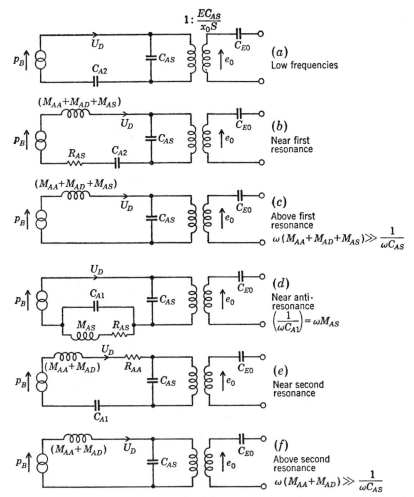

FIG. 6.23. Capacitor microphone. Simplified circuits for six frequency regions (impedance analogy). The excess pressure produced by the sound wave at the diaphragm of the microphone with the microphone held motionless is p_B, and the open-circuit voltage is e_o.

Above the resonance frequency, the circuit becomes that of Fig. 6.23c. The volume velocity is controlled entirely by the mass reactance. Hence,

$$e_o = \frac{p_B E}{\omega^2 (M_{AA} + M_{AD} + M_{AS}) x_0 S} \tag{6.36}$$

In this frequency region the response decreases at the rate of 12 db per octave [see region (c) of Fig. 6.24].

At a high frequency, the parallel circuit antiresonates, and the response drops very low. This condition can be seen by reference to Fig. 6.23d. Just above the point of antiresonance, a resonance occurs as shown in (e).

The magnitude of the resonance peak is here limited by the radiation resistance R_{AA}, which is no longer negligible compared with $j\omega M_{AA}$ [see (e) of Fig. 6.24]. Finally, above this resonance frequency, the open-circuit voltage again drops off at the rate of 12 db per octave [see (f)].

6.6. Piezoelectric Microphones. Piezoelectric microphones employ crystals or dielectrics which, when acted on by suitable forces, produce electrical potentials linearly related to the deformation of the substance.

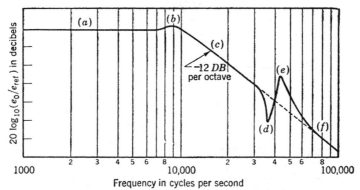

Fig. 6.24. Open-circuit-voltage response characteristic of a capacitor microphone of the type shown in Fig. 6.18b. Normally e_{ref} is taken as 1 volt.

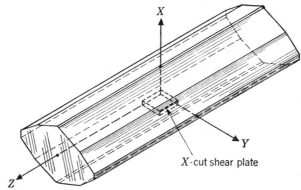

Fig. 6.25. Typical form of a large Rochelle salt crystal. The coordinate axes and the way in which an X-cut shear plate is cut from the crystal are indicated. (*Courtesy of Brush Electronics Company.*)

Piezoelectric substances have been used extensively in microphones because of their low cost and ruggedness.

Four common types of piezoelectric substances are discussed in this chapter, *viz.*, Rochelle salt, ammonium dihydrogen phosphate (ADP) and lithium sulfate crystals,† and barium titanate ceramic plates.

---

† ADP crystals are sold under the trade name PN, and lithium sulfate crystals under the trade name LH.

Crystal Microphones. Crystal microphones are used primarily in public-address systems, sound-level meters, and hearing aids. They have satisfactory frequency response for these applications and are high in sensitivity and low in cost. A diaphragm type can be purchased at low cost to cover the frequency range of 20 to 8000 cps with a maximum variation of 6 db from the average. Sound levels as low as 20 db and as high as 160 db *re* 0.0002 microbar may be measured.

The electrical impedance is that of a pure capacitance, which for some crystals varies with temperature, as we shall see later. The mechanical impedance for microphones using a diaphragm is not usually high enough

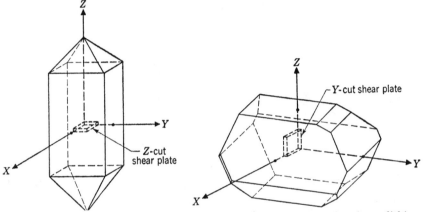

FIG. 6.26. Typical form of a large ammonium dihydrogen phosphate (ADP or PN) crystal. (*Courtesy of Brush Electronics Company.*)

FIG. 6.27. Typical form of a large lithium sulfate (LH) crystal. (*Courtesy of Brush Electronics Company.*)

so that the microphone may be used in a closed cavity without seriously changing the sound pressure therein.

Typical forms of the whole crystals of Rochelle salt, ADP, and lithium sulfate are shown in Figs. 6.25 to 6.27. Transducing elements are obtained by cutting slabs of material from these whole crystals. Usually these slabs are thin and either square or rectangular in shape. If the X axis of the crystal is perpendicular to the flat face of the slab, the crystal is said to be X-cut (see Fig. 6.25).

Two other common cuts are the Y cut and the Z cut. Reference to Table 6.1 indicates that each crystal material can best be used only in certain cuts.

SHEAR PLATES. If two edges of an X-cut Rochelle salt crystal are parallel to the Y and Z axes, a *shear plate* is obtained (see Fig. 6.28). When a shear plate is used as a transducer, a metal foil is cemented to each side of it, as shown in Fig. 6.28a. The two foils and the crystal itself form an electrical capacitor of the solid dielectric type. When the crystal is deformed as shown in Fig. 6.28b, the resulting shearing effect in the crystal causes a charge to appear on the capacitor, thereby producing a potential difference between the two plates, with a certain polarity. Deformation of the crystal in the opposite direction, as shown in Fig. 6.28c, reverses the polarity. Similarly, if a potential is applied between the two faces of the slab, shearing stresses are produced in the plane of the plate and a deformation takes place like that of (b) or (c) depending on the polarity.

Microphones generally use either X or Y cuts of Rochelle salt and Z cuts of ADP.

EXPANDER PLATES. Another important form of crystal transducer, viz., the 45° *expander plate*, is obtained if a cut is taken from a shear plate

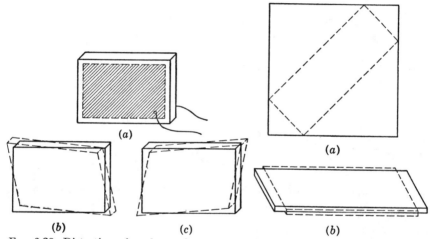

(a)

(b)   (c)

(a)

(b)

FIG. 6.28. Distortion of a shear plate occurs when a potential is applied between two foils cemented to its faces. (a) Placement of foils. (b) Distortion of shear plate with one polarity. (c) Distortion of shear plate with opposite polarity. (*Courtesy of Brush Electronics Company.*)

FIG. 6.29. (a) Method of cutting a 45° expander bar from a shear plate. (b) Sketch showing expansion of sides and compression of length of such a bar when a potential is applied between the two faces. (*Courtesy of Brush Electronics Company.*)

in the manner shown in Fig. 6.29. From Fig. 6.28, it is apparent that when a voltage is applied to a shear plate, one diagonal lengthens and the other shortens. Hence, if a crystal is made by cutting a piece from a shear plate in the manner shown in Fig. 6.29a, potentials will be generated by squeezing or extending the crystal along its length, as shown in Fig. 6.29b.

Most piezoelectric crystals of commercial interest, including Rochelle salt and ADP, yield no significant output when subjected to hydrostatic pressure. This happens because the algebraic sum of the potentials developed by deformations along the three axes simultaneously is zero.

TABLE 6.1.   Properties of Piezoelectric Substances

Property	Crystals				Ceramic
	Rochelle salt	ADP	LH	Quartz	Barium titanate
Type of strain obtainable	Transverse shear; and expander	Transverse shear; and expander	Transverse and longitudinal shear; and expander and hydrostatic	Transverse and longitudinal shear	Transverse and longitudinal shear; and expander; and hydrostatic
Common cuts.....	X; $45°X$; Y; $45°Y$	Z; $45°Z$; L	Y	X, Y, AT	
Density, kg/m³....	1.77×10^3	1.80×10^3	2.06×10^3	2.65×10^3	5.7×10^3
Low-frequency cutoff at 25°C	0.1 cps	AAA1, 9 cps AAA, 14 cps	Less than 0.2 cps	Less than 0.0001 cps	Less than 0.0001 cps
Volume resistivity		See Fig. 6.37	10^{10} ohm-m	Very high	More than 10^{11} ohm-m
Temperature, for complete destruction	55°C (131°F)	†	†	Loses piezoelectric properties at 576°C	Loses piezoelectric properties at 120°C
Maximum safe temperature	45°C (113°F)	125°C (257°F)	76°C (167°F)	Above 250°C slow decrease of piezoelectric activity	90°C (194°F)
Temperature for appreciable leakage............	50°C (122°F)	40°C (104°F)	†		
Maximum safe humidity (unprotected element)	70 %	94 %	95 %	(Not a critical factor except for external short circuits of electrodes)	Moisture absorption 0.1 %
Protected with Metalseal.......	100 %				
Humidity above which surface leakage becomes appreciable (unprotected element)	50 %	50 %	50 %	Usually much higher than other piezoelectric materials, except that value depends on surface finish and condition	95 %
Minimum safe relative humidity (unprotected element)	40 %	0 %	0 %	0 %	0 %
Dielectric constant at 30°C.........	350 (X cut) 9.4 (Y cut)	15.3	10.3	4.5	1700
Maximum breaking stress (alternating compression-tension)	14.7×10^6 newtons/m²	20.13×10^6 newtons/m²	†	76×10^6 newtons/m²	45×10^6 newtons/m²

† No data available.

However, there is a sizable response of the lithium sulfate crystal to hydrostatic pressure.

The same electromechanical relationships exist for a crystal microphone as for an electrostatic microphone (see the preceding section of this part), $viz.$,

$$e_0 = -\tau\xi$$
$$f = \tau q \tag{6.37}$$

where e_0 = open-circuit voltage in volts produced by a deformation ξ of the crystal in meters, f = force acting to deform the crystal produced by an electrical charge q, and τ is a coupling coefficient defined after Eq. (3.27b). As was described in Part VIII (pages 71 to 75), such a

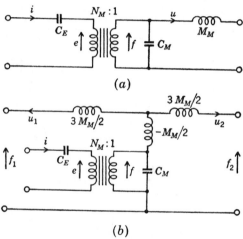

FIG. 6.30. Electromechanical equivalent circuits for piezoelectric microphones (impedance analogy). (a) Circuit for one face of piezoelectric element blocked, $e.g.$, cemented to a rigid surface. (b) Circuit for both faces free to move.

device may be represented by the equivalent circuit of Fig. 6.30a, provided one of the two faces of the crystal across which the force is being produced is held stationary. If both faces are permitted to move, the equivalent circuit must be modified as shown in Fig. 6.30b.

In these circuits, C_E is the electrical capacitance, in farads, measured at low frequencies with the crystal in a vacuum (or, for all practical purposes, in air). M_M is the effective mechanical mass, in kilograms; N_M is the electromechanical transformation ratio in volts per newton (or meters per coulomb); and C_M is the mechanical compliance, in meters per newton, measured at low frequencies with the electrical terminals open-circuited.

For Rochelle salt X cut crystals, the electrical capacitance varies as a function of temperature. The factor of proportionality μ for unmounted and unrestrained crystals is shown in Fig. 6.31. Two discontinuities in this curve occur at $-18°C$ and $+22°C$ and are known as the Curie points.[1] At these points, the capacitance varies radically with temperature and is influenced considerably by the manner in which the crystal is mounted. For example, a mounted crystal of the Bimorph† type has a value of C_E at 22°C, which is a little over three times that at 10°C. This anomalous behavior is usually described as a violent variation of the free dielectric constant. An obvious way to avoid variations in the terminal voltage corresponding to variations in capacitance is to operate the microphone into an open circuit.

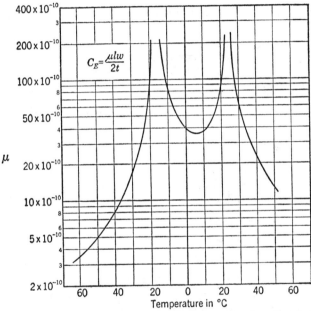

Fig. 6.31. Factor of proportionality μ in the expression for electrical capacitance $C_e = \mu lw/2t$ of an unmounted X-cut Rochelle salt expander bar as a function of temperature. The discontinuities at $-18°C$ and $+22°C$ are known as the Curie points. (*Courtesy of Brush Electronics Company.*)

The electromechanical transformation ratio N_M for this type of crystal in effect does not vary with temperature. For an unmounted uncoated crystal the variation amounts to about 0.05 db per degree Fahrenheit. For a coated crystal mounted in a microphone this variation may be as little as 0.02 db per degree Fahrenheit. Again, this holds for a crystal working into an open circuit.

[1] A discussion of the detailed performance of piezoelectric dielectrics and their resemblance to ferromagnetic substances appears in a text now in press: T. F. Hueter and R. H. Bolt, "Sonics," John Wiley & Sons, Inc., New York.

† Bimorph is a registered trade-mark of Brush Electronics Company, Cleveland, Ohio.

The element sizes (Fig. 6.30) for a *Rochelle salt X-cut length expander bar* are $N_M = 0.093/w$, $C_M = 31.4 \times 10^{-12} l/wt$, $M_M = 715lwt$, and $C_E = \mu lw/2t$. The dimensions l, w, and t in meters are as shown in Fig. 6.32a, and μ is given in Fig. 6.31.

The element sizes for an *ADP Z-cut expander bar* are $N_M = 0.185/w$, $C_M = 47.4 \times 10^{-12} l/wt$, $M_M = 737lwt$, and $C_E = 128 \times 10^{-12} lw/t$. The dimensions l, w, and t in meters are as shown in Fig. 6.32a.

The element sizes for a *Y-cut LH thickness expander bar* (the electrical field is parallel to the mechanical deformation of the crystal, see Fig. 6.32b) are $N_M = 0.175t/lw$, $C_M = 16.3 \times 10^{-12} t/lw$, $N_M = 832lwt$, and $C_E = 91 \times 10^{-12} lw/t$. The dimensions l, w, and t are in meters.

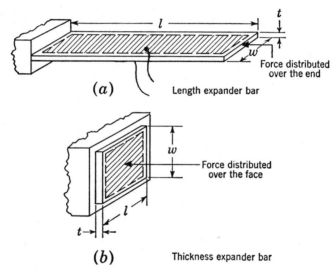

(a) Length expander bar

(b) Thickness expander bar

Fig. 6.32. Dimensions, method of mounting, and points of application of force for (a) length expander bar; (b) thickness expander bar.

The element sizes, in *mks acoustic-impedance* units, for an *LH block operated in the hydrostatic mode* are $N_A = 0.148t$, $C_A = 32 \times 10^{-12} lwt$, and $C_E = 91 \times 10^{-12} lw/t$. The value of M_A has little significance because of the various resonances that can be excited. The dimensions l, w, and t in meters are as shown in Fig. 6.32b except that the crystal is free of a support.

BIMORPHS. An important disadvantage of the expander-bar type of structure in some applications is its large mechanical impedance. For use in liquids, such an impedance is desirable, but in air, because of the large mechano-acoustic impedance mismatch, very small output voltages are obtained for normal sound pressures. To reduce this mechanical impedance appreciably without lowering the output voltage, two plates or bars may be combined to produce a Bimorph. In a sense, a Bimorph is a mechanical transformer operating on a principle resembling that of a bimetallic strip. The flat faces of two crystals are cemented together in such a way that when a potential is applied to them one expands and the other contracts. Conversely, a force acting perpendicular to the face of the "bimetallic" strip causes a sizable compressive force in one plate and a tensile force in the other.

Examples of a bender Bimorph and a torque Bimorph are shown in Figs. 6.33 and 6.34, respectively. The former makes use of two expander bars and the latter of two shear plates. Generally, the bender Bimorph

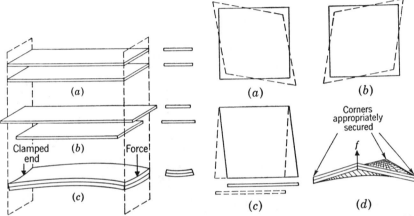

FIG. 6.33. Bender Bimorph. (a) Two expander-bar crystals ready to be cemented together to form a bender Bimorph. (b) Application of potentials of opposite polarity to the crystals causes the upper to lengthen and to contract in width and the other to shorten and to expand in width. (c) When these two crystals are cemented together and either a force f or electrical potentials are applied, the crystal assumes the shape shown. (Courtesy of Brush Electronics Company.)

FIG. 6.34. Torque Bimorph. (a), (b), and (c) Distortion of shear plates when potentials of opposite polarity are applied across their faces. (d) Two shear plates are cemented together to form a torque Bimorph. When either a force f or electrical potentials are applied, the Bimorph assumes the shape shown. (Courtesy of Brush Electronics Company.)

is supported at both ends and the force applied in the middle, although it can be clamped at one end and the force applied at the other. A torque Bimorph is generally held at three corners and the force applied at the fourth. With these types of structures, a force applied as shown in Figs. 6.33c and 6.34d is equivalent to a greater force acting on the end of a single expander bar or along one end of a single shear plate.

Electrodes may be applied to a Bimorph in two ways to form either a series or a parallel arrangement of the two plates. For a series connection, the electrical terminals are the two outer foils. For a parallel connection, the foil between the crystals forms one terminal, and the two outside foils connected together form the other.

A series Bimorph has one-quarter the capacitance and twice the output voltage of a parallel Bimorph.

Refer to the circuit of Fig. 6.30a (the circuit of Fig. 6.30b cannot be used for Bimorphs mounted as described above). For *square-torque Bimorphs made from X-cut Rochelle salt shear plates* as shown in Fig. 6.35a† connected electrically in *parallel*, the element sizes at 30°C are $N_M = 0.143/t$, $C_M = 2.87 \times 10^{-10}l^2/t^3$, $M_M = 262l^2t$, and $C_E = \beta \times 1.18l^2/t \times 10^{-8}$. The factor β is given in Fig. 6.36 and equals 1.0 at 30°C. The dimensions l and t are in meters. When the plates are connected electrically in *series*,

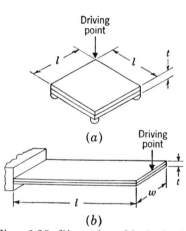

(a)

(b)

FIG. 6.35. Bimorphs. Method of mounting, dimensions, and points of application of force for (a) square torque Bimorph and (b) rectangular bender Bimorph.

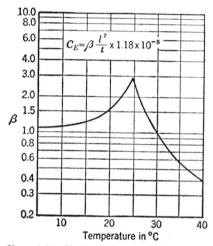

$$C_E = \beta \frac{l^2}{t} \times 1.18 \times 10^{-8}$$

Temperature in °C

FIG. 6.36. Factor of proportionality β in the expression for electrical capacitances C_E of an X-cut Rochelle salt square torque Bimorph mounted as shown in Fig. 6.35a. (*Courtesy of Brush Electronics Company.*)

the value of N_M is doubled and that of C_E is reduced to one-fourth its parallel value.

For *ADP square-torque Bimorphs*† (see Fig. 6.35a) connected electrically in *parallel*, the element sizes are $N_M = 0.250/t$, $C_M = 4.45 \times 10^{-10}l^2/t^3$, $M_M = 228l^2t$, and $C_E = 507l^2/t \times 10^{-12}$.

For *Rochelle salt bender Bimorphs* (see Fig. 6.35b) connected electrically in *parallel*, the element sizes are $N_M = 0.0735l/wt$, $C_M = 179 \times 10^{-12}l^3/wt^3$, $M_M = 407lwt$, and $C_E = 1.18 \times 10^{-8}lw/t$. The dimensions l, w, and t are in meters. When the plates are connected electrically in *series*, the value of N_M is doubled and that of C_E is reduced to one-fourth its parallel value. The circuit of Fig. 6.30a must be used.

For *ADP bender Bimorphs* (see Fig. 6.35b) connected electrically in

† Element sizes for rectangular-torque Bimorphs of Rochelle salt or ADP are available from the Brush Electronics Company, Cleveland, Ohio.

parallel, the element sizes are $N_M = 0.1523l/wt$, $C_M = 199 \times 10^{-12}l^3/wt^3$, $M_M = 435lwt$, and $C_E = 5.09 \times 10^{-10}lw/t$.

The nonlinear distortion produced by ADP and LH crystals is very small. Rochelle salt is an exception, however. In the temperature range from $-18°C$ to $+24°C$, it exhibits hysteresis effects in the relation between the applied force and the voltage produced across a small load resistance. This hysteresis effect arises in the electrical capacitance C_E and is of negligible importance if the microphone operates in a near-open-circuit condition, such as into the input of a cathode-follower vacuum-tube stage. In microphones working at ordinary sound levels, the

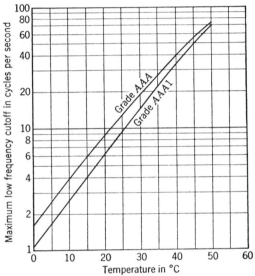

FIG. 6.37. Maximum expected low-frequency cutoff in cycles for ADP (PN) crystals as a function of ambient temperature in centigrade degrees. The cutoff frequency is defined as that frequency where the capacitive reactance and the resistance (each measured between the same two faces) become equal. (*Courtesy of Brush Electronics Company.*)

hysteresis effect in Rochelle salt is negligible even when the crystal is loaded.

All crystals have a bulk resistivity† which appears as a shunt resistance across the electrical terminals of the crystal. This factor is not of practical importance at ordinary temperatures except in ADP crystals.

For ADP crystals, the bulk resistivity increases rapidly with temperature, as can be seen from Fig. 6.37. The ordinate of this graph is the cutoff frequency, equal to the frequency at which the reactance and resistance measured across two faces of a crystal plate are equal.

† Bulk resistivity is the resistance in ohms measured between two faces of a one-unit cube.

The self-noise produced by an electrostatic microphone is that produced by the d-c resistance of the crystal or dielectric shunted by the capacitance C_E. In addition, the load resistance will contribute to the noise in that it is in parallel with the internal resistance. The spectrum level of the self-noise drops off at 6 db per octave as soon as the capacitive reactance becomes less than the total shunt resistance (external and internal). A crystal that has been damaged by heat or humidity is likely to have a high internal noise level.

Temperature not only affects the electrical capacitance of Rochelle salt crystals and the resistivity of the ADP crystals but also may permanently damage the crystals. Data showing the approximate maximum safe temperature, temperatures for appreciable leakage, and the temperatures for complete destruction are given in Table 6.1.

Crystals are affected by humidity, as can be seen from Table 6.1. This is particularly true of the Rochelle salt type. Rochelle salt chemically is sodium potassium tartrate with four molecules of water of crystallization. If the humidity is too low (less than about 30 per cent), the crystal gradually dehydrates and becomes a powder. If the humidity is too high (above 84 per cent), the crystal gradually dissolves. Neither result is reversible. Protective coatings of recent vintage give nearly complete protection from the influence of humidity extremes.

To test a crystal for normalcy, measure its resistivity and electrical capacitance, being careful not to apply so much voltage to the crystal that it will fracture. If the capacitance is lower than normal, the crystal has dehydrated. If the resistivity is too low and the capacitance is too high, the crystal has partially or entirely dissolved. If the resistivity is low and the shunt capacitance has not changed, surface leakage is taking place.

Barium Titanate Ceramic Microphones. A recent and useful form of piezoelectric material is a ceramic made from barium titanate. The ceramic is rendered piezoelectric by permanently polarizing it with a high electrostatic potential of about 40,000 to 60,000 volts/in. for a period of several minutes. Lower voltage is satisfactory if polarization is carried out at elevated temperatures. Pure barium titanate has a Curie point like that of Rochelle salt, except that it occurs at a temperature of 120°C, well above the normal operating range for a microphone. Dielectric anomalies also exist near 12°C.

Barium titanate microphones may be used interchangeably with crystal microphones, except that their sensitivity is about 6 to 18 db below that of Rochelle salt or ADP types. Temperature changes do not affect the dielectric constant appreciably.

The advantages of barium titanate over the crystal materials described above are that it has a higher dielectric constant than that for Rochelle salt, and one that changes very little with temperature in the range between $+10$ and $+80°C$. The mechanical compliance C_M and the transducer ratio N_M are lower than for Rochelle salt. Two length-expander units which are made from Rochelle salt and barium titanate, respectively, and which are designed to have the same capacitance and mechanical compliance at $15°C$ will differ in sensitivity by 12 to 16 db, the barium titanate unit being the less sensitive of the two.

Barium titanate ceramics exhibit a slow decrease in dielectric constant and electromechanical response with age. The decrease of each quantity is believed to be less than 10 per cent, that is, 1.0 db, during the first year after manufacture and only a few per cent per year thereafter.

The free unmounted capacity C_E as a function of temperature of a barium titanate ceramic unit relative to the free capacity for a temperature of $25°C$ is shown in Fig. 6.38.

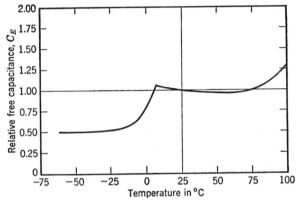

Fig. 6.38. Barium titanate ceramic. Relative free capacitance C_E as a function of temperature. (*Courtesy of Brush Electronics Company.*) The exact nature of this curve depends upon the manner in which the dielectric is rendered piezoelectric.

Referring to the circuits of Fig. 6.30, for a *barium titanate length-expander bar* like that shown in Fig. 6.32a, the element sizes are $N_M = 0.0051/w$, $C_M = 10.2 \times 10^{-12}l/wt$, $M_M = 2260lwt$, and C_E (at $25°C$) $= 1.5 \times 10^{-8}lw/t$. The dimensions l, w, and t are in meters.

For a *barium titanate thickness-expander bar* like that in Fig. 6.32b, the element sizes are $N_M = 0.0127t/lw$, $C_M = 8.13 \times 10^{-12}t/lw$, $M_M = 2260lwt$, and C_E (at $25°C$) $= 1.5 \times 10^{-8}lw/t$. The dimensions l, w, and t are in meters.

Referring to the circuit of Fig. 6.30a, the element sizes for a *barium titanate bender Bimorph* like that shown in Fig. 6.35b, with the plates connected electrically in series, are $N_M = 0.00068l/wt$, $C_M = 40.6 \times 10^{-12}l^3/wt^3$, $M_M = 1530lwt$, and C_E (at $25°C$) $= 1.28 \times 10^{-8}lw/t$.

The barium titanate transducer also responds to hydrostatic pressures such as would be obtained if a ceramic plate were suspended freely in air or in a liquid. The circuit of Fig. 6.30a is likewise used here except that

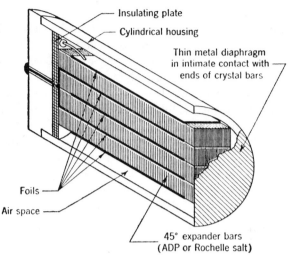

FIG. 6.39. Arrangement for mounting expander bars to form a pressure microphone. The foils are connected so that all capacitances are in parallel. The sound field acts on the thin metal diaphragm and causes expansion or compression of the bars along their length. (*Prepared from rough sketch by F. Massa of Massa Labs.*)

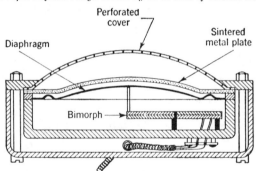

FIG. 6.40. Diaphragm type of crystal microphone using a torque Bimorph. The sintered metal plate is an acoustic damping element.

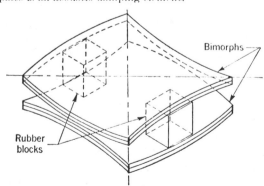

FIG. 6.41. Sound-cell crystal microphone. The required compressive and tensile forces are produced in each of the four crystal plates when a force acts on the centers and the sides of the two bender Bimorph elements.

acoustical elements are substituted for mechanical elements. In this case the transducer is called a *volume expander*, and the element sizes are $N_A = 0.0022t$ volt/newton/m^2 (or m^3/coulomb), $C_A = 12.2 \times 10^{-12}lwt$ m^5/newton, and $C_E = 1.5 \times 10^{-8}lw/t$ farads. The dimensions l, w, and t are in meters.

Uses of Piezoelectric Transducers. Ways in which piezoelectric transducers are incorporated into microphones are shown in Figs. 6.39 to 6.42.

In the first of these figures, a microphone is sketched using ADP or Rochelle salt 45°-cut expander bars or barium titanate length-expander bars. A sound pressure causes a force to be exerted on the end of the piezoelectric plates, and a voltage is generated because of changes in their lengths. The sides of the plates are free to expand into the air space.

In Fig. 6.40 a diaphragm is connected to a torque Bimorph element by a short rod. The Bimorph is mounted securely on three corners so that the force is applied in the manner shown in Fig. 6.35a. The sintered metal plate in front of the diaphragm is an acoustic resistance that damps the resonance of the mechanical system.

Another form of mounting Bimorph plates is shown in Fig. 6.41. Here two square bender Bimorph elements are held apart by two rubber blocks. The entire assembly is encased in a thin wax-impregnated paper jacket, thereby forming two "diaphragms" with an air space between. An increase in sound pressure causes a distortion of the plates as shown in Fig. 6.41. This distortion is of such a nature as to produce a potential difference across the electrodes fastened to the piezoelectric plates.

A fourth form of mounting suitable for use with piezoelectric plates that respond to hydrostatic changes in pressure is shown in Fig. 6.42. Here a block of four lithium sulfate plates is shown immersed in a castor oil bath. The outer flexible housing serves both as a diaphragm and as a retainer for the castor oil. A pressure over the outside surface is transferred through the housing and the castor oil bath to the element inside.

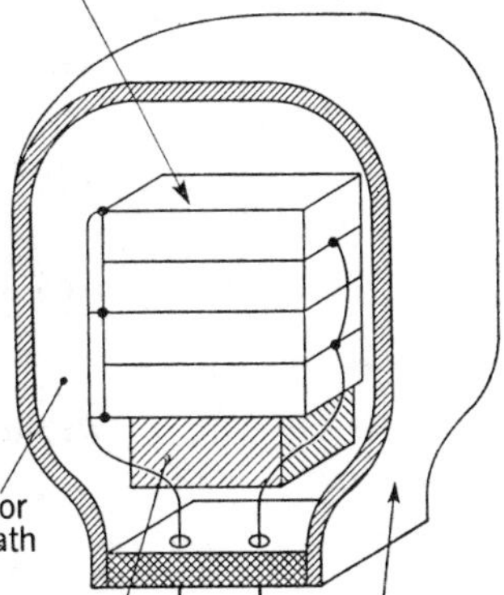

FIG. 6.42. Sketch of a hydrostatically actuated crystal microphone using lithium sulfate (LH) crystals as the active element. The castor oil bath is provided to convey the sound pressure uniformly to all surfaces of the crystal. Data taken on a cubical stack of six plates connected in parallel, ¼ in. on a side with each plate 0.04 in. thick, show that the capacitance is about 23×10^{-12} farad and the open-circuit output voltage is about -96 db *re* 1 volt per dyne/cm^2. (*Courtesy of Brush Electronics Company.*)

PART XVI *Gradient and Combination Microphones*

6.7. Pressure-gradient Microphones. *General Features.* The ribbon microphone has approximately the same sensitivity and impedance as a moving-coil microphone when used with a suitable impedance-matching transformer. Because of its figure 8 directivity pattern it is extensively used in broadcast and public-address applications to eliminate unwanted sounds that are situated in space, relative to the microphone, about 90° from those sounds which are wanted. It is also used by singers to

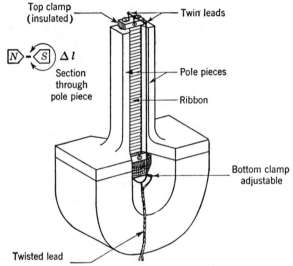

Fig. 6.43. Sketch of the ribbon and magnetic structure for a velocity microphone. (*After Olson, Elements of Acoustical Engineering, 2d ed., p. 239, D. Van Nostrand Company, Inc., 1947.*)

introduce a "throaty" or "bassy" quality into their voice. A disadvantage of the ribbon microphone is that, unless elaborate wind screening is resorted to, it is often very noisy when used outdoors.

Construction. A typical form of pressure-gradient microphone is that represented by Fig. 6.43. It consists of a ribbon with a very low resonant frequency hung in a slot in a baffle. A magnetic field transverses the slot so that a movement of the ribbon causes a potential difference to appear across its ends. In this way, the moving conductor also serves as the diaphragm. In modern design, the ribbon element might be 1 in. long, $\frac{1}{16}$ in. wide, and 0.0001 in. thick with a clearance of 0.003 in. at each side.

From Eq. (6.11) we see that the pressure difference acting to move the diaphragm is

$$p_R = f_R/S = u\omega\rho_0 \,\Delta l \cos \theta \qquad (6.38)$$

where f_R = rms net force acting to move the ribbon
    u = rms particle velocity in the wave in the direction of propagation of the wave
    S = effective area of ribbon
    Δl = effective distance between the two sides
    θ = angle the normal to the ribbon makes with the direction of travel of the wave

This equation is valid as long as the height of the baffle is less than approximately one-half wavelength.

Analogous Circuit. The equivalent circuit (impedance analogy) for this type of microphone is shown in Fig. 6.44. There p_R is the pressure difference which would exist between the two sides of the ribbon if it were held rigid and no air could leak around it; Z_{AA} is the acoustic impedance of the medium viewed from one side of the ribbon; $U_R = Su_r$ = volume velocity of the ribbon; u_R = linear velocity of the ribbon; M_{AR}, C_{AR},

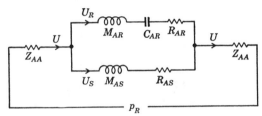

FIG. 6.44. Analogous acoustical circuit for a ribbon microphone (impedance analogy).

and R_{AR} are the acoustic constants of the ribbon itself (for example, $M_{AR} = M_{MR}/S^2$, where M_{MR} is the mass of the ribbon); M_{AS} and R_{AS} are the acoustic mass and resistance, respectively, of the slots at either edge of the ribbon; and U_S is the volume velocity of movement of the air through the slot on the two sides of the ribbon.

Over nearly all the frequency range, the radiation impedance Z_{AA} is a pure mass reactance corresponding to an acoustic mass M_{AA} [see Eq. (5.31)]. In a properly designed microphone, $U_S \ll U_R$. Also, the microphone is operated above the resonance frequency so that $\omega M_{AR} \gg 1/(\omega C_{AR})$. Usually, also, $\omega M_{AR} \gg R_{AR}$. Hence, the circuit of Fig. 6.44 simplifies into a single acoustic mass of magnitude $2M_{AA} + M_{AR}$.

When the mobility analogy is used and the electrical circuit is considered, we get the complete circuit of Fig. 6.45. Here, $M_{MA} = M_{AA}S^2$, $M_{MR} = M_{AR}S^2$, B = flux density, l = length of the ribbon, and $f_r = p_rS$.

Performance. The open-circuit voltage e_o of the microphone is found from solution of Fig. 6.45 to be

$$e_o = \frac{Blf_R}{j\omega(2M_{MA} + M_{MR})} \tag{6.39}$$

Substitution of (6.38) in (6.39) yields

$$|e_o| = |u| \frac{(Bl)\rho_0 \, \Delta l}{2M_{MA} + M_{MR}} S \cos \theta \qquad (6.40)$$

The open-circuit voltage is directly proportional to the component of the particle velocity perpendicular to the plane of the ribbon. In a well-designed ribbon microphone, this relation holds true over the frequency range from 50 to 10,000 cps. The lower resonance frequency is usually about 15 to 25 cps. The effects of diffraction begin at frequencies of about 2000 cps but are counterbalanced by appropriate shaping of the magnetic pole pieces.

6.8. Combination Pressure and Pressure-gradient Microphones.
Electrical Combination of Pressure and Pressure-gradient Transducers.
One possible way of producing a directivity pattern that has a single maximum (so-called unidirectional characteristic) is to combine electrically the outputs of a pressure and a pressure-gradient microphone.

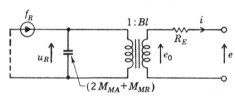

Fɪɢ. 6.45. Simplified electromechanical analogous circuit for a ribbon microphone (mobility analogy).

The two units must be located as near to each other in space as possible so that the resulting directional characteristic will be substantially independent of frequency.

Microphones with unidirectional, or cardioid, characteristics are used primarily in broadcast or public-address applications where it is desired to suppress unwanted sounds that are situated, with respect to the microphone, about 180° from wanted sounds. In respect to impedance and sensitivity this type of cardioid microphone is similar to a ribbon or to a moving-coil microphone when suitable impedance-matching transformers are used.

The equation for the magnitude of the open-circuit output voltage of a pressure microphone in the frequency range where its response is "flat" is

$$e_o = Ap \qquad (6.41)$$

The equation for the open-circuit output voltage of a magnetic or ribbon type of pressure-gradient microphone in the same frequency range is

$$e_o' = Cp \cos \theta \qquad (6.42)$$

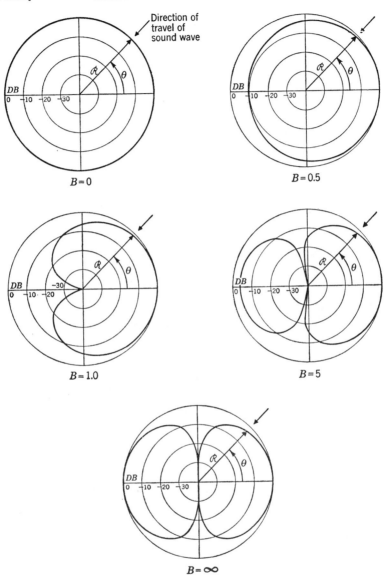

FIG. 6.46. Graphs of the expression $\Re = 20 \log [(1 + B \cos \theta)/(1 + B)]$ as a function of θ for $B = 0, 0.5, 1.0, 5.0,$ and ∞.

Adding (6.41) and (6.42) and letting $C/A = B$ gives

$$e_o = Ap(1 + B \cos \theta) \tag{6.43}$$

B will be a real positive number only if e_o and e'_o have the same phase. The directional characteristic for a microphone obeying Eq. (6.43) will depend on the value of B. For $B = 0$, the microphone is a nondirectional type; for $B = 1$, the microphone is a cardioid type; for $B = \infty$, the microphone is a figure 8 type. In Fig. 6.46 directional characteristics for five values of B are shown.

The voltage e'_o is a function of kr, as we discussed in Par. 6.3, so that the voltage e_o as given by Eq. (6.43) will vary as a function of frequency for small values of $\omega r/c$, where r is the distance between the microphone and a small source of sound. Here, as is the case for a pressure-gradient microphone, a "bassy" quality is imparted to a person's voice if he stands very near the microphone.

Acoustical Combination of Pressure and Pressure-gradient Microphones. One example of an acoustical design responding to both pressure and pressure gradient in a sound wave was described earlier in Par. 6.3 (pp. 149 to 150). The directional patterns for this type of design are the same as those shown for Fig. 6.46.

In order that this type of microphone have a flat response as a function of frequency for p_{rms} constant (*i.e.*, constant sound pressure at all frequencies in the sound wave), a transducer must be chosen whose output voltage for a constant differential force acting on the diaphragm is inversely proportional to the quantity $\mathfrak{a}$ defined in Eq. (6.19), *i.e.*,

$$e_o \propto \frac{1}{|\mathfrak{a}|} = \left| \frac{Z_{AD} - j[(R_A + Z_{AD})/\omega C_A R_A]}{Z_{AD}} \right| \tag{6.44}$$

As an example, let us take the case of a microphone for which $Z_{AD} \gg R_A$ and $1/\omega C_A R_A \gg 1$. In this case the response of the transducer must be proportional to

$$\frac{1}{|\mathfrak{a}|} = \frac{1}{\omega C_A R_A} = \frac{cB}{\Delta l \omega} \tag{6.45}$$

where B is given by Eq. (6.17).

Restated, the transducer must have an output voltage for a constant net force acting on the diaphragm that is inversely proportional to frequency, if a flat frequency response is desired. This is the case for a moving-coil or ribbon transducer above the natural resonance frequency of the diaphragm.

CHAPTER 7

DIRECT-RADIATOR LOUDSPEAKERS
ELECTROMAGNETIC TRANSDUCER.

PART XVII *Basic Theory of Direct-radiator Loudspeakers*

7.1. Introduction. A loudspeaker is an electromagnetic transducer for converting electrical signals into sounds. There are two principal types of loudspeakers: those in which the vibrating surface (called the diaphragm) radiates sound directly into the air, and those in which a horn is interposed between the diaphragm and the air. The direct-radiator type is used in most home radio receiving sets, in phonographs, and in small public-address systems. The horn type is used in high-fidelity reproducing systems, in large sound systems in theaters and auditoriums, and in music and outdoor-announcing systems.

The principal advantages of the direct-radiator type are (1) small size, (2) low cost, and (3) a satisfactory response over a comparatively wide frequency range. The principal disadvantages are (1) low efficiency, (2) narrow directivity pattern at high frequencies, and (3) frequently, irregular response curve at high frequencies. For use in home radio receiving sets where little acoustic power is necessary and where the listeners are generally not very critical, the advantages far outweigh the disadvantages. In theater and outdoor sound systems where large amounts of acoustic power are necessary and where space is not important, the more efficient horn-type loudspeaker is generally used.

All the types of transduction discussed in the previous chapter on Microphones might be used for loudspeakers. In this text, however, we shall limit ourselves to moving-coil loudspeakers, the type commonly used in radios and home music systems.

7.2. Construction.[1] A cross-sectional sketch of a typical direct-radiator loudspeaker is shown in Fig. 7.1. The diaphragm is a cone,

[1] For supplemental reading, the student will find the following publications valuable: H. F. Olson, "Elements of Acoustical Engineering," 2d ed., Chap. VI, D. Van Nostrand Company, Inc., New York, 1947; M. S. Corrington, Amplitudes and Phase Measurements on Loudspeaker Cones, *Proc. IRE*, **39**: 1021–1026 (1951); Transient Testing of Loudspeakers, *Audio Engineering*, **34**: 9–13 (August, 1950).

generally made of paper or aluminum, which is supported at the outer edge and near the voice coil so that it is free to move only in an axial direction. Current through the voice coil creates a magnetomotive force which interacts with the air-gap flux of the permanent magnet and causes a translatory movement of the voice coil and, hence, of the cone to which it is attached. Usually the cone is sufficiently stiff at low frequencies to move as a whole. At high frequencies, however, vibrations from the center travel outward toward the edge in the form of waves. The results

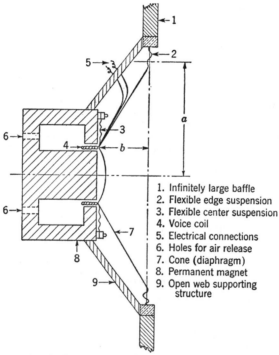

1. Infinitely large baffle
2. Flexible edge suspension
3. Flexible center suspension
4. Voice coil
5. Electrical connections
6. Holes for air release
7. Cone (diaphragm)
8. Permanent magnet
9. Open web supporting
   structure

Fig. 7.1. Cross-sectional sketch of a direct-radiator loudspeaker assumed to be mounted in an infinite baffle.

of these traveling waves and of resonances in the cone itself are to produce irregularities in the frequency-response curve at the higher frequencies and to influence the relative amounts of sound radiated in different directions.

In Fig. 7.1, the loudspeaker is shown mounted in a flat baffle assumed to be of infinite extent. By definition, a baffle is any means for acoustically isolating the front side of the diaphragm from the rear side. For purposes of analysis, the loudspeaker diaphragm may be considered at low frequencies to be a piston of radius a moving with uniform velocity over its entire surface. This is a fair approximation at frequencies for which the distance b on Fig. 7.1 is less than about one-tenth wavelength.

7.3. Electro-mechano-acoustical Circuit. Before drawing a circuit diagram for a loudspeaker, we must identify the various elements involved. The voice coil has inductance and resistance, which we shall call L and R_E, respectively. The diaphragm and the wire on the voice coil have a total mass M_{MD}. The diaphragm is mounted on flexible suspensions at the center and at the edge. The total effect of these suspensions may be represented by a mechanical compliance C_{MS} and a mechanical resistance $R_{MS} = 1/r_{MS}$, where r_{MS} is the mechanical responsiveness.

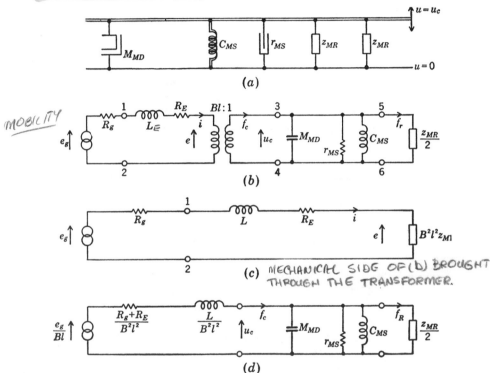

FIG. 7.2. (a) Mechanical circuit of direct-radiator loudspeaker; (b) electromechanical analogous circuit of the mobility type; (c) electrical circuit showing motional electrical impedance; (d) analogous circuit of the mobility type with electrical quantities referred to the mechanical side.

The air cavity and the holes at the rear of the center portion of the diaphragm form an acoustic network which, in most loudspeakers, can be neglected in analysis because they have no appreciable influence on the performance of the loudspeaker. However, both the rear and the front side of the main part of the diaphragm radiate sound into the open air.

A radiation impedance is assigned to each side and is designated as $Z_{MR} = 1/z_{MR}$, where z_{MR} is the radiation mobility.

We observe that one side of each flexible suspension is at zero velocity. For the mechanical resistance this also must be true because it is contained in the suspensions.   We already know from earlier chapters that one side of the mass and one side of the radiation mobility must be considered as having zero velocity.   Similarly, we note that the other sides of the masses, the compliance, the responsiveness, and the radiation mobilities all have the same velocity, *viz.*, that of the voice coil.

From inspection we are able to draw a mechanical circuit and the electromechanical analogous circuit using the mobility analogy.   These are shown in Fig. 7.2a and b, respectively.   The symbols have the following meanings:

e_g = open-circuit voltage of the generator (audio amplifier) in volts.

R_g = generator resistance in electrical ohms

L = inductance of voice coil in henrys, measured with the voice-coil movement blocked, *i.e.*, for $u_c = 0$.

R_E = resistance of voice coil in electrical ohms, measured in the same manner as L

B = steady air-gap flux density in webers per square meter.

l = length of wire in meters on the voice-coil winding.

i = electric current in amperes through the voice-coil winding.

f_c = force in newtons generated by interaction between the alternating and steady mmfs, that is, $f_c = Bli$.

u_c = voice-coil velocity in meters per second, that is, $u_c = e/Bl$, where e is the so-called counter emf.

a = radius of diaphragm in meters.

M_{MD} = mass of the diaphragm and the voice coil in kilograms.

C_{MS} = total mechanical compliance of the suspensions in meters per newton.

r_{MS} = $1/R_{MS}$ = mechanical responsiveness of the suspension in meters per newton-second (mks mechanical mohms†).

R_{MS} = mechanical resistance of the suspensions in newton-seconds per meter (mks mechanical ohms).

z_{MR} = $1/Z_{MR}$ = $r_{MR} + jx_{MR}$ = mechanical radiation mobility in mks mechanical mohms from one side of the diaphragm (see Fig. 5.4).   The German $\mathfrak{r}$ indicates that $\mathfrak{r}_{MR}$ varies with frequency.

Z_{MR} = $\mathfrak{R}_{MR} + jX_{MR}$ = mechanical radiation impedance in newton-seconds per meter (mks mechanical ohms) from one side of a piston of radius a mounted in an infinite baffle (see Fig. 5.3).   The German $\mathfrak{R}$ indicates that $\mathfrak{R}_{MR}$ varies with frequency.

The circuit of Fig. 7.2b with the mechanical side brought through the transformer to the electrical side is shown in Fig. 7.2c.   The mechanical mobility $z_{M1} = u_c/f_c$ is zero if the diaphragm is blocked so that there is no

† A mohm is a mobility ohm.   See Par. 3.3 for discussion.

motion ($u_c = 0$) but has a value different from zero whenever there is motion. For this reason the quantity $B^2l^2z_{M1}$ is usually called the *motional electrical impedance*. When the electrical side is brought over to the mechanical side, we have the circuit of Fig. 7.2d.

The circuit of Fig. 7.2d will be easier to solve if its form is modified. First we recognize the equivalence of the two circuits shown in Fig. 7.3a

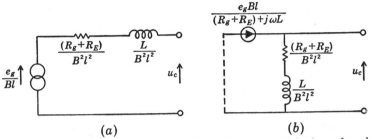

(a)   (b)

F$_{\text{IG}}$. 7.3. The electrical circuit (referred to the mechanical side) is shown here in two equivalent forms. The circuits are of the mobility type.

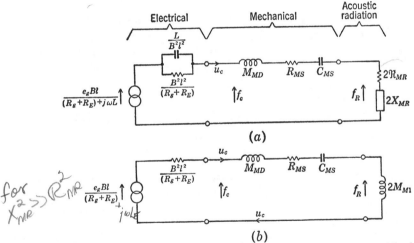

(a)

(b)

F$_{\text{IG}}$. 7.4. (a) Low-frequency analogous circuit of the impedance type with electrical quantities referred to mechanical side. Z_{MR} is given by Fig. 5.3. The quantity f_c represents the total force acting in the equivalent circuit to produce the voice-coil velocity u_c. (b) Single-loop approximation to Fig. 7.4a valid for $X_{MR}^2 \gg \Re_{MR}^2$.

and b. Next we substitute Fig. 7.3b for its equivalent in Fig. 7.2d. Then we take the dual of Fig. 7.2d to obtain Fig. 7.4a.

The performance of a direct-radiator loudspeaker is directly related to the diaphragm velocity. Having solved for it, we may compute the acoustic power radiated and the sound pressure produced at any given distance from the loudspeaker in the far-field.

Voice-coil Velocity at Medium and Low Frequencies. The voice-coil velocity u_c, neglecting ω^2L^2 compared with $(R_g + R_E)^2$, is found from

Fig. 7.4a,

$$u_c \doteq \frac{e_g B l}{(R_g + R_E)(R_M + j X_M)} \tag{7.1}$$

where

$$R_M = \frac{B^2 l^2}{R_g + R_E} + R_{MS} + 2\Re_{MR} \tag{7.2}$$

$$X_M = \omega M_M = \omega M_{MD} + 2X_{MR} - \frac{1}{\omega C_{MS}} \tag{7.3}$$

Voice-coil Velocity at Low Frequencies. At low frequencies, assuming in addition that $X_{MR}^2 \gg \Re_{MR}^2$, we have from Fig. 7.4b that

$$(X_M)_{\text{low } f} = \omega(M_{MD} + 2M_{M1}) - \frac{1}{\omega C_{MS}} \tag{7.4}$$

where $M_{M1} = 2.67a^3\rho_0$ = mass in kilograms contributed by the air load on one side of the piston for the frequency range in which $ka < 0.5$. The quantity ka equals the ratio of the circumference of the diaphragm to the wavelength.

The voice-coil velocity is found from Eq. (7.1), using Eqs. (7.2) and (7.4) for R_M and X_M, respectively.

7.4. Power Output. The acoustic power radiated in watts from *both* the rear and the front sides of the loudspeaker is

$$W = |u_c|^2(2\Re_{MR}) \tag{7.5}$$

Hence, assuming $\omega^2 L^2 \ll (R_g + R_E)^2$,

$$W = \frac{2e_g^2 B^2 l^2 \Re_{MR}}{(R_g + R_E)^2 (R_M^2 + X_M^2)} \tag{7.6}$$

7.5. Sound Pressure Produced at Distance r. *Low Frequencies.* In Chap. 4 we showed that a piston whose diameter is less than one-third wavelength ($ka < 1.0$) is essentially nondirectional at low frequencies. Hence, we can approximate it by a hemisphere whose rms volume velocity equals $U_c = S_D u_c$, where S_D is the projected area of the loudspeaker cone. By the projected area, we mean πa^2 of Fig. 7.1.

From Eq. (4.3) we see that the magnitude of the rms pressure at a point in free space a distance r from *either* side of the loudspeaker in an infinite baffle is

$$|p(r)| \doteq \frac{|U_c| f \rho_0}{r} \tag{7.7}$$

It is assumed in writing this equation that the distance r is great enough so that it is situated in the "far-field." Hence, the pressure at r is

$$|p(r)| = \frac{e_g B l S_D f \rho_0}{r(R_g + R_E)\sqrt{R_M^2 + X_M^2}} \tag{7.8}$$

Equation (7.8) is also readily derived from Eq. (7.6) by observing from Table 5.1 (page 124) that, at low frequencies,

$$\Re_{MR} = \frac{\omega^2 S_D{}^2 \rho_0}{2\pi c} \tag{7.9}$$

and

$$W = 4\pi r^2 I = \frac{4\pi r^2 |p(r)|^2}{\rho_0 c} \tag{7.10}$$

where I is the intensity at distance r in watts per square meter.

Medium Frequencies. At medium frequencies, where the radiation from the diaphragm becomes directional but yet where the diaphragm vibrates as one unit, *i.e.*, as a rigid piston, the pressure produced at a distance r depends on the power radiated and the directivity factor Q.

The directivity factor Q was defined in Chap. 4 as the ratio of the intensity on a designated axis of a sound radiator to the intensity that would be produced at the same position by a point source radiating the same acoustic power.

From Eq. (7.10) we see for a *point source radiating to both sides* of an infinite baffle that

$$|p(r)| = \sqrt{\frac{W \rho_0 c}{4\pi r^2}} \tag{7.11}$$

For a directional source in an infinite baffle such as we are considering here,

$$|p(r)| = \sqrt{\frac{W_1 Q \rho_0 c}{4\pi r^2}} \tag{7.12}$$

where W_1 = acoustic power in watts radiated from *one* side of the loudspeaker.

Q = directivity factor for *one* side of a piston in an infinite plane baffle. Values of Q are found from Fig. 4.20. Note that W_1 equals $W/2$ and, at low frequencies where there is no directionality, $Q = 2$, so that Eq. (7.12) reduces to Eq. (7.11) at low frequencies.

The sound pressure is found by substituting Eq. (7.6) divided by 2 into (7.12), giving,

$$|p(r)| = \frac{e_g Bl \sqrt{Q \rho_0 c \Re_{MR}}}{2r \sqrt{\pi} (R_g + R_E) \sqrt{R_M{}^2 + X_M{}^2}} \tag{7.13}$$

7.6. Frequency-response Curves. A frequency-response curve of a loudspeaker is defined as the variation in sound pressure or acoustic power as a function of frequency, with some quantity such as voltage or electrical power held constant. Inspection of Eqs. (7.6), (7.8), and (7.13) shows that the quantity $R_M{}^2 + X_M{}^2$ in the denominator, $\Re_{MR}$ in the numerator, and the directivity factor Q are terms that vary with frequency. The variation in sound pressure or diaphragm velocity due to the variation of the denominator is exactly the same as the variation of electric current as a function of frequency in a series RLC electrical resonant circuit. A plot of this variation as a function of normalized frequency is called a universal resonance curve.

At low frequencies, the quantity $\mathfrak{R}_{MR}$ varies with the square of the frequency. So, if Q is constant, the numerator of (7.13) varies in direct proportion to frequency. If Q is not constant but increases with frequency, the variation is more rapid. In other words, neglecting the directivity (*i.e.*, let Q = constant), the curve of sound pressure as a function of frequency is identical in shape to a universal electrical resonance curve multiplied by frequency.

When an electrical universal resonance curve is expressed in decibels, below the resonance frequency it has a slope of $+6$ db per octave of frequency. Above the resonance frequency it has a slope of -6 db per octave. In the case of a sound-pressure or an acoustic-power *vs.* frequency-response curve, the slopes are further increased by the linear frequency term, so that below the resonance frequency the slope is $+12$ db per octave and above it the slope is 0 db (flat).

Whether one considers sound pressure or power radiated, the frequency-response curve when expressed in decibels has the same shape, provided the directivity factor is constant. We shall now study one of the quantities, the power radiated, as a function of the quantity $f^2/(R_M^2 + X_M^2)$.

7.7. Maximum Power Available Efficiency (PAE). Often, the response of a loudspeaker is stated in terms of its *maximum power available efficiency*, which is 100 times the ratio of the acoustic power radiated to the maximum power that the electrical generator can supply. The maximum power is available from the electrical generator when the load resistance equals the generator resistance.

$$W_E = \frac{e_g^2}{4R_g} \qquad (7.14)$$

Medium and Low Frequencies. The *maximum power available efficiency* (*PAE*) at medium and low frequencies $[\omega^2 L^2 \ll (R_g + R_E)^2]$ is Eq. (7.6) divided by the maximum power available.

$$\text{PAE} \equiv \frac{W}{W_E} \times 100 = \frac{800B^2l^2R_g\mathfrak{R}_{MR}}{(R_g + R_E)^2(R_M^2 + X_M^2)} \qquad (7.15)$$

where R_M and X_M are given by Eqs. (7.2) and (7.3). At the lower frequencies we may replace $\mathfrak{R}_{MR}$ by Eq. (7.9) and X_M by Eq. (7.4).

Let us divide the frequency region into five parts and treat each part separately by simplifying the circuit of Fig. 7.4a to correspond to that part alone. Reference is made to Figs. 7.5 and 7.6 for the breakdown.

In region A, where the loudspeaker is stiffness-controlled, the power output increases as the fourth power of frequency or 12 db per octave.

In region B, at the resonance frequency ω_0, the power output is determined by the total resistance in the circuit because X_M passes through zero.

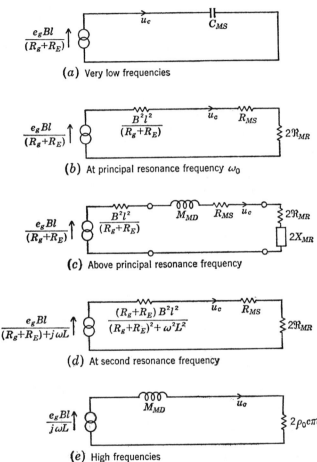

(a) Very low frequencies

(b) At principal resonance frequency ω_0

(c) Above principal resonance frequency

(d) At second resonance frequency

(e) High frequencies

Fig. 7.5. Simplified forms of the circuit of Fig. 7.4a valid over limited frequency ranges.

In region C, above the first resonance frequency, the power output (and the sound pressure) approaches a constant value, provided the circuit impedance approaches being a pure mass reactance. That is to say, $\Re_{MR}$ increases with the square of the frequency, and $X_M{}^2$ also increases as the square of the frequency, and so the frequency variation cancels out.

For small values of amplifier resistance R_g, the total mechanical resistance R_M becomes quite large in some loudspeakers so that the resonance is more than critically damped. Reference to Eq. (7.8) shows that if

$R_M{}^2 \gg X_M{}^2$, the sound pressure increases linearly with frequency f.   This condition is shown in Fig. 7.6 by the dashed line.

High Frequencies.   Referring back to Fig. 7.4a, we see that there is a possibility of a second resonance taking place involving L/B^2l^2 and the masses $M_{MD} + X_{MR}/\omega$.   The voice-coil velocity at this resonance can be determined from the circuit of Fig. 7.5d.   The resonance frequency will occur when

$$\frac{\omega L(B^2l^2)}{\omega^2 L^2 + (R_g + R_E)^2} = \omega M_{MD} + 2X_{MR} \tag{7.16}$$

We must note, however, that if $(R_g + R_E)^2$ is large compared with $L^2\omega^2$, the reactance of the capacitance L/B^2l^2 and resistance $B^2l^2/(R_g + R_E)$

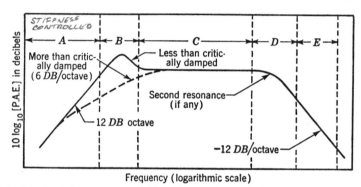

Frequency (logarithmic scale)

Fig. 7.6. Graph of the power available efficiency in decibels of a hypothetical direct-radiator loudspeaker in an infinite baffle.   It is assumed that the diaphragm acts like a rigid piston over the entire frequency range.   The power is the total radiated from *both* sides of the diaphragm.   Zero decibels is the reference power available efficiency level.   The solid curve is for a loudspeaker with a Q_T of about 2.   The dashed curve is for a Q_T equal to about 0.5.

in parallel becomes that of a negative inductance equal to $-B^2l^2L/(R_g + R_E)^2$.   In this case, no resonance can occur.

The solution of Fig. 7.5d applies to the peak of the region marked D in Fig. 7.6.

At frequencies above the second resonance frequency, the radiation resistance on each side of the diaphragm becomes approximately equal to $\pi a^2 \rho_0 c$, where a is the effective radius of the loudspeaker.   Also, $\omega^2 M_{MD}$ becomes large compared with the resistance in the circuit, and $\omega^2 L^2$ becomes large compared with $(R_g + R_E)^2$.   The voice-coil velocity is determined from Fig. 7.5e.   The power available efficiency is

$$\text{PAE} = \frac{800 R_g B^2 l^2 \rho_0 c \pi a^2}{\omega^4 L^2 M_{MD}{}^2} \tag{7.17}$$

This region is marked E in Fig. 7.6.   Here, the power output decreases by 12 db for each doubling of frequency.

The response curve given in Fig. 7.6 is for a typical loudspeaker used for the reproduction of music in the home. For this application, the mass of the cone is made as light as possible and the compliance of the suspension as high as possible consistent with mechanical stability. For special applications, C_{MS} can be small so that the resonance frequency is high. Also, it is common in practice to make R_M so large that the velocity u_c is nearly constant as a function of frequency through regions B and C. In this case, the sound pressure increases linearly with frequency, and there is no flat region C.

7.8. Reference Efficiency. It is convenient to define a reference efficiency which permits one to plot the shape of the frequency-response curve without showing the actual acoustic power that is being radiated at the time. The reference power available efficiency (both sides of the diaphragm) is defined as,

$$\text{PAE}_{\text{ref}} \equiv \frac{800R_g B^2 l^2 \Re_{MR}}{(R_g + R_E)^2 \omega^2 (M_{MD} + 2M_{M1})^2} \tag{7.18}$$

or, with the help of Eq. (7.9),

$$\text{PAE}_{\text{ref}} \equiv \frac{800R_g B^2 l^2 S_D^2 \rho_0}{2\pi c (R_g + R_E)^2 (M_{MD} + 2M_{M1})^2} \tag{7.19}$$

If the loudspeaker is less than critically damped, Eq. (7.19) gives the actual response in frequency region C, which lies above the first resonance frequency. Even for loudspeakers that are highly damped so that there is no flat region C, Eq. (7.19) forms a convenient reference to which the rest of the curve is compared.

Expressed as a ratio, the PAE response at medium and low frequencies where the radiation is nondirectional [see Eq. (7.15)] is

$$\frac{\text{PAE}}{\text{PAE}_{\text{ref}}} = \frac{\omega^2 (M_{MD} + 2M_{M1})^2}{R_M^2 + X_M^2} \tag{7.20}$$

At the resonance frequency ω_0, where $X_M = 0$,

$$\frac{\text{PAE}}{\text{PAE}_{\text{ref}}} = \frac{\omega_0^2 (M_{MD} + 2M_{M1})^2}{R_M^2} \equiv Q_T^2 \tag{7.21}$$

where Q_T is analogous to the Q of electrical circuits. Equations (7.20) and (7.21) may be expressed in decibels by taking $10 \log_{10}$ of both sides of the equations.

In Chap. 8 of this book, on Loudspeaker Enclosures, design charts are presented from which it is possible to determine, without laborious computation, the sound pressure from a direct-radiator loudspeaker as a function of frequency including the directivity characteristics. Methods for determining the constants of loudspeakers and of box and bass-reflex enclosures are also presented. If the reader is interested only in learning how to choose a baffle for a loudspeaker, he may proceed directly to Chap. 8. The next part deals with the factors in design that determine the over-all response and efficiency of the loudspeaker.

7.9. Examples of Loudspeaker Calculations

Example 7.1. Given the reference power available efficiency of Eq. (7.19) for a loudspeaker in an infinite baffle, determine the reference sound pressure equivalent to the reference power available efficiency assuming that the directivity factor Q (for radiation to one side) equals 2.

Solution. The sound pressure at distance r, assuming no directivity, is related to the acoustic power radiated to one side as follows (see Eq. 7.12):

$$p = \sqrt{\rho_0 c I} = \sqrt{\frac{\rho_0 c W_1}{2\pi r^2}}$$

where I = intensity at distance r

$W_1 = W/2$ total acoustic power radiated from *one* side of the diaphragm

The equivalent reference sound pressure is

$$p_{ref} = \sqrt{\frac{\rho_0 c}{2\pi r^2}} \sqrt{\frac{PAE_{ref}}{200}} W_E$$

$$= \frac{e_g \rho_0 Bl S_D}{2\pi r (R_g + R_E)(M_{MD} + 2M_{M1})}$$

Example 7.2. As an example of the power available efficiency to be expected from a direct-radiator loudspeaker of conventional design mounted in an infinite baffle and radiating from both sides of the baffle, let us calculate the reference power available efficiency (PAE$_{ref}$) from Eq. (7.18) for the case of a commercial loudspeaker with an advertised diameter of 12 in. Also, let us calculate the ratio of the PAE to the reference power available efficiency at the first resonance frequency. Typical values of the constants are

$B = 10,000$ gauss $= 1.0$ weber/m^2
$l = 9$ m
$L = 7 \times 10^{-4}$ henry
$R_E = 8$ ohms
$R_g = 2$ ohms
a = effective radius of diaphragm $= 0.13$ m
S_D = effective area of diaphragm $= 0.0531$ m^2
ρ_0 = density of air $= 1.18$ kg/m^3
c = speed of sound $= 344.8$ m/sec
$\Re_{MR} = 1.57\omega^2 a^4 \rho_0/c$ (see Table 5.1) $= 1.53 \times 10^{-6}\omega^2$ newton-sec/m
$M_{M1} = 2.67\rho_0 a^3$ (see Table 5.1) $= 0.00694$ kg
$M_{MD} = 0.011$ kg
$R_{MS} = 0.5$ mks mechanical ohm
$C_{MS} = 1.79 \times 10^{-4}$

Solution. From Eq. (7.18), we obtain

$$PAE_{ref} = \frac{800 \times 2 \times 1^2 \times 9^2 \times 1.53 \times 10^{-6}}{10^2 \times (0.025)^2} \doteq 3.2\%$$

For radiation from one side of the loudspeaker only, divide this figure by 2.

Only 1.6 per cent of the available electrical power in region C of Fig. 7.6 is radiated to one side of the diaphragm. This illustrates the statement made at the beginning of the chapter that the efficiency of this type of loudspeaker is usually low.

The upper resonance frequency, if such exists, is determined from Eq. (7.16), *i.e.*,

$$\omega^2 = \cfrac{1}{\left(M_{MD} + \cfrac{2X_{MR}}{\omega} \right) \left[\cfrac{L}{B^2 l^2} + \cfrac{(R_g + R_E)^2}{\omega^2 L B^2 l^2} \right]}$$

For our example, the $(R_E + R_g)^2$ is so large compared with $L^2\omega^2$ that the shunt resistance and capacitance act like a series resistance and a negative inductance, and no second resonance occurs in the frequency range for which a lumped-element circuit holds.

The boundary between regions C and D of Fig. 7.5 occurs when ka lies approximately between 1 and 2. For our example $ka = 1$ corresponds to a frequency of

$$f = \frac{c}{2\pi a} = \frac{344.8}{2\pi \times 0.13} = 424 \text{ cps}$$

Obviously, a smaller diaphragm of lighter weight would result in region C extending to a higher frequency. However, a reduction in the mass M_{MD} occasioned by a smaller diaphragm will cause an increase in the first resonance frequency with a resulting loss in bass response. A further disadvantage of a smaller diaphragm is that, for a given sound pressure, a greater voice-coil velocity u_c is needed. A longer air gap and a larger magnet structure must therefore be provided.

The first resonance frequency equals

$$f_0 = \frac{1}{2\pi \sqrt{M_M C_{MS}}}$$

where

$$M_M = M_{MD} + 2M_{M1} = 0.025 \text{ kg}$$

so that

$$f_0 = \frac{100}{2\pi \sqrt{(0.025)(1.79)}} = 75 \text{ cps}$$

$$R_M = {}^8\!\!\!\;\frac{1}{10} + 0.5 + 0.3 \doteq 8.9 \text{ mks mechanical ohms}$$

$$Q_T = \frac{\omega_0 M_M}{R_M} = \frac{(471)(0.025)}{8.9} = 1.32$$

From Eq. (7.21) we see that the ratio of the PAE at ω_0 to the reference PAE is equal to Q_T^2. Hence, this ratio equals 1.74, and PAE at ω_0 equals 5.6 per cent (both sides).

PART XVIII *Design Factors Affecting Direct-radiator Loud-speaker Performance*

A loudspeaker generally is designed to provide an efficient transfer of electric power into acoustic power and to effect this transfer uniformly over as wide a frequency range as possible. To accomplish this, the voice coil, diaphragm, and amplifier must be properly chosen. The choice of the elements and their effect on efficiency, directivity, and transient response are discussed here.

7.10. Voice-coil Design. Inspection of Fig. 7.6 reveals that region C is a very important part of the response curve, because the average efficiency is governed by it.   From Eq. (7.19), which is valid for this region, if the indicated approximations hold, we see that the maximum power available efficiency at a given frequency is proportional to

$$\text{PAE} \propto \frac{l^2 R_g}{[(R_g + R_E)(M_{MD} + 2M_{M1})]^2} \tag{7.22}$$

Now, the resistance R_E can be expressed in terms of the mass of the voice-coil winding M_{MC} by writing

$$R_E = \frac{\kappa l}{\pi a_w^2} \tag{7.23}$$

where κ = resistivity of voice-coil conductor in units of ohm-meters.   The value of κ for different materials is given in Table 7.1.

a_w = radius of wire in meters.

l = length of voice-coil winding in meters.

Also,

$$M_{MC} = \pi a_w^2 l \rho_w \tag{7.24}$$

where ρ_w = density of the voice-coil wire in kilograms per cubic meter (see Table 7.1).   Combining (7.23) and (7.24), we get

$$R_E = \frac{\kappa l^2 \rho_w}{M_{MC}} \tag{7.25}$$

Substituting (7.24) in (7.22) yields

$$\text{PAE} \propto \frac{l^2 R_g}{\left[\left(R_g + \dfrac{\kappa l^2 \rho_w}{M_{MC}}\right)(M_{MC} + M'_{MD} + 2M_{M1})\right]^2} \tag{7.26}$$

where $M'_{MD} = M_{MD} - M_{MC}$.

Differentiation of this equation with respect to M_{MC} and equating the result to zero gives the value of M_{MC} necessary for maximum power output from a generator of impedance R_g provided we assume that the coil length is already predetermined.   Hence, M_{MC} for maximum PAE is found from

$$M_{MC}^2 = \frac{\kappa l^2 \rho_w}{R_g}(M'_{MD} + 2M_{M1}) \tag{7.27}$$

Further, substituting (7.25) in (7.27), we get

$$M_{MC} = \frac{R_E}{R_g}(M'_{MD} + 2M_{M1}) \tag{7.28}$$

As an alternate possibility, we assume that the resistance of the coil, R_E, is to be constant.   Allow M_{MC} and l to vary.   Then determine M_{MC} for maximum PAE.

TABLE 7.1.　Resistivity and Density of Various Metals

Metal element	Resistivity, ohm-m	Density, kg/m³
Aluminum.............	0.0283×10^{-6}	2.70×10^3
Antimony.............	0.417	6.6
Bismuth..............	1.190	9.8
Cadmium.............	0.075	8.7
Calcium..............	0.046	1.54
Carbon...............	8.0	2.25
Cesium...............	0.22	1.9
Chromium............	0.026	6.92
Cobalt...............	0.097	8.71
Copper...............	0.0172	8.7
Gold.................	0.0244	19.3
Iridium..............	0.061	22.4
Iron.................	0.1	7.9
Lead.................	0.220	11.0
Lithium..............	0.094	0.534
Magnesium...........	0.046	1.74
Manganese...........	0.050	7.42
Mercury.............	0.958	13.5
Molybdenum..........	0.057	10.2
Nickel...............	0.078	8.8
Platinum.............	0.10	21.4
Potassium............	0.071	0.87
Silver...............	0.0163	10.5
Sodium..............	0.046	0.97
Tin..................	0.115	7.3
Titanium.............	0.032	4.5
Tungsten.............	0.055	19.0
Zinc.................	0.059	7.1

From Eq. (7.25), we have

$$l^2 = \frac{R_E M_{MC}}{\kappa \rho_w} \tag{7.29}$$

Putting this in (7.22) yields

$$\text{PAE} \propto \frac{R_E R_g M_{MC}}{(R_g + R_E)^2 (M_{MC} + M'_{MD} + 2M_{M1})^2} \tag{7.30}$$

where, as above, $M'_{MD} = M_{MD} - M_{MC}$. Maximizing, we get

$$M_{MC} = M'_{MD} + 2M_{M1} \tag{7.31}$$

Finally, let us assume that M_{MC} of Eq. (7.30) is a constant and that we wish to let R_E and l vary. Then determine R_E for maximum PAE. Maximizing Eq. (7.30), we get

$$R_E = R_g \tag{7.32}$$

Hence, for the optimum value of power available efficiency, we see

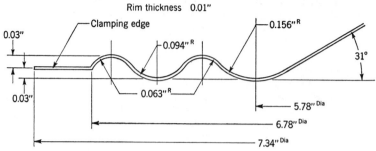

FIG. 7.7. Detail of the edge of a felted-paper loudspeaker cone from an 8-in. loud-speaker. [*After Corrington, Amplitude and Phase Measurements on Loudspeaker Cones, Proc. IRE,* **39**: 1021–1026 (1951).]

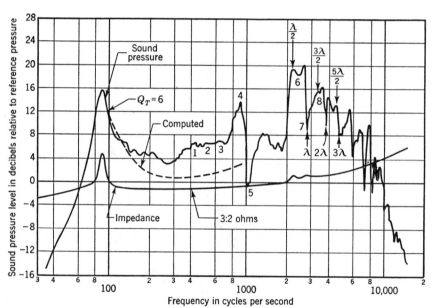

FIG. 7.8. Relative power-available response of an 8-in.-diameter loudspeaker mounted in an infinite baffle. The dashed curve was computed from Figs. 8.12 and 8.13 for $Q_T = 6$. [*After Corrington, Amplitude and Phase Measurements on Loudspeaker Cones, Proc. IRE,* **39**: 1021–1026 (1951).]

from Eqs. (7.28), (7.31), and (7.32), that $R_E = R_g$ and

$$M_{MC} = M'_{MD} + 2M_{M1}.$$

It is not usual, however, that the voice coil should be this massive, for the reason that a large voice coil demands a correspondingly large magnet structure.

Values of voice-coil resistances and masses for typical American loud-speakers are given in Table 8.1 of the next chapter.

7.11. Diaphragm Behavior. The simple theory using the method of equivalent circuits, which we have just derived, is not valid above some frequency between 300 and 1000 cps. In the higher frequency range the cone no longer moves as a single unit, and the diaphragm mass M_{MD} and also the radiation impedance change. These changes may occur with great rapidity as a function of frequency. As a result, no tractable mathematical treatment is available by which the exact performance of a loudspeaker can be predicted in the higher frequency range.

A detailed study of one particular loudspeaker is reported here as an example of the behavior of the diaphragm.[2] The diaphragm is a felted paper cone, about 6.7 in. in effective diameter (see Fig. 7.7), having an included angle of 118°.

The sound-pressure-level response curve for this loudspeaker measured on the principal axis is shown in Fig. 7.8. This particular loudspeaker has, in addition to its fundamental resonance, other peaks and dips in the response at points 1 to 8 as indicated on the curve.

The major resonance at 90 cps is the principal resonance and has the relative amplitude given by Eq. (7.21). Above that is the fairly flat region that we have called region C. At point 1, which is located at 420 cps, the cone breaks up into a resonance of the form shown by the first sketch in Fig. 7.9. Here, there are four nodal lines on the cone extending radially, and four regions of maximum movement. As in-

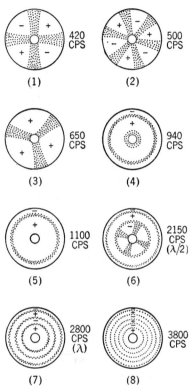

FIG. 7.9. Nodal pattern of the cone of the loudspeaker whose response curve is given in Fig. 7.8. The shaded and dashed lines indicate lines of small amplitude of vibration. The $(+)$ and $(-)$ signs indicate regions moving in opposite directions, i.e., opposite phases. [*After Corrington with changes.*]

dicated by the plus and minus signs, two regions move outward while two regions move inward. The net effect is a pumping of air back and forth across the nodal lines. The cone is also vibrating as a whole in and out of the page. The net change in the output is an increase of about 5 db relative to that computed. A similar situation exists at point 2 at

[2] M. S. Corrington, Amplitude and Phase Measurements on Loudspeaker Cones, *Proc. IRE*, **39**: 1021–1026 (1951).

500 cps, except that the number of nodal lines is increased from 4 to 6. At point 3, 650 cps, the vibration becomes more complex. Nodal lines are no longer well defined, and the speaker vibrates in such a way that the increase in pressure level is about 4.5 db, exclusive of 0.5 db increase due to directivity.

For point 4 at 940 cps, a new type of vibration has become quite apparent. The diaphragm moves in phase everywhere except at the rim. Looking at the rim construction shown in Fig. 7.7 and at the vibration pattern of Fig. 7.9(4), we can deduce what happens. The center part of the cone vibrates at a fairly small amplitude while the main part of the cone has a larger amplitude. At the 5.78-in. diameter the amplitude of vibration is very small. At this point the corrugation has a large radius (0.156 in.). As the cone moves to and fro, the paper tends to roll around this curve, and this excites the 0.094-in. corrugation that follows into violent oscillation at its resonance frequency. The rim resonance is 180° out of phase with respect to the main part of the cone. However, the main part of the cone has a high amplitude produced by the rocking motion around the 5.78-in. diameter, and because of its greater area, only part of its effect in producing a high sound level is canceled out by the rim motion. The net result is a peak in output (see point 4 of Fig. 7.8).

At point 5, 1100 cps, a sharp decrease in response is observed. The decrease seems to be the result of a movement of the nodal line toward the apex of the cone, and a reduction of the amplitude of the (+) portion. Here, the effect is a pumping of air back and forth across the nodal line, with a cancellation in output. This vibration is very characteristic, and at the time such motion occurs, the response drops vigorously.

As frequency is increased, the loudspeaker breaks up into still different characteristic modes of vibration. As shown in Fig. 7.9, case 6, several nodal lines appear concentric to the rim of the loudspeaker. When these occur, a large increase in output is obtained, as shown at point 6 of Fig. 7.8. As frequency is increased, other such resonances occur, with more nodal lines becoming apparent. These nodal lines are the result of waves traveling from the voice coil out to the edge of the cone and being reflected back again. These outwardly and inwardly traveling waves combine to produce a standing-wave pattern that will radiate a maximum of power at some particular angle with the principal axis of the loudspeaker.

In order to reduce standing-wave patterns of the type shown in cases 6, 7, and 8 of Fig. 7.9, it is necessary that a termination of proper mechanical impedance be placed at the outer edge of the diaphragm. This termination must be one that absorbs waves traveling outward from the center of the cone so that no wave is reflected back. In practical design, a leather supporting edge is frequently employed. Such a leather might be a very soft sheepskin, having a weight of approximately 3 oz/yd², and carefully tanned.　A leather supporting edge is also effective in reducing the rim resonance.　The resulting effect is to produce a more uniform response in the frequency region between 700 and 1500 cps of Fig. 7.8.

7.12. Divided-cone Driving Unit.　We have seen that important deterrents to a flat response at high frequencies are large radius and large mass of the diaphragm.　A logical means, therefore, for improving the

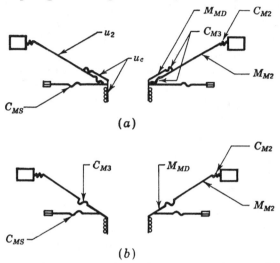

(a)

(b)

Fig. 7.10. Two methods for effectively having two sizes of cones with a single voice coil.　(a) Two separate cones joined by a compliant element.　(b) One cone with a compliant element molded into it at a fraction of its radius.

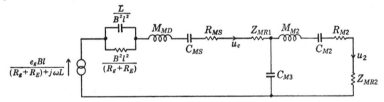

Fig. 7.11. Mechano-acoustical circuit for loudspeakers with the cones shown in Fig. 7.10.

high-frequency response would be to design the diaphragm so that at the high frequencies only the portion of it near the voice coil will move.　Two means for doing this are shown in Fig. 7.10.　The performance of these diaphragms is indicated by the equivalent circuit of Fig. 7.11.　Here, the quantities M_{MD}, C_{MS}, and R_{MS} are the mechanical constants of the smaller (inner) cone, and M_{M2}, C_{M2}, and R_{M2} are the mechanical constants of the outer cone.　The two are connected mechanically with each other through a compliance C_{M3}.　The radiation impedance of the inner cone is Z_{MR1}, and that of the outer cone is Z_{MR2}.　The other constants are the same as in Fig. 7.4.　At high frequencies $1/\omega C_{M3}$ is essentially a short circuit, and only the constants of the smaller cone are involved.

In practice the design of **Fig. 7.10b** is more commonly used because of its lower cost.

Another but more expensive means for separating the inner portion of the cone from the outer is to use a leather strip for the compliance C_{M3}. A material sometimes used for this is goatskin, having a weight of about 2.5 oz/yd². At the highest frequencies at which the inner cone of the loudspeaker is to operate, the goatskin acts to absorb waves traveling in the cone outward from the voice coil. A smoother response curve is thereby obtained.

7.13. Multiple Driving Units. Another means of accomplishing the equivalent of several sizes of cones is to mount two or more loudspeakers of different diameters near each other. An electrical network, called a *crossover network*, is used to supply electrical power to one loudspeaker at

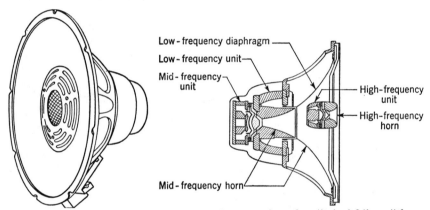

Low-frequency diaphragm

Low-frequency unit

Mid-frequency unit

High-frequency unit

High-frequency horn

Mid-frequency horn

FIG. 7.12. A "coaxial," or "two-way," loudspeaker.

FIG. 7.13. Cross section of a "triaxial," or "three-way" loudspeaker. (*Courtesy of Jensen Mfg. Co., Chicago, Ill.*)

low frequencies and to the other, or others, at higher frequencies. A difficulty with this arrangement is that if the loudspeakers are mounted side by side, the path that the sound has to travel from each of the loud-speakers to a listener will be different in different parts of the listening room. Hence, in the vicinity of the crossover frequencies cancellation of the sound will result at some parts of the room, and addition will occur at others.

To avoid this effect, the loudspeakers are often mounted *concentrically i.e.*, the smaller loudspeakers are placed in the front of and on the axis of the larger loudspeaker (see Figs. 7.12 and 7.13). In the vicinity of the crossover frequency there is usually some shielding of the radiation from the larger loudspeakers by the smaller ones, with resulting irregularity in the response curve.

7.14. Directivity Characteristics. The response curve of Fig. 7.6 and the information of the previous three paragraphs reveal that, above the frequency where $ka = 2$ (usually between 800 to 2000 cps), a direct-radiator loudspeaker can be expected to radiate less and less power. The rate at which the radiated power would decrease, if the cone were a rigid piston, is between 6 and 12 db for each doubling of frequency. This decrease in power output is not as apparent directly in front of the loudspeaker as at the sides because of directivity. That is to say, at high frequencies, the cone directs a larger proportion of the power along the axis than in other directions. Also, the decrease in power is overcome in part by the resonances that occur in the diaphragm, as we have seen from Fig. 7.8.

Directivity Patterns for Typical Loudspeakers. Typical directivity patterns for a 12-in.-diameter direct-radiator loudspeaker, mounted in one of the two largest sides of a closed box having the dimensions 27 by 20 by 12 in., were shown in Fig. 4.23. These data are approximately correct for loudspeakers of other diameters if the frequencies beneath the graphs are multiplied by the ratio of 12 in. to the diameter of the loudspeaker in inches.

Comparison with the directivity patterns for a flat rigid piston in the end of a long tube, as shown in Fig. 4.12, reveals that the directivity patterns for a flat piston are different from those for an actual loudspeaker. This difference results from the cone angle, the speed of propagation of sound in the cone relative to that in the air, and the resonances in the cone. In this connection, it is interesting to see how the speed varies with frequency in an actual cone.

Speed of Propagation of Sound in Cone. Let us define the average speed of propagation of sound in the cone as the distance between the apex and the rim, divided by the number of wavelengths in that distance, multiplied by the frequency in cycles per second. For the particular 8-in. loudspeaker of Figs. 7.7 to 7.9, the phase shift and the average speed of propagation of the sound wave from the apex to the rim of the cone are given in Fig. 7.14. At low frequencies the cone moves in phase so that the speed can be considered infinite. At high frequencies the speed asymptotically approaches that in a flat sheet of the same material, infinite in size.

Intensity Level on Designated Axis. We have stated already that at high frequencies a loudspeaker diaphragm becomes directional. In order to calculate the enhancement of the sound pressure on the axis of the loudspeaker as compared with that indicated by the equations for maximum available power efficiency, it is convenient to use the concepts of directivity factor and of directivity index as defined in Part XI (pages 109 to 115). For example, we might wish to know the intensity (or the sound pressure level) on the axis of the loudspeaker, given the power-available response and the directivity factor. This is done as follows:

The intensity as a function of frequency on the axis of symmetry of the loudspeaker divided by the electrical power available is equal to the product of (1) the power-available-efficiency response characteristic, (2) the directivity factor, and (3) $1/4\pi r^2$, where r is the distance at which the intensity is being measured.   In decibels, we have

$$10 \log_{10} \frac{I_{ax}}{W_E} = 10 \log_{10} \text{PAE}_1 + \text{DI} - 10 \log_{10} 4\pi r^2 \qquad (7.33)$$

where $I_{ax} = p_{ax}^2/\rho_0 c$ = intensity in watts per square meter on the desig-
    nated axis at a particular frequency

  p_{ax} = sound pressure level in newtons per square meter measured
    on the designated axis at a particular frequency

$\text{PAE}_1 = W_1/W_E$ = ratio of total acoustic power in watts radiated by
    the front side of the loudspeaker to the maximum electrical
    power in watts available from the source [see Eq. (7.14)]

and where DI is given by Eq. (4.19) and Fig. 4.20.   Note that, for the piston in an infinite baffle, the DI at low frequencies is 3 db because the

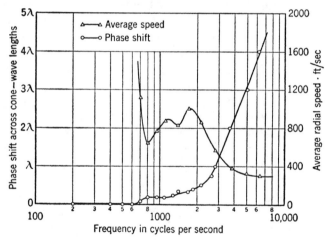

Fig. 7.14. Phase shift and average wave speed in the cone of an 8-in. loudspeaker. [*After Corrington, Amplitude and Phase Measurements on Loudspeaker Cones, Proc. IRE,* **39**: 1021–1026 (1951).]

power is radiated into a hemisphere, and that the last term of Eq. (7.32) is the area of a sphere, in decibels.

Expressed in terms of the sound pressure level on the designated axis *re* 0.0002 microbar (0.00002 newton/m²), Eq. (7.33) becomes

$$\text{SPL } re \text{ 0.0002 microbar} = 20 \log_{10} \frac{p_{ax}}{0.00002} = 10 \log_{10} W_E$$
$$+ 10 \log_{10} \text{PAE}_1 + \text{DI} - 10 \log_{10} 4\pi r^2$$
$$+ 10 \log_{10} \rho_0 c + 94 \text{ db} \qquad (7.34)$$

7.15. Transient Response. The design of a loudspeaker enclosure and the choice of amplifier impedance eventually must be based on subjective judgments as to what constitutes "quality" or perhaps simply on listening "satisfaction." It is believed by many observers that a flat sound-pressure-level response over at least the frequency range between 70 and 7000 cps is found desirable by most listeners. Some observers believe that the response should be flat below 1000 cps but that between 1000 and 4000 cps it should be about 5 db higher than its below-1000-cps value. Above 4000 cps, the response should return to its low-frequency value. It is also believed by some observers that those loudspeakers which sound best generally reproduce tone bursts† well, although this requirement is better substantiated in the literature for the high frequencies than for the low.

An important factor determining the transient response of the circuits of Fig. 7.4 is the amount of damping of the motion of the loudspeaker diaphragm that is present. For a given loudspeaker, the damping may be changed (1) by choice of the amplifier impedance R_g, or (2) by adjustment of the resistive component of the impedance of the enclosure for the loudspeaker, or (3) by both. Generally, the damping is adjusted by choice of amplifier impedance because this is easier to do.

The instantaneous velocity u_c for both steady-state and suddenly applied sine waves is

$$u_c = \frac{e_g Bl}{(R_g + R_E)|Z_M|} \left[\sin(\omega t - \theta) + \frac{\sin \theta}{\sin \psi_0} e^{-R_M t/2M_M} \sin(\omega_0 t + \psi_0) \right] \quad (7.35)$$

where

$$Z_M = |Z_M|e^{j\theta} = R_M + jX_M \quad (7.36)$$

where R_M and X_M are given by Eqs. (7.2) and (7.4).

$$\sin \psi_0 = \frac{1}{\sqrt{1 - \left[\frac{(R_M^2/4M_M^2) + \omega_0^2}{\omega \omega_0} \frac{R_M}{X_M} + \frac{R_M}{2M_M \omega_0} \right]^2}} \quad (7.37)$$

$$\omega_0 = \sqrt{\frac{1}{M_M C_{MS}} - \frac{R_M^2}{4M_M^2}} = 2\pi \text{ times the resonance frequency} \quad (7.38)$$

From Eq. (7.1) we see that the first fraction on the right side of (7.35) and the first term within the brackets is the steady-state term. The second term within the brackets is the transient term, which dies out at the rate of $\exp(-R_M t/2M_M)$.

It is known that the reverberation time in the average living room is

---

† A tone burst is a wave-train pulse that contains a number of waves of a certain frequency.

about 0.5 sec, which corresponds to a decay constant of 13.8 sec^{-1}. Psychological studies also indicate that if a transient sound in a room has decreased to less than 0.1 of its initial value within 0.1 sec, most listeners are not disturbed by the "overhang" of the sound.   This corresponds to a decay constant of 23 sec^{-1}, which is a more rapid decay than occurs in the average living room.   Although criteria for acceptable transient distortion have not been established for loudspeakers, it seems reasonable to assume that if the decay constant for a loudspeaker is greater than four times this quantity, *i.e.*, greater than 92 sec^{-1}, no serious objection will be met from most listeners to the transient occurring with a tone burst. Accordingly, the criterion that is suggested here as representing satisfactory transient performance is

$$\frac{R_M}{2M_M} > 92 \text{ sec}^{-1} \tag{7.39}$$

Equation (7.35) reveals that, the greater $R_M/2M_M$, the shorter the transient.   Equation (7.39) should be construed as setting a lower limit on the amount of damping that must be introduced into the system.   It is not known how much damping ought to be introduced beyond this minimum amount.

In the next chapter we shall discuss the relation between the criterion of Eq. (7.39) and the response curve with baffle.

Each of the diaphragm resonances (*e.g.*, points 1 to 8 in Fig. 7.8) has associated with it a transient decay time determined from an equation like Eq. (7.35).   In order to fulfill the criterion of Eq. (7.39), it is usually necessary to damp the loudspeaker cone and to terminate the edges so that a response curve smoother than that shown in Fig. 7.8 is obtained. With the very best direct-radiator loudspeakers much smoother response curves are obtained.   The engineering steps and the production control necessary to achieve low transient distortion and a smooth response curve may result in a high cost for the completed loudspeaker.

Example 7.3.   If the circular gap in the permanent magnet has a radial length of 0.2 cm, a circumference of 8 cm, and an axial length of 1.0 cm, determine the energy stored in the air gap if the flux density is 10,000 gauss.
   Solution.

Volume of air gap $= (0.002)(0.08)(0.01) = 1.6 \times 10^{-6}$ m^3
Flux density $= 1$ weber/m^2

From books on magnetic devices, we find that the energy stored is

$$W = \frac{B^2 V}{2\mu}$$

where the permeability μ for air is $\mu_0 = 4\pi \times 10^{-7}$ weber/(amp-turn m).   Hence, the air-gap energy is

$$W = \frac{(1)(1.6 \times 10^{-6})}{(2)(4\pi \times 10^{-7})} = \frac{2}{\pi} = 0.636 \text{ joule}$$

Example 7.4. A 12-in. loudspeaker is mounted in one of the two largest sides of a closed box having the dimensions 27 by 20 by 12 in. Determine and plot the relative power available efficiency and the relative sound pressure level on the principal axis.

Solution. Typical directivity patterns for this loudspeaker are shown in Fig. 4.23. The directivity index on the principal axis as a function of frequency is shown in Fig. 4.24. It is interesting to note that the transition frequency from low directivity to high directivity is about 500 cps. Since the effective radius of the radiating cone for this loudspeaker is about 0.13 m, ka at this transition frequency is

$$ka = \frac{2\pi f a}{c} = \frac{1000\pi \times 0.13}{344.8} = 1.18$$

or nearly unity, as would be expected from our previous studies. The transition from region C [where we assumed that $\omega^2 M_{M1}^2 \gg \Re_{MR}^2$ and $\omega^2 L^2 \ll (R_g + R_E)^2$] to region E of Fig. 7.6 also occurs at about $ka = 1$.

In the frequency region between $ka = 0.5$ and $ka = 3$, the loudspeaker can be represented by the circuit of Fig. 7.4a. Let us assume that it is mounted in an infinite baffle and that one-half the power is radiated to each side. Also, let us assume that the amplifier impedance is very low.

The power available efficiency, *from one side of the loudspeaker*, is

$$\text{PAE}_1 = \frac{W_1}{W_E} = \frac{400 B^2 l^2 R_g \Re_{MR}}{(R_g + R_E)^2 (R_{MT}^2 + X_{MT}^2)} \tag{7.40}$$

where

$$R_{MT} = \frac{B^2 l^2}{\sqrt{(R_g + R_E)^2 + \omega^2 L^2}} + (R_{MS} + 2\Re_{MR}) \sqrt{1 + \frac{\omega^2 L^2}{R_g + R_E}} \tag{7.41}$$

$$X_{MT} = (\omega M_{MD} + 2X_{MR}) \sqrt{1 + \frac{\omega^2 L^2}{(R_g + R_E)^2}} - \frac{\omega L B^2 l^2}{(R_g + R_E)^2 \sqrt{1 + [\omega^2 L^2/(R_g + R_E)^2]}} \tag{7.42}$$

If we assume the constants for Example 7.2, we obtain the solid curve of Fig. 7.15a. It is seen that, above $f = 1000$ cps, the power available efficiency drops off.

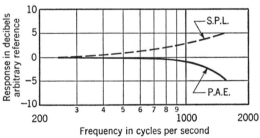

Fig. 7.15. Graphs of the relative power available efficiency and the sound pressure level measured on the principal axis of a typical 12-in.-diameter loudspeaker in a closed-box baffle. The reference level is chosen arbitrarily.

Now, let us determine the sound pressure level on the principal axis of the loudspeaker, using Eq. (7.34). The directivity index for a piston in a long tube is found from Fig. 4.20. The results are given by the dashed curve in Fig. 7.15. Obviously, the directivity index is of great value in maintaining the frequency response on the principal axis out to higher frequencies. At still higher frequencies, cone resonances occur, as we said before, and the typical response curve of Fig. 7.8 is obtained.

CHAPTER 8

LOUDSPEAKER ENCLOSURES

PART XIX *Simple Enclosures*

Loudspeaker enclosures are the subject of more controversy than any other item connected with modern high-fidelity music reproduction. Because the behavior of enclosures has not been clearly understood, and because no single authoritative reference has existed on the subject, opinions and pseudo theories as to the effects of enclosures on loudspeaker response have been many and conflicting. The problem is complicated further because the design of an enclosure should be undertaken only with full knowledge of the characteristics of the loudspeaker and of the amplifier available, and these data are not ordinarily supplied by the manufacturer.

A large part of the difficulty of selecting a loudspeaker and its enclosure arises from the fact that the psychoacoustic factors involved in the reproduction of speech and music are not understood. Listeners will rank-order differently four apparently identical loudspeakers placed in four identical enclosures. It has been remarked that if one selects his own components, builds his own enclosure, and is convinced he has made a wise choice of design, then his own loudspeaker sounds better to him than does anyone else's loudspeaker. In this case, the frequency response of the loudspeaker seems to play only a minor part in forming a person's opinion.

In this chapter, we shall discuss only the physics of the problem. The designer should be able to achieve, from this information, any reasonable frequency-response curve that he may desire. Further than that, he will have to seek information elsewhere or to decide for himself which shape of frequency-response curve will give greatest pleasure to himself and to other listeners.

With the information of this chapter, the high-fidelity enthusiast should be able to calculate, if he understands a-c circuit theory, the frequency-response curve for his amplifier-loudspeaker-baffle combination. Design graphs are presented to simplify the calculations, and three complete examples are worked out in detail. Unfortunately, the calculations are sometimes tedious, but there is no short cut to the answer.

As we have stated earlier, all calculations are based on the mks system. A conversion table is given in Appendix II that permits ready conversion from English units. The advantage of working with meters and kilograms is that all electrical quantities may be expressed in ordinary watts, volts, ohms, and amperes. It is believed that use of the mks system leads to less confusion than use of the cgs system[1] where powers are in ergs per second, electrical potentials are volts $\times 10^8$, electrical currents are amperes $\times 10^{-1}$, and impedances are ohms $\times 10^9$.

8.1. Unbaffled Direct-radiator Loudspeaker. A baffle is a structure for shielding the front-side radiation of a loudspeaker diaphragm from the rear-side radiation. The necessity for shielding the front side from the rear side can be understood if we consider that an unbaffled loudspeaker at low frequencies is the equivalent of a pair of simple spherical sources of equal strength located near each other and pulsing out of phase (see Fig. 8.1). The rear side of the diaphragm of the loudspeaker is equivalent to one of these sources, and the front side is equivalent to the other.

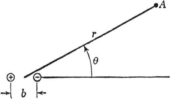

Fig. 8.1. Doublet sound source equivalent at low frequencies to an unbaffled vibrating diaphragm. The point A is located a distance r and at an angle θ with respect to the axis of the loudspeaker.

If we measure, as a function of frequency f, the magnitude of the rms sound pressure p at a point A, fairly well removed from these two sources, and if we hold the volume velocity of each constant, we find from Eq. (4.15) that

$$|p| = \frac{\rho_0 f^2 U_0 b \pi}{rc} \cos \theta \tag{8.1}$$

where U_0 = rms strength of each simple source in cubic meters per second.

b = separation between the simple sources in meters.

ρ_0 = density of air in kilograms per cubic meter (1.18 kg/m³ for ordinary temperature and pressure).

r = distance in meters from the sources to the point A. It is assumed that $r \gg b$.

θ = angle shown in Fig. 8.1.

c = speed of sound in meters per second (344.8m/sec, normally).

[1] H. F. Olson, "Elements of Acoustical Engineering," 2d ed., pp. 84–85, Table 4.3, D. Van Nostrand Company, Inc., New York, 1947. For a discussion of simple loudspeaker enclosures, see pp. 144–154.

In other words, for a constant-volume velocity of the loudspeaker diaphragm, the pressure p measured at a distance r is proportional to the square of the frequency f and to the cosine of the angle θ and is inversely proportional to r. In terms of decibels, the sound pressure p increases at the rate of 12 db for each octave (doubling) in frequency.

In the case of an actual unbaffled loudspeaker, below the first resonance frequency where the system is stiffness-controlled, the velocity of the diaphragm is not constant but doubles with each doubling of frequency. This is an increase in velocity of 6 db per octave. Hence, the pressure p from a loudspeaker *without* a baffle increases $12 + 6 = 18$ db for each octave increase in frequency. Above the first resonance frequency, where the system is mass-controlled, the velocity of the diaphragm decreases 6 db for each octave in frequency. Hence, in that region, the pressure p increases $12 - 6 = 6$ db for each octave increase in frequency.

8.2. Infinite Baffle. In the previous chapter we talked about direct-radiator loudspeakers in infinite baffles. Reference to Fig. 7.6 reveals that with an infinite baffle, the response of a direct-radiator loudspeaker is enhanced over that just indicated for no baffle. It was shown that if one is above the first resonance frequency usually the response is flat with frequency unless the Bl product is large (region C) and that if one is below the first resonance frequency the response decreases at the rate of 12 db per octave instead of 18 db per octave. Hence, the isolation of the front side from the back side by an infinite baffle is definitely advantageous.

In practice, the equivalent of an infinite baffle is a very large enclosure, well damped by absorbing material. One practical example is to mount the loudspeaker in one side of a closet filled with clothing, allowing the front side of the loudspeaker to radiate into the adjoining listening room.

Design charts covering the performance of a direct-radiator loudspeaker in an infinite baffle are identical to those for a closed-box. We shall present these charts in Par. 8.5.

8.3. Finite-sized Flat Baffle. The discussion above indicated that it is advisable to shield completely one side of the loudspeaker from the other, as by mounting the loudspeaker in a closet. Another possible alternative is to mount the loudspeaker in a flat baffle of finite size, free to stand at one end of the listening room.

The performance of a loudspeaker in a free-standing flat baffle leaves much to be desired, however. If the wavelength of a tone being radiated is greater than twice the smallest lateral dimension of the baffle, the loudspeaker will act according to Eq. (8.1). This means that for a finite flat baffle to act approximately like an infinite baffle at 50 cps, its smallest lateral dimension must be about 3.5 m (11.5 ft). However, even above this frequency, sound waves traveling from behind the loudspeaker reflect off walls and meet with those from the front and cause alternate cancellations and reinforcements of the sound as the two waves come into phase or out of phase at particular frequencies in particular parts of the room. This effect can be reduced by locating the loudspeaker off center in the baffle, but it cannot be eliminated because of reflections from the walls of the room behind the loudspeaker. Also, a flat baffle makes the loudspeaker more directional than is desirable because in the plane of the baffle the sound pressure tends to reduce to zero regardless of the baffle size.

8.4. Open-back Cabinets. An open-back cabinet is simply a box with one side missing and with the loudspeaker mounted in the side opposite the open back. Many home radios are of this type. Such a cabinet performs nearly the same as a flat baffle that provides the same path length between the front and back of the loudspeaker. One additional

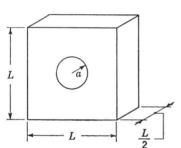

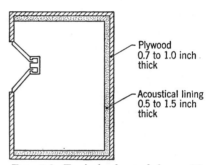

Plywood
0.7 to 1.0 inch
thick

Acoustical lining
0.5 to 1.5 inch
thick

Fig. 8.2. Loudspeaker of radius a mounted in an unlined box with dimensions $L \times L \times L/2$. While this type of box is convenient for analysis, the construction shown in Fig. 8.3 is more commonly used.

Fig. 8.3. Typical plywood box with loudspeaker mounted off center in one side and lined with a layer of soft absorbent acoustical material.

effect, usually undesirable, occurs at the frequency where the depth of the box approaches a quarter wavelength. At this frequency, the box acts as a resonant tube, and more power is radiated from the rear side of the loudspeaker than at other frequencies. Furthermore, the sound from the rear may combine in phase with that from the front at about this same frequency, and an abnormally large peak in the response may be obtained.

8.5. Closed-box Baffle.[1,2] The most commonly used type of loudspeaker baffle is a closed box in one side of which the loudspeaker is mounted. In this type, discussed here in considerable detail, the back side of the loudspeaker is completely isolated from the front. Customary types of closed-box baffles are shown in Figs. 8.2 and 8.3. The sides are made as rigid as possible using some material like 5-ply plywood, 0.75 to 1.0 in. thick and braced to prevent resonance. A slow air leak must be

[2] D. J. Plach and P. B. Williams, Loudspeaker Enclosures, *Audio Engineering*, **35**: 12*ff.* (July, 1951).

SUMMARY OF CLOSED-BOX BAFFLE DESIGN

1. *To determine the volume of the closed box:*

   a. Find the values of f_0 (without baffle) and C_{MS} from Par. 8.7 (pages 229 to 230). Approximate values may be determined from Fig. 8.5b and d.

   b. Determine S_D from Fig. 8.5a, and calculate $C_{AS} = C_{MS}S_D^2$.

   c. Decide what percentage shift upward in resonance frequency due to the addition of the box you will tolerate, and, from the lower curve of Fig. 8.11, determine the values of C_{AB}/C_{AS} and, hence, C_{AB}.

   d. Having C_{AB}, determine the volume of the box from Eq. (8.7).

   e. Shape and line the box according to Pars. 8.6 (page 227) and 8.5 (page 217).

2. *To determine the response of the loudspeaker at frequencies below the rim resonance frequency (about 500 cps):*

   a. Find the values of M_{MD}, R_{MS}, C_{MS}, Bl, R_E, and S_D from Par. 8.7 (pp. 228 to 232). Approximate values may be obtained from Fig. 8.5, Table 8.1, and the sentence preceding Table 8.1.

   b. Determine $M_{AD} = M_{MD}/S_D^2$, $C_{AS} = C_{MS}S_D^2$, and $R_{AS} = R_{MS}/S_D^2$.

   c. Determine $\Re_{AR}$, X_{AR}, M_{A1}, C_{AB}, and M_{AB} from Eqs. (8.4) to (8.8).

   d. If the flow resistance and volume of the acoustical lining are known, determine R_{AB} from Fig. 8.8. Otherwise, neglect R_{AB} to a first approximation.

   e. Determine the actual (not the rated) output resistance of the power amplifier, R_g. All the constants for solving the circuit of Fig. 8.4 are now available.

   f. Calculate the total resistance R_A, total mass M_A, and total compliance C_A from Eqs. (8.19) to (8.21). Determine ω_0 and Q_T from Eqs. (8.22) and (8.23).

   g. Determine the reference sound pressure at distance r from the loudspeaker by Eq. (8.27).

   h. Determine the ratios of the driving frequencies at which the response is desired to the resonance frequency ω_0, that is, ω/ω_0. Determine the ratio of M_A/R_A

   i. Obtain the frequency response in decibels relative to the reference sound pressure directly from Figs. 8.12 and 8.13.

---

provided in the box so that changes in atmospheric pressure do not displace the neutral position of the diaphragm.

Analogous Circuit. A closed box reacts on the back side of the loudspeaker diaphragm. This reaction may be represented by an acoustic impedance which at low frequencies is a compliance operating to stiffen the motion of the diaphragm and to raise the resonance frequency. At high frequencies, the reaction of the box, if unlined, is that of a multiresonant circuit. This is equivalent to an impedance that varies cyclically with frequency from zero to infinity to zero to infinity, and so on. This varying impedance causes the frequency-response curve to have corresponding peaks and dips.

If the box is lined with a sound-absorbing material, these resonances are damped and at high frequencies the rear side of the diaphragm is loaded with an impedance equal to that for the diaphragm in an infinite baffle radiating into free space.

At low frequencies, where the diaphragm vibrates as one unit so that it can be treated as a rigid piston, a complete electro-mechano-acoustical circuit can be drawn that describes the behavior of the box-enclosed loudspeaker. This circuit is shown in Fig. 8.4 and was developed by procedures given in Part XVII.

Some interesting facts about loudspeakers are apparent from this circuit. First, the electrical generator (power amplifier) resistance R_g

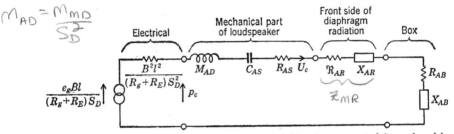

$$M_{AD} = \frac{M_{MD}}{S_D^2}$$

Fig. 8.4. Circuit diagram for a direct-radiator loudspeaker mounted in a closed-box baffle. This circuit is valid for frequencies below about 400 cps. The volume velocity of the diaphragm = U_c; e_g = open-circuit voltage of generator; R_g = generator resistance; R_E = voice-coil resistance; B = air-gap flux density; l = length of wire on voice-coil winding; S_D = effective area of the diaphragm; M_{AD} = acoustic mass of diaphragm and voice coil; C_{AS} = total acoustic compliance of the suspensions; R_{AS} = acoustic resistance in the suspensions; $\mathfrak{R}_{AR}$, X_{AR} = acoustic-radiation imped-ance from the front side of the diaphragm; R_{AB}, X_{AB} = acoustic-loading impedance of the box on the rear side of the diaphragm.

and the voice-coil resistance R_E appear in the denominator of one of the resistances shown. This means that if one desires a highly damped or an overdamped system, it is possible to achieve this by using a power amplifier with very low output impedance. Second, the circuit is of the simple resonant type so that we can solve for the voice-coil volume velocity (equal to the linear velocity times the effective area of the dia-phragm) by the use of universal resonance curves. Our problem becomes, therefore, one of evaluating the circuit elements and then determining the performance by using standard theory for electrical series LRC circuits.

Values of Electrical-circuit Elements. All the elements shown in Fig. 8.4 are in units that yield acoustic impedances in mks acoustic ohms (newton-seconds per meter[5]), which means that all elements are trans-formed to the acoustical side of the circuit. This accounts for the effec-tive area of the diaphragm S_D appearing in the electrical part of the circuit. The quantities shown are

e_g = open-circuit voltage in volts of the audio amplifier driving the loudspeaker

B = flux density in the air gap in webers per square meter (1 weber/m^2 = 10^4 gauss)

l = length of the wire wound on the voice coil in meters (1 m = 39.37 in.)

R_g = output electrical impedance (assumed resistive) in ohms of the audio amplifier

R_E = electrical resistance of the wire on the voice coil in ohms

a = effective radius in meters of the diaphragm

$S_D = \pi a^2$ = effective area in square meters of the diaphragm

Values of the Mechanical-circuit Elements. The elements for the mechanical part of the circuit differ here from those of Part XVII in that they are transformed over to the acoustical part of the circuit so that they yield acoustic impedances in mks acoustic ohms.

$M_{AD} = M_{MD}/S_D{}^2$ = acoustic mass of the diaphragm and voice coil in kilograms per meter⁴

M_{MD} = mass of the diaphragm and voice coil in kilograms

$C_{AS} = C_{MS}S_D{}^2$ = acoustic compliance of the diaphragm suspensions in meters⁵ per newton (1 newton = 10^5 dynes)

C_{MS} = mechanical compliance in meters per newton

$R_{AS} = R_{MS}/S_D{}^2$ = acoustic resistance of the suspensions in mks acoustic ohms

R_{MS} = mechanical resistance of the suspensions in mks mechanical ohms

As we shall demonstrate in an example shortly, these quantities may readily be measured with a simple setup in the laboratory. It is helpful, however, to have typical values of loudspeaker constants available for rough computations, and these are shown in Fig. 8.5 and in Table 8.1. The magnitude of the air-gap flux density B varies from 0.6 to 1.4 webers/m^2 depending on the cost and size of the loudspeaker.

TABLE 8.1.   Typical Values of l, R_E, and M_{MC} for Various Advertised Diameters of Loudspeakers

Advertised diam, in.	Nominal impedance, ohms	l, m	R_E, ohms	M_{MC}, mass of voice coil, g
4–5	3.2	2.7	3.0	0.35–0.4
6–8	3.2	3.4	3.0	0.5 –0.7
10–12	3.2	4.4	3.0	1.0 –1.5
12	8.0	8.0	7.0	3
15–16	16.0	. . .	. . .	12

Values of Radiation (Front-side) Impedance. Acoustical elements always give the newcomer to the field of acoustics some difficulty because they are

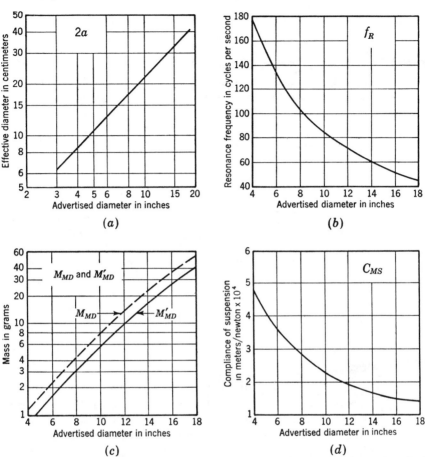

FIG. 8.5a. Relation between effective diameter of a loudspeaker and its advertised diameter.

FIG. 8.5b. Average resonance frequencies of direct-radiator loudspeakers when mounted in infinite baffles vs. the advertised diameters.

FIG. 8.5c. Average mass of voice coils and diaphragms of loudspeakers as a function of advertised diameters. M_{MD} is the mass of the diaphragm including the mass of the voice-coil wire, and M'_{MD} is the mass of the diaphragm excluding the mass of the voice-coil wire.

FIG. 8.5d. Average compliances of suspensions of loudspeakers as a function of advertised diameters. Note, for example, that 3 on the ordinate means 3×10^{-4} m/newton.

not well behaved. That is to say, the resistances vary with frequency, and, when the wavelengths are short, so do the masses.

The radiation impedance for the radiation from the front side of the diaphragm is simply a way of indicating schematically that the air has mass, that its inertia must be overcome by the movement of the diaphragm, and that it is able to accept power from the loudspeaker. The magnitude of the front-side radiation impedance depends on whether the box is very large so that it approaches being an infinite baffle or whether the box has dimensions of less than about 0.6 by 0.6 by 0.6 m (7.6 ft³), in which case the behavior is quite different.

VERY-LARGE-SIZED BOX (APPROXIMATE INFINITE BAFFLE)

$\Re_{AR}$ = radiation resistance for a piston in an infinite baffle in mks acoustic ohms. This resistance is determined from the ordinate of Fig. 5.3 multiplied by $407/S_D$. If the frequency is low so that the effective circumference of the diaphragm ($2\pi a$) is less than λ, that is, $ka < 1$ (where $k = 2\pi/\lambda$) $\Re_{AR}$ may be computed from

$$\Re_{AR} \doteq \frac{0.159\omega^2\rho_0}{c} \doteq 0.0215f^2 \qquad (8.2)$$

X_{AR} = radiation reactance for a piston in an infinite baffle. Determine from the ordinate of Fig. 5.3, multiplied by $407/S_D$. For $ka < 1$, X_{AR} is given by

$$X_{AR} = \omega M_{A1} = \frac{0.270\omega\rho_0}{a} \doteq \frac{2.0f}{a} \qquad (8.3a)$$

and

$$M_{A1} = \frac{0.270\rho_0}{a} \doteq \frac{0.318}{a} \qquad (8.3b)$$

MEDIUM-SIZED BOX (LESS THAN 8 FT³)

$\Re_{AR}$ = approximately the radiation impedance for a piston in the end of a long tube. This resistance is determined from the ordinate of Fig. 5.7 multiplied by $407/S_D$. If the frequency is low so that the effective circumference of the diaphragm ($2\pi a$) is less than λ, $\Re_{AR}$ may be computed from

$$\Re_{AR} = \frac{\pi f^2 \rho_0}{c} \doteq 0.01076f^2 \qquad (8.4)$$

X_{AR} = approximately the radiation reactance for a piston in the end of a long tube. Determine from the ordinate of Fig. 5.7 multiplied by $407/S_D$. For $ka < 1$, X_{AR} is given by

$$X_{AR} = \omega M_{A1} = \frac{\omega(0.1952)\rho_0}{a} \doteq \frac{1.45f}{a} \qquad (8.5a)$$

and

$$M_{A1} = \frac{(0.1952)\rho_0}{a} \doteq \frac{0.23}{a} \qquad (8.5b)$$

Closed-box (*Rear-side*) *Impedance.* The acoustic impedance Z_{AB} of a closed box in which the loudspeaker is mounted is a reactance X_{AB} in series with a resistance $\Re_{AB}$. As we shall see below, neither X_{AB} nor $\Re_{AB}$ is well behaved for wavelengths shorter than 8 times the smallest dimension of the box.† If the dimension behind the loudspeaker is less than about $\lambda/4$, the reactance is negative. If that dimension is greater than $\lambda/4$, the reactance is usually positive if there is absorbing material in the box so that the loading on the back side of the loudspeaker is approximately that for an infinite baffle.

MEDIUM-SIZED BOX. For those frequencies where the wavelength of sound is greater than *eight times* the smallest dimension of the box ($4L < \lambda$ for the box of Fig. 8.2), the mechanical reactance presented to the rear side of the loudspeaker is a series mass and compliance,

$$X_{AB} = \omega M_{AB} - \frac{1}{\omega C_{AB}} \tag{8.6}$$

where

$$C_{AB} = \frac{V_B}{\gamma P_0} \tag{8.7}$$

is the acoustic compliance of the box in meters⁵ per newton and

$$M_{AB} = \frac{B\rho_0}{\pi a} \tag{8.8}$$

is the acoustic mass in kilograms of the air load on the rear side of the diaphragm due to the box; and where

V_B = volume of box in cubic meters. The volume of the loudspeaker should be subtracted from the actual volume of the box in order to obtain this number. To a first approximation, the volume of the speaker in meters³ equals 0.4 × the fourth power of the advertised diameter in meters.

γ = 1.4 for air for adiabatic compressions.

P_0 = atmospheric pressure in newtons per square meter (about 10^5 on normal days).

$\pi a = \sqrt{S_D \pi}$ if the loudspeaker is not circular.

B = a constant, given in Fig. 8.6, which is dependent upon the ratio of the effective area of the loudspeaker diaphragm S_D to the area L^2 of the side of the box in which it is mounted.

As an example, assume that the depth of the box $L/2$ is 1 ft. Then, since Eq. (8.6) is restricted to the frequency region where $\lambda/8 > L/2$, the maximum frequency for it is 140 cps.

LARGE-SIZED BOX. If the box is large so that its smallest dimension is greater than one-eighth wavelength, and if it is unlined, the mechanical reactance is determined from Fig. 8.7.

Impedance of Closed Box with Absorptive Lining. The type of reactance function shown in Fig. 8.7 is not particularly desirable because of the very

† At 1000 cps, a wavelength at 72°F is about 13.5 in.; at 500 cps, 27 in.; at 2000 cps, 6.8 in.; and so on.

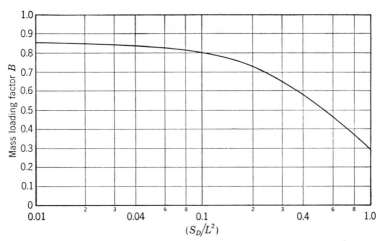

FIG. 8.6. End-correction factor B for the reactance term of the impedance at the rear side of the loudspeaker diaphragm mounted in a box of the type shown in Fig. 8.2. The acoustic reactance of the box on the diaphragm is given by $X_{AB} = -\gamma P_0/\omega V_B + \omega B\rho_0/\pi a$. For a noncircular diaphragm of area S_D, $\pi a = \sqrt{S_D\pi}$.

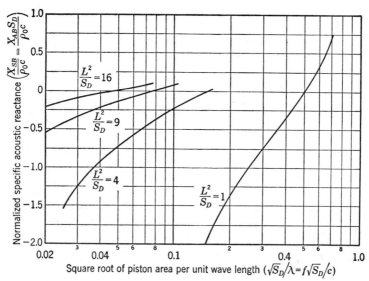

FIG. 8.7. Specific acoustic reactance of a closed box $L \times L \times L/2$ with a diaphragm of area S_D at center of $L \times L$ face. The position of the first normal mode of vibration occurs when $L/2 = \lambda/2$, that is, it occurs at $\sqrt{S_D}/\lambda = 0.25$ for $L^2/S_D = 16$; at 0.333 for $L^2/S_D = 9$; at 0.5 for $L^2/S_D = 4$; and at $\sqrt{S_D}/\lambda = 1$ for $L^2/S_D = 1$.

high value that X_{AB} reaches at the first normal mode of vibration (resonance) for the box, which occurs when the depth of the box equals one-half wavelength. A high reactance reduces the power radiated to a very small value. To reduce the magnitude of X_{AB} at the first normal mode of vibration, an acoustical lining is placed in the box. This lining should be highly absorptive at the frequency of this mode of vibration and at all higher frequencies. For normal-sized boxes, a satisfactory lining is a 1-in.-thick layer of bonded mineral wool, bonded Fiberglas, bonded hair felt, Cellufoam (bonded wood fibers), etc. For small cabinets, where the largest dimension is less than 18 in., a ½-in.-thick layer of absorbing material may be satisfactory.

At low frequencies, where the thickness of the lining is less than 0.05 wavelength, the impedance of the box presented to the rear side of the diaphragm equals

$$Z_{AB} = R_{AB} + jX_{AB} \qquad (8.9)$$

where X_{AB} is given in Eqs. (8.6) to (8.8) and

$$R_{AB} \doteq \frac{R_{AM}}{\omega^2 C_{AB}{}^2 R_{AM}{}^2 + \left(1 + \dfrac{V_B}{\gamma V_M} + \dfrac{V_B{}^2}{\gamma^2 V_M{}^2}\right)} \qquad (8.10)$$

$R_{AM} = R_f/3S_M$ = one-third of the total flow resistance of a layer of the acoustical material that lines the box divided by the area of the acoustical material S_M. The units are mks acoustic ohms. The flow resistance equals the ratio of the pressure drop across the sample of the material to the linear air velocity through it. For lightweight materials the flow resistance R_f is about 100 mks rayls for each inch of thickness. For dense materials like PF Fiberglas board or rockwool duct liner, the flow resistance may be as high as 2000 mks rayls for each inch of thickness of the material. For example, if the flow resistance per inch of material is 500 mks rayls, the thickness 3 in., and the area 0.2m², then $R_{AM} = 1500/(3)(0.2) = 2500$ mks acoustic ohms. It is assumed in writing this equation that the material does not occupy more than 10 per cent of the volume of the box.

V_B = volume of the box in cubic meters including the volume of the acoustical lining material.

V_M = volume of the acoustical lining material in cubic meters.

Graphs of Eq. (8.10) are given in Fig. 8.8.

At all frequencies where the absorption coefficient† of the lining is high (say, greater than 0.8), the impedance of the box presented to the back side of the diaphragm will be the same as that presented to a piston in an

† Tables and graphs of absorption coefficients for common materials are given in Chap. 10.

infinite baffle radiating into free space, so that Z_{AB} is found from Fig. 5.3 or Eqs. (8.2) and (8.3).

Acoustical material may also be used to enlarge effectively the volume of enclosed air. Gaseous compressions in a sound wave are normally adiabatic. If the air space is completely filled with a soft, lightweight

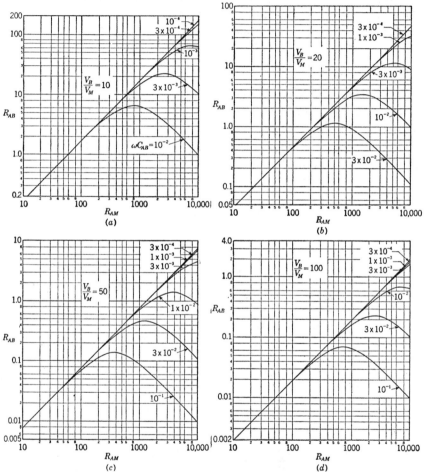

FIG. 8.8. Graphs for determining the acoustic resistance R_{AB} of the closed-box baffle as a function of R_{AM} with ωC_{AB} as the parameter. The four charts are for different ratios of the box volume to the volume of the acoustical material.

material such as kapok or Cellufoam (foamed wood fibers), the compressions become isothermal. This means that the speed of sound decreases from $c = 344.8$ m/sec to $c = 292$ m/sec. Reference to Fig. 8.7 shows that this lowers the reactance at low frequencies just as does an increase in box dimension L. This also means that in Eq. (8.7) the value of γ is 1.0 instead of 1.4.

One special type of closed box with an absorptive lining is shown in Fig. 8.9.[3] Here, the loudspeaker is mounted near the end of a rectangular box of length L and of cross-sectional area approximately equal to πa^2, the effective area of the loudspeaker diaphragm. A glass fiber wedge whose length is $L/2$ is used to terminate the box. The specific acoustic resistance R_{AB} and reactance X_{AB} (multiplied by $S_D/\rho_0 c$) for such a box with and without the wedge are given in Fig. 8.10.

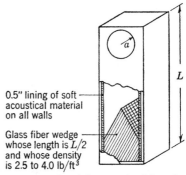

0.5" lining of soft acoustical material on all walls

Glass fiber wedge whose length is $L/2$ and whose density is 2.5 to 4.0 lb/ft³

FIG. 8.9. Special type of cabinet for minimizing the shift of resonance frequency of a loudspeaker mounted in a closed-box baffle.

Three things of importance are observed about the impedance: (1) for a given volume of box, at low frequencies, the reactance X_{AB} is smaller than that for the box of Fig. 8.2; (2) at high frequencies, the box resonances (normal modes of vibration) are damped out so that R_{AB} approaches $\rho_0 c/S_D$ and X_{AB} approaches zero; and (3) between these frequency regions [that is, $0.2 < L/\lambda < 0.5$], the reactance X_{AB} is positive.

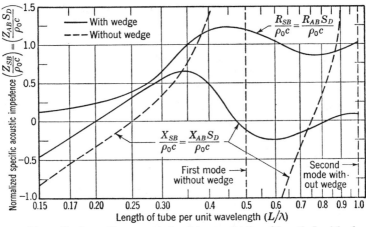

FIG. 8.10. Normalized specific acoustic impedance of tube of length L with absorbing wedge mounted in end. Without the wedge, $R_{SB} = 0$.

Effect of Box Compliance on Resonance Frequency. Let us analyze the effect of the closed-box baffle on the lowest resonance frequency of a direct-radiator loudspeaker. For a loudspeaker mounted in an infinite

[3] D. A. Dobson, "Closed-box Loudspeaker Enclosures," master's thesis, Electrical Engineering Department, Massachusetts Institute of Technology, Cambridge, Mass., 1951.

baffle, the frequency for zero reactance is

$$f_1 = \frac{1}{2\pi\sqrt{C_{AS}(M_{AD} + 2M'_{A1})}} \tag{8.11}$$

where we have assumed that the radiation reactance X_{AR} from each side of the diaphragm equals $\omega M'_{A1}$ and that $M'_{A1} = 0.27\rho_0/a$.

From Fig. 8.4 we see that the resonance frequency for the loudspeaker in a closed-box baffle with a volume less than about 8 ft^3 is

$$f_2 = \frac{1}{2\pi}\sqrt{\frac{C_{AS} + C_{AB}}{C_{AS}C_{AB}(M_{AD} + M_{A1} + M_{AB})}} \tag{8.12}$$

where C_{AB} and M_{AB} are given by Eqs. (8.7) and (8.8) and $M_{A1} \doteq \dfrac{0.2\rho_0}{a}$.

The ratio of (8.12) to (8.11) is equal to the ratio of the resonance frequency with the box to the resonance frequency with an infinite baffle. This ratio is

$$\frac{f_2}{f_1} = \sqrt{\left(1 + \frac{C_{AS}}{C_{AB}}\right)\left(1 + \frac{1.25M'_{A1} - M_{AB}}{M_{AD} + M_{A1} + M_{AB}}\right)} \tag{8.13}$$

If the loudspeaker occupies less than one-third of the area of the side of the box in which it is mounted, Eq. (8.13) is approximately

$$\frac{f_2}{f_1} = \sqrt{1.13\left(1 + \frac{C_{AS}}{C_{AB}}\right)} \tag{8.14}$$

This equation is plotted in Fig. 8.11.

Often, it is difficult to find an "infinite" baffle in which to determine the resonance frequency. If the loudspeaker is held in free space without a baffle, the mass loading M''_{A1} on the diaphragm will be exactly one-half its value in an infinite baffle, that is, $M''_{A1} = 0.135\rho_0/a$. Hence, the ratio of the resonance frequency in the closed box f_2 to the resonance frequency without baffle f_3 is, approximately,

$$\frac{f_2}{f_3} = \sqrt{0.87\left(1 + \frac{C_{AS}}{C_{AB}}\right)} \tag{8.15}$$

This equation also is plotted in Fig. 8.11.

Radiation Equation. At very low frequencies where the diaphragm has not yet become a directional radiator (*i.e.*, its circumference is less than about a wavelength), the loudspeaker in a closed-box baffle may be treated as though it were a simple spherical source of sound. We find from Eq. (4.3) that the sound pressure a distance r away from such a source in a free field is given by

$$|p| \doteq \frac{f\rho_0|U_c|}{2r} \tag{8.16}$$

where $|p|$ = magnitude of the rms sound pressure in newtons per square meter at a distance r from the loudspeaker

$|U_c| = |u_c|S_D$ = magnitude of the rms volume velocity of the diaphragm in cubic meters per second

ρ_0 = density of air in kilograms per cubic meter (about 1.18 kg/m³ for normal room conditions)

r = distance r from the loudspeaker in meters (1 m = 3.28 ft)

f = frequency in cycles per second

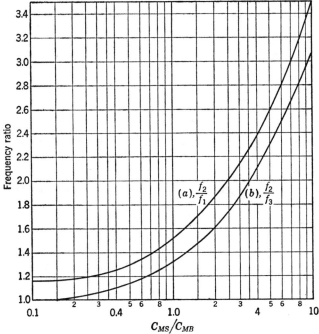

FIG. 8.11. (a) f_2/f_1 = ratio of the resonance frequency for a loudspeaker in a closed-box baffle to the resonance frequency for the same loudspeaker in an infinite baffle. (b) f_2/f_3 = ratio of the resonance frequency for a loudspeaker in a closed-box baffle to the resonance frequency for the same loudspeaker unbaffled.

At medium frequencies, where the diaphragm is becoming directive but yet is still vibrating substantially as a rigid piston, the pressure at a distance r in a free field is

$$|p| = \sqrt{\frac{|U_c|^2 \Re_{AR} Q \rho_0 c}{4\pi r^2}} \tag{8.17}$$

where $\Re_{AR}$ = radiation resistance from the front side of the diaphragm in mks mechanical ohms [see Eq. (8.4)]

Q = approximately the directivity factor for a piston in the end of a tube (see Fig. 4.20)

$\rho_0 c$ = characteristic impedance of air in mks rayls (about 407 mks rayls for normal room conditions)

c = speed of sound in meters per second (about 344.8 m/sec for normal room conditions)

At low frequencies, Eq. (8.17) reduces to Eq. (8.16) because $Q \doteq 1$ and $\Re_{AR} = \pi f^2 \rho_0 / c$.

Diaphragm Volume Velocity U_c. We determine the volume velocity U_c from Fig. 8.4.

$$|U_c| = \frac{e_g B l}{S_D(R_g + R_E)\sqrt{R_A{}^2 + [\omega M_A - (1/\omega C_A)]^2}} \qquad (8.18)$$

where, from Fig. 8.4,

$$R_A = \frac{B^2 l^2}{(R_g + R_E)S_D{}^2} + R_{AS} + R_{AB} + \Re_{AR} \qquad (8.19)$$

$$M_A = M_{AD} + M_{A1} + M_{AB} \qquad (8.20)$$

$$C_A = \frac{C_{AS}C_{AB}}{C_{AS} + C_{AB}} \qquad (8.21)$$

The radiation mass and resistance $\Re_{AR}$ and M_{A1} are generally given by Eqs. (8.4) and (8.5) but for very large boxes or for infinite baffles are given by (8.2) and (8.3).

In an effort to simplify Eq. (8.18), let us define a Q_T in the same manner as we do for electrical circuits. First, let us set

$$\omega_0{}^2 \equiv \frac{1}{M_A C_A} \qquad (8.22)$$

where ω_0 = angular resonance frequency for zero reactance. Then,

$$Q_T \equiv \frac{\omega_0 M_A}{R_A} \qquad (8.23)$$

and

$$\left(\omega M_A - \frac{1}{\omega C_A}\right)^2 \equiv \omega_0{}^2 M_A{}^2 \left(\frac{\omega}{\omega_0} - \frac{\omega_0}{\omega}\right)^2 \qquad (8.24)$$

Substitution of Eqs. (8.22) to (8.24) in Eq. (8.18) yields

$$|U_c| = \frac{e_g B l}{S_D(R_g + R_E)\omega_0 M_A \sqrt{\dfrac{1}{Q_T{}^2} + \left(\dfrac{\omega}{\omega_0} - \dfrac{\omega_0}{\omega}\right)^2}} \qquad (8.25)$$

When the right-hand side is so normalized that its value is unity when $\omega = \omega_0$, we have the equation for the universal resonance curves from which the exact resonance curve may be obtained without calculation when Q_T is known.[4]

[4] F. E. Terman, "Radio Engineers' Handbook," pp. 136–138, McGraw-Hill Book Company, Inc., New York, 1943.

Reference Volume Velocity and Sound Pressure. A reference diaphragm volume velocity is arbitrarily defined here by the equation

$$|U_c|_C = \frac{e_g Bl}{(R_g + R_E)\omega M_A S_D} \tag{8.26}$$

This reference volume velocity is equal to the actual volume velocity above the resonance frequency under the special condition that $R_A{}^2$ of Eq. (8.19) is small compared with $\omega^2 M_A{}^2$. This reference volume velocity is consistent with the reference power available efficiency defined in Par. 7.8.

The reference sound pressure at low frequencies, where it can be assumed that there is unity directivity factor, is found from Eqs. (8.16) and (8.26).†

$$|p_C| = \frac{e_g Bl\rho_0}{(R_g + R_E)M_A 4\pi r S_D} \tag{8.27}$$

It is emphasized that the reference sound pressure will not be the actual sound pressure in the region above the resonance frequency unless the motion of the diaphragm is mass-controlled and unless the directivity factor is nearly unity. The reference pressure is, however, a convenient way of locating "zero" decibels on a relative sound-pressure-level response curve, and this is the reason for defining it here.

Radiated Sound Pressure for $ka < 1$. The radiated sound pressure in the frequency region where the circumference of the diaphragm $(2\pi a)$ is less than a wavelength (*i.e.*, where there is negligible directivity) is found by taking the ratio of Eq. (8.16) to Eq. (8.27), using Eq. (8.25) for $|U_c|$.

$$\left| \frac{p}{p_c} \right| = \frac{\omega/\omega_0}{\left[\dfrac{1}{Q_T{}^2} + \left(\dfrac{\omega}{\omega_0} - \dfrac{\omega_0}{\omega} \right)^2 \right]^{1/2}} \tag{8.28}$$

The ratio, in decibels, of the sound pressure at the resonance frequency ω_0 to the reference sound pressure is

$$20 \log_{10} \left| \frac{p}{p_c} \right|_0 = 20 \log_{10} Q_T \tag{8.29}$$

For *flat* response down to the lowest frequency possible, Q_T should approximately equal unity. (Note that for critical damping $Q_T = 0.5$.)

Referring back to Eq. (7.39), we find that we suggested for satisfactory transient response that $R_M/M_M = R_A/M_A > 184 \text{ sec}^{-1}$. Let us see what this means in terms of Q_T.

---

† Equation (8.27) is the same as that derived in Example 7.1, p. 194 except for a factor of 2 in the denominator. This factor of 2 expresses the difference between radiation into full space as compared to radiation into half space (infinite baffle case).

In terms of Q_T the suggested criterion for satisfactory transient response is

$$Q_T = \frac{\omega_0 M_A}{R_A} < \frac{\omega_0}{184} \qquad (8.30)$$

As an example, if $\omega_0 = 2\pi f_0 = 2\pi 40 = 250$ radians/sec, then Q_T should be less than 1.36. This would mean that the peak in the response curve

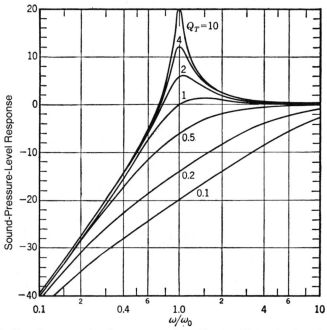

Fig. 8.12. Sound-pressure-level response of a direct-radiator loudspeaker at low frequencies. An infinite baffle or a closed-box enclosure is assumed. Zero decibel is referred to the reference sound pressure defined by Eq. (8.27). Q_T is the same as Q_T of Eq. (8.23) and ω_0 is found from Eq. (8.22). The graph applies only to the frequency range where the wavelengths are greater than about three times the advertised diameter of the diaphragm.

must be less than 3 db. Methods for achieving desired Q_T's will be discussed as part of the example below.

For ease in design of direct-radiator loudspeakers, Eq. (8.28) is given in decibels by Eq. (8.31) below.

$$20 \log_{10} \left| \frac{p}{p_c} \right| = 20 \log_{10} \frac{\omega}{\omega_0} - 10 \log_{10} \left[\frac{1}{Q_T^2} + \left(\frac{\omega}{\omega_0} - \frac{\omega_0}{\omega} \right)^2 \right] \quad \text{db} \quad (8.31)$$

Equation (8.31) is plotted in Fig. 8.12. It should be remembered that Eqs. (8.31) to (8.33) and Fig. 8.12 are valid for either the small closed box or for the infinite baffle, depending on how $\mathfrak{R}_{AR}$ and M_{A1} are chosen.

We should observe that, even in the frequency range where the diaphragm diameter is less than one-third wavelength, the value of Q_T is not strictly constant because $\mathfrak{R}_{AR}$ increases with the square of the frequency. In using Eq. (8.31) and Fig. 8.12, therefore, R_A in Q_T probably ought to be calculated as a function of ω/ω_0. Usually, however, the value of R_A at ω_0 is the only case for which calculation is necessary.

Radiated Sound Pressure at Medium and High Frequencies. When the frequency becomes large enough so that the diameter of the diaphragm is greater than one-third wavelength, Eq. (8.28) may no longer be used. If the diaphragm still vibrates as a rigid piston, we may find the ratio of $|p|$ to $|p_c|$ by taking the ratio of Eq. (8.17) to Eq. (8.27), using Eq. (8.25) for $|U_c|$ and assuming that we are above the first resonance frequency so that $\omega/\omega_0 \gg \omega_0/\omega$. This operation yields

$$\left|\frac{p}{p_c}\right| = \sqrt{\frac{4\mathfrak{R}_{AR}Qc\pi}{\rho_0(R_A^2/M_A^2 + \omega^2)}} \tag{8.32}$$

where ω_0 is given in (8.22) and

$$M_A = M_{AD} + \frac{2X_{AR}}{\omega} \tag{8.33}$$

$$R_M = \frac{B^2l^2}{(R_g + R_E)S_D^2} + R_{AS} + 2\mathfrak{R}_{AR} \tag{8.34}$$

Q = directivity index for a piston in the end of a tube (see Fig. 4.20)

Equation (8.32) expressed in decibels becomes,

$$20 \log \left|\frac{p}{p_c}\right| = 10 \log_{10} \mathfrak{R}_{AR} + \text{DI}$$
$$+ 10 \log_{10} \frac{4\pi c}{\rho_0} - 10 \log_{10} \left(\frac{R_A^2}{M_A^2} + \omega^2\right) \tag{8.35}$$

where DI = $10 \log_{10} Q$ = directivity index for a piston in the end of a long tube (see Fig. 4.20)

$10 \log_{10} \dfrac{4\pi c}{\rho_0}$ = 35.7 db at normal room conditions

Inspection of Eq. (8.35) shows that the sound-pressure response $|p|$ of the loudspeaker relative to its reference response $|p_c|$ is a function of frequency, effective radius of the diaphragm, and the ratio R_A/M_A.

For convenience in design, Eq. (8.35) is plotted in Fig. 8.13 for four values of advertised diameters of loudspeakers having effective radii of 0.0875, 0.11, 0.13, and 0.16 m, respectively. These graphs join on to Fig. 8.12 at frequencies where ω/ω_0 is large compared with ω_0/ω.

8.6. Location of Loudspeaker in Box. The results shown in Fig. 8.7 for the reactance of the closed box apply to a loudspeaker mounted in the center of one of the L by L sides. This location of the loudspeaker leaves something to be desired, because waves traveling outward from the diaphragm reach the outside edges of the box simultaneously and in combination set up a strong diffracted wave in the listening space. To reduce the magnitude of the diffracted wave, the loudspeaker should be moved off center by several inches—preferably in the direction of one corner.

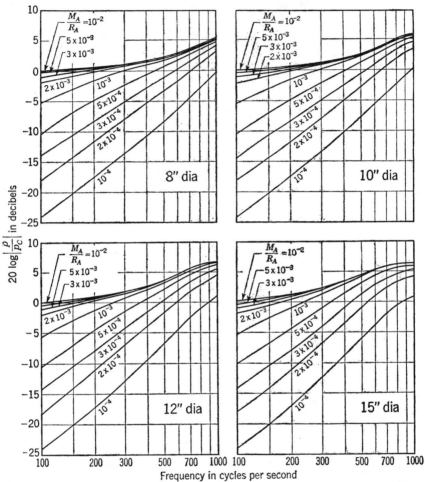

Fig. 8.13. Graphs for calculating the relative frequency response in the frequency region where the loudspeaker is no longer a simple source, but below the frequency of the rim resonance. The parameter is M_A/R_A. These charts cover the frequency range just above that covered by Fig. 8.12.

The front face of the box of Fig. 8.2 need not be square. It is possible to make the ratio of the two front edges vary between 1 and 3 without destroying the validity of the charts, for the same total volume.

8.7. Measurement of Loudspeaker Constants. The constants of a loudspeaker are not difficult to determine if a-c and d-c voltmeters and a calibrated variable-frequency oscillator are available.

During the tests, the direct-radiator loudspeaker is hung *without baffle* in an anechoic chamber, or outdoors, or in a large room that is not too reverberant. The various constants are then determined as follows:

Measurement of ω_0. A variable-frequency source of sound with an output impedance greater than 100 times the nominal impedance of the loudspeaker is connected to the loudspeaker terminals. A voltmeter is then connected across the terminals, and the frequency is varied until a maximum meter reading is obtained (see Fig. 8.14). From Fig. 7.2, we see that maximum electrical loudspeaker impedance corresponds to maximum mechanical mobility, which in turn occurs at the resonance frequency f_0 or ω_0.

Measurement of R_E. The electrical resistance of the voice coil is measured with a d-c bridge.

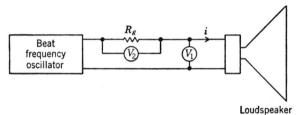

FIG. 8.14. Circuit for determining the resonance frequency of a loudspeaker. The voltage V_2 is held constant, and the oscillator frequency is varied until V_1 is maximum.

Measurement of Q. The Q of the loudspeaker is found by determining the width of the diaphragm-velocity curve u_c plotted as a function of frequency for a constant current i (see Fig. 8.14). Because $e = Blu_c$, we may determine the shape of the velocity curve by plotting against frequency the quantity

$$e = V_1 - R_E i = V_1 - \frac{R_E V_2}{R_g}$$

The width, Δf cps, is measured between the points of the curve on either side of the resonance peak where the voltage is 3 db down (0.707) from the peak voltage.

The value of Q for a constant current i is given by

$$Q = \frac{f_0}{\Delta f} \tag{8.36}$$

Measurement of M_{MD}. To find M_{MD}, we must determine the total mass of the diaphragm, including the air load, and subtract from it the air-load mass. A known mass is added to the diaphragm, and a new resonance frequency is determined. The added mass is usually a lead rod or a roll of solder bent into a circle of 3 to 6 in. diameter and taped to the diaphragm so that it does not bounce. Lead is used because it is nonmagnetic.

If the original resonance frequency was ω_0 and the resonance frequency after addition of a mass M' kg is ω_0', then

$$\omega_0 = \frac{1}{\sqrt{(M_{MD} + M_{M1})C_{MS}}} \qquad (8.37)$$

and

$$\omega_0' = \frac{1}{\sqrt{(M_{MD} + M_{M1} + M')C_{MS}}} \qquad (8.38)$$

where C_{MS} = mechanical compliance of the suspension in meters per newton

$M_{M1} = 2.67a^3\rho_0 = 3.15a^3$ = air-load mass in kilograms on the *two* sides of the diaphragm

Simultaneous solution of (8.37) and (8.38) yields

$$M_{MD} + M_{M1} = \frac{M'}{(\omega_0/\omega_0')^2 - 1} \qquad (8.39)$$

Determination of R_{MS}. The mechanical resistance R_{MS} of the diaphragm suspension plus the air-load resistance is determined from the data above. From circuit theory we know that

$$R_{MS} + \Re_{MR} = \frac{\omega_0(M_{MD} + M_{M1})}{Q} \qquad (8.40)$$

where $\Re_{MR} = 0.1886a^6\rho_0\omega^4/c^3 = (8.45 \times 10^{-6})a^6f^4$ mechanical air-load resistance in mks mechanical ohms for *both* sides of an unbaffled loudspeaker. This resistance generally can be neglected.

Measurement of C_{MS}. The mechanical compliance of the suspension may be determined by adding known masses to the cone and measuring the steady displacement when the axis of the cone is vertical. A depth micrometer is mounted rigidly with respect to the loudspeaker frame to accomplish this. That is to say,

$$C_{MS} = \frac{\text{displacement (m)}}{\text{added weight (newtons)}} \qquad (8.41)$$

The added weight is, of course, the added mass in kilograms M' times the acceleration due to gravity, *i.e.*,

$$\text{Added force (weight)} = f_1 = M'g = 9.8M' \qquad \text{newtons} \quad (8.42)$$

The compliance C_{MS} may also be determined from Eq. (8.37) if the masses $M_{MD} + M_{M1}$ and the angular resonance frequency ω_0 are known. In this case the depth micrometer is not needed.

Measurement of Bl. The value of the electromagnetic coupling constant Bl is determined by using an added mass and the depth micrometer described above or a bent piece of wire fastened to the frame of the loudspeaker to mark the initial position of the diaphragm. When a mass M'

is added to the diaphragm, a downward displacement occurs.   To restore
the diaphragm to its original position, a steady force f_s is needed.

If a direct current i is passed through the voice-coil winding with the
proper direction and magnitude, the required force f_s will be produced as
follows:

$$f_s = Bli \qquad (8.43)$$

The displacement of the diaphragm was produced by the weight of the
added mass [see Eq. (8.42)].   The diaphragm will return to its zero position when

$$f_1 = f_s \qquad (8.44)$$

Solution of Eqs. (8.42) and (8.43) simultaneously gives

$$Bl = \frac{M'g}{i} = \frac{9.8M'}{i} \qquad (8.45)$$

Measurement of S_D.   The effective area of the diaphragm can be
determined only by coupling its front side to a closed box.   The volume of
air V_0 enclosed in the space bounded by the diaphragm and the sides of
the box must be determined accurately.   Then a slant manometer for
measuring air pressure is connected to the airspace.   The cone is then
displaced a known distance ξ meters, the manometer is read, and the
incremental pressure p is determined.   Then,

$$p = \frac{P_0}{V_0} \xi S_D$$

or

$$S_D = \frac{V_0}{P_0} \frac{p}{\xi} \qquad \text{m}^2 \qquad (8.46)$$

where P_0 is the ambient pressure.   The pressures P_0 and p both must be
measured in the same units, and V_0/ξ should be determined in square
meters.

Usually, S_D can be determined accurately enough for most calculations
from Fig. 8.5a, that is, $S_D = \pi a^2$.

8.8. Measurement of Baffle Constants.   The constants of the baffle
may be measured after the loudspeaker constants are known.   Refer to
Fig. 8.4.   The quantities $\Re_{AR}$ and X_{AR} are determined from Eqs. (8.4)
and (8.5).   The electrical and mechanical quantities are measured
directly.

Measurement of X_{AB}.   Using the same procedure as for measuring ω_0
in Par. 8.7, determine a new ω_0 and solve for X_{AB} from Fig. 8.4.

Measurement of R_{AB}.   Using the same procedure as for measuring Q
in Par. 8.7, determine a new Q and solve for R_{AB} from

$$R_{AB} = \frac{\omega_0(M_{AD} + M_{A1})}{Q} - (R_{AS} + \Re_{AR})$$

where $M_{AD} = M_{MD}/S_D{}^2$; M_{A1} is given by Eq. (8.5), and

$$R_{AS} + \Re_{AR} = \frac{R_{MS} + \Re_{MR}}{S_D{}^2}.$$

Example 8.1. Direct-radiator Loudspeaker in Infinite Baffle. Determine Q_T and R_{MS} for the loudspeaker of Fig. 7.8.

Solution. The advertised diameter of that loudspeaker is 8 in., and from Fig. 8.5a we find that this corresponds to an effective diaphragm radius of $a = 0.088$ m and an effective area $S_D = 0.0244$ m². Hence, Fig. 8.12 is valid for this loudspeaker for wavelengths down to 0.6 m, or frequencies as high as about 550 cps. The diaphragm on this loudspeaker ceases to vibrate as a rigid piston at about 400 cps so that theory and measurements would be expected to agree well only up to about 200 cps.

By superimposing the measured data plotted in Fig. 7.8 on the contours of Fig. 8.12, we find that $Q_T \doteq 6$. We also observe that $\omega_0 = 565$ radians/sec.

From Fig. 8.5c we find that $M_{MD} \doteq 0.0042$ kg and

$$M_{AD} = \frac{0.0042}{(0.0244)^2} = 7.05 \text{ kg/m}^4$$

The loudspeaker was in an infinite baffle so that, from Eq. (8.3),

$$2M_{A1} = \frac{(2)(0.27)(1.18)}{0.088} = 7.25 \text{ kg/m}^4$$

Hence,

$$M_A = M_{AD} + 2M_{A1} = 14.3 \text{ kg/m}^4$$

From Eq. (8.23) we have

$$R_A = \frac{\omega_0 M_A}{Q_T} = \frac{(565)(14.3)}{6}$$
$$= 1346 \text{ mks acoustic ohms}$$

The radiation resistance $2\Re_{AR}$ at $\omega_0 = 565$ radians/sec ($f_0 = 90$ cps) should equal [see Eq. (8.2)]

$$2\Re_{AR} = (2)(0.0215)(90)^2 = 348 \text{ mks acoustic ohms}$$

The damping contributed by the suspension at the resonance frequency is

$$R_{AS} = R_A - 2\Re_{AR} - \frac{B^2 l^2}{(R_g + R_E)S_D{}^2}$$

From Table 8.1, R_E was approximately 3 ohms. The loudspeaker was operated from an amplifier with an output resistance of 12 ohms, and Bl was approximately 2.5 webers/m. So

$$\frac{B^2 l^2}{(R_g + R_E)S_D{}^2} = \frac{6.25}{(15)(0.0244)^2} = 700 \text{ mks acoustic ohms}$$

Hence,

$$R_{AS} = 1346 - 348 - 700 = 298 \text{ mks acoustic ohms}$$

It is interesting to observe that Q_T is a function of frequency because R_A equals

$$R_A = 298 + 730 + 0.043 f_0{}^2$$

For example, Q_T at 300 cps equals

$$Q_T = \frac{\omega_0 M_A}{R_A} = \frac{(565)(14.3)}{5000} = 1.62$$

Reference to Fig. 8.12 shows that the response at 300 cps essentially is the same for $Q_T = 1.62$ as for its value of 6 at 90 cps.

Example 8.2. Wide-frequency-range Loudspeaker. A loudspeaker designed to be operated without a tweeter unit has a nominal diameter of 12 in. and the physical constants given below:

$$a = 13 \text{ cm (effective radius)}$$
$$C_{MS} = 2 \times 10^{-4} \text{ m/newton}$$
$$M_{MD} = 12 \text{ g}$$
$$R_g = 2 \text{ ohms}$$
$$R_E = 8 \text{ ohms}$$
$$B = 1 \text{ weber/m}^2$$
$$l = 10 \text{ m}$$
$$R_{MS} = 1 \text{ newton-sec/m}$$
$$S_D = \pi a^2 = 0.053 \text{ m}^2$$
$$S_D{}^2 = 0.00281 \text{ m}^4$$

   a. Determine the percentage shift in the first resonance frequency of the loudspeaker from the value for an infinite baffle if a box having dimensions 1.732 by 1.732 by 1 ft $= 3 \text{ ft}^3$ is used.

   b. Determine the height of the resonance peak, assuming $R_{AB} = 0$.

   c. Determine the dimensions for a baffle of the type shown in Fig. 8.9 that will cause a shift in infinite-baffle resonance frequency of only 10 per cent.

   d. Determine the height of the resonance peak for the baffle of (*c*).

   Solution. *a.* First, let us determine the value of the acoustic-radiation masses for the loudspeaker in an infinite baffle. From Eq. (8.2),

$$2M_{A1} \doteq \frac{0.54\rho_0}{a} = \frac{(0.54)(1.18)}{0.13}$$
$$= 4.9 \text{ kg/m}^4$$

Also,

$$C_{AS} = C_{MS}S_D{}^2 = (2 \times 10^{-4})(2.81 \times 10^{-3})$$
$$= 5.62 \times 10^{-7} \text{ m}^5/\text{newton}$$
$$M_{AD} = \frac{M_{MD}}{S_D{}^2} = \frac{0.012}{0.00281} = 4.26 \text{ kg/m}^4$$

The resonance frequency for the speaker in an infinite baffle is found from Eq. (8.11).

$$f_1 = \frac{1}{2\pi \sqrt{5.62 \times 10^{-7}(4.9 + 4.26)}}$$
$$= \frac{10^2}{6.28 \sqrt{0.0516}} \doteq 70 \text{ cps}$$

From Eq. (8.7) we find the formula for C_{AB}.

$$C_{AB} = \frac{V}{\gamma P_0} = \frac{(3)(2.832 \times 10^{-2})}{1.4 \times 10^5}$$
$$\doteq 6.06 \times 10^{-7} \text{ m}^5/\text{newton}$$

The ratio of C_{AS} to C_{AB} is

$$\frac{C_{AS}}{C_{AB}} = \frac{5.62 \times 10^{-7}}{6.06 \times 10^{-7}} = 0.93$$

From Fig. 8.11 we find the ratio of the resonance frequency with a closed-box baffle to the resonance frequency in an infinite baffle.

$$\frac{f_2}{f_1} = 1.48$$

That is to say, the new resonance frequency is 48 per cent higher; so

$$f_2 = (1.48)(70) = 103.5 \text{ cps}$$

b. The height of the resonance peak is to be found from Eqs. (8.23) and (8.29). The total resistance R_M is given by Eq. (8.19) for low frequencies. Using Eq. (8.4) and $f_0 = 103.5$ cps, we determine the radiation resistance at the resonance frequency.

$$\Re_{AR} = 0.01076f^2 = (0.01076)(103.5)^2 = 115 \text{ mks acoustic ohms}$$
$$R_{AS} = \frac{R_{MS}}{S_D{}^2} = \frac{1}{0.00281} = 356 \text{ mks acoustic ohms}$$

So

$$R_A = \frac{10^2}{(10)(0.00281)} + 356 + 115 = 4030 \text{ mks acoustic ohms}$$

The value of M_A is given by Eq. (8.20). From Eq. (8.5),

$$M_{A1} \doteq \frac{0.1952\rho_0}{a} = \frac{(0.1952)(1.18)}{0.13} = 1.772 \text{ kg/m}^4$$

From Eq. (8.8),

$$M_{AB} = \frac{(0.85)(1.18)}{(3.14)(0.13)} = 2.46 \text{ kg/m}^4$$

So

$$M_A = 4.26 + 2.46 + 1.77 = 8.49 \text{ kg/m}^4$$
$$Q_T = \frac{\omega_0 M_A}{R_A} = \frac{(650)(8.49)}{4030} \doteq 1.37$$
$$20 \log_{10} \frac{p}{p_c} = 20 \log_{10} 1.37 \doteq 2.7 \text{ db}$$

The resonance peak lies about 3 db above the reference pressure level. Because Q_T is greater than unity, the reference pressure is the pressure in the flat region C above the resonance frequency, as we can see from Fig. 8.12.

c. For only a 10 per cent shift in resonance frequency, $f_2 = 70 \times 1.1 = 77$ cps. From Eq. (8.14) we find that

$$\frac{C_{AS}}{C_{AB}} = \frac{(1.1)^2}{1.13} - 1 = 0.07$$

or

$$C_{AB} = \frac{5.62 \times 10^{-7}}{0.07} = 8.04 \times 10^{-6}$$

To achieve this ratio, the volume of the box either must be quite large, or else we might use the arrangement of Fig. 8.9 as was asked for in the statement of the problem. The desired normalized reactance is

$$\frac{X_{AB}S_D}{\rho_0 c} = \frac{-S_D}{\omega C_{AB}\rho_0 c} = \frac{-(10^6 \times 0.053)}{484 \times 8.04 \times 407} = -0.0334$$

From Fig. 8.10 we see that for this kind of baffle a vertical length is required of

$$L = 0.2\lambda = \frac{0.2 \times 344.8}{77} = 0.9 \text{ m, or } 35 \text{ in.}$$

If the 12-in.-wide loudspeaker is to fit into a cross-sectional area of 0.053 m², the side that it is mounted in will need to be about 13 in. = 0.33 m wide. This gives us, for the other side, 0.053/0.33 = 0.16 m ≐ 6.34 in.

d. The height of the resonance peak is found from Q_T. To determine Q_T, we need R_A and X_A.

$$R_A = 4030 + R_{AB}$$

We find R_{AB} from Fig. 8.10.

$$R_{AB} = \frac{\rho_0 c (0.25)}{S_D} = \frac{(410)(0.25)}{0.053}$$
$$= 1935 \text{ mks acoustic ohms}$$

which gives us

$$R_A = 4030 + 1935 \doteq 6000 \text{ mks acoustic ohms}$$
$$M_A = 4.26 + 1.77 + 0 = 6.03 \text{ kg}$$
$$Q_T = \frac{\omega_0 W_A}{R_A} = \frac{(484)(6.72)}{6000} \doteq 0.54$$

In other words, this arrangement is nearly critically damped. At 77 cps, the response will be down about 5 db from the reference response.

Example 8.3. Low-frequency Loudspeaker (Woofer). Many high-fidelity sound systems employ two loudspeakers. One covers the low-frequency range, and the other covers the high-frequency range. An electrical network is used to divide the output energy from the amplifier into two frequency regions centered on the "crossover" frequency. Common crossover frequencies are 500, 800, 1000, and 1500 cps.

In this example, we have chosen a 15-in. commercial low-frequency unit, designed to be used with a 500 cps crossover network. As no information was available from the manufacturer on the constants of this loudspeaker, they were determined experimentally. The loudspeaker was then put in a closed-box baffle, and its frequency response was measured for comparison with calculations. The complete procedures used in carrying out the experiments are described here, and the results are compared with the computations in order that the reader may get an indication of the reliability of the method.

The loudspeaker was placed in a closed box with dimensions

$$V = (30 \text{ in.} \times 35 \text{ in.} \times 18 \text{ in.})(1.639 \times 10^{-5}) = 0.31 \text{ m}^3.$$

The other constants were determined by the procedures of Par. 8.7.

$$R_E = 5.5 \text{ ohms}$$
$$C_{MS} = 2.82 \times 10^{-4} \text{ m/newton}$$
$$a = 0.16 \text{ m}$$
$$S_D = 8.03 \times 10^{-2} \text{ m}^2$$
$$S_D^2 = 6.45 \times 10^{-3} \text{ m}^4$$
$$R_{MS} = 2.3 \text{ mks mechanical ohms}$$
$$M_{MD} = 0.045 \text{ kg}$$
$$f_0 = 39 \text{ cps}$$
$$\omega_0 = 247 \text{ radians/sec}$$
$$Bl = 25 \text{ webers/m}$$

The volume of the box occupied by the loudspeaker was about 0.01 m³, so that

$$V_B = 0.3 \text{ m}^3$$

From these quantities we determine the sizes of the elements in Fig. 8.4.

$$C_{AS} = C_{MS} S_D^2 = 1.82 \times 10^{-6} \text{ m}^5/\text{newton}$$
$$R_{AS} = \frac{R_{MS}}{S_D^2} = 356 \text{ mks acoustic ohms}$$
$$M_{AD} = \frac{M_{MD}}{S_D^2} = 6.98 \text{ kg/m}^4$$
$$\frac{B^2 l^2}{S_D^2} = 9.7 \times 10^4 \text{ webers}^2/\text{m}^6$$

A 3-in.-thick sound-absorbing blanket with a flow resistance of 2000 mks rayls/in. of thickness is placed on the 18- by 35-in. wall of the box to reduce the effects of standing waves at higher frequencies. This material yields a value of R_{AB} that varies between the values given by Eq. (8.10) at low frequencies to $\Re_{AR}$ for an infinite baffle at 500 cps. If the amplifier resistance is small, R_{AB} may be neglected at low frequencies.

Values of R_{AB}, calculated from Eq. (8.10) at low frequencies and estimated at higher frequencies, are shown in Table 8.2. At very low frequencies, for example, 40 cps, the plywood box absorbs sound because of the frictional losses in the wood as it vibrates.

TABLE 8.2. Determination of R_A

f, cps	ka	R_{AS}	$\Re_{AR}$†	R_{AB}	$R_g = 3$ ohms		$R_g = 130$ ohms	
					$\dfrac{B^2 l^2}{(R_g + R_E)S_D^2}$	R_A	$\dfrac{B^2 l^2}{(R_g + R_E)S_D^2}$	R_A
	$ka < 1$:							
20	0.0582	356	4	300	11,400	12,000	716	1380
100	0.291	356	108	93	11,400	12,000	716	1270
150	0.437	356	242	155	11,400	12,100	716	1470
200	0.582	356	430	295	11,400	12,500	716	1800
250	0.728	356	672	465	11,400	12,900	716	2210
300	0.874	356	968	775	11,400	13,500	716	2820
	$ka > 1$:							
400	1.164	356	1622	1470	11,400	14,800	716	4160
500	1.455	356	2640	2640	11,400	17,000	716	6350

† For $ka < 1$, $\Re_{AR} = 0.01076 f^2$.

For example, at 40 cps, the box might be expected to absorb one-half as much energy per square foot of area as the acoustical material placed inside. By 80 cps, however, the acoustical material will absorb many times more than the box. Hence, for the case of the closed box we may neglect the losses in the wood except at the resonance frequency (near 50 cps), where we may estimate in the calculations the effect by tripling the actual amount of acoustical material.

The quantities $\Re_{AR}$ and M_{A1} (see Fig. 8.4) are found from Fig. 5.7 or, for $ka < 1$, Eqs. (8.4) and (8.5).

The quantity M_{AB} for $ka < 1$ is found from Eq. (8.8) and Fig. 8.6. For our box, the quantity $S_D/L^2 = 0.0803/0.618 = 0.118$. From Fig. 8.6, we see that $B = 0.788$. The value of M_{AB} for $ka > 1$ is equal to $0.92\rho_0 c/S_D\omega$ times the ordinate of Fig. 5.3.

DETERMINATION OF R_A, M_A, AND C_A. These quantities are given by Eqs. (8.19) to (8.21). For this example, calculated values of R_A and M_A are given in Tables 8.2 and 8.3, respectively. C_A is found as follows:

$$C_{AB} = \frac{V}{\gamma P_0} = \frac{0.3}{1.4 \times 10^5} = 2.14 \times 10^{-6} \text{ m}^5/\text{newton}$$

$$C_A = \frac{C_{AS}C_{AB}}{C_{AS} + C_{AB}} = \frac{1.82 \times 2.14 \times 10^{-12}}{(1.82 + 2.14) \times 10^{-6}}$$
$$= 9.84 \times 10^{-7} \text{ m}^5/\text{newton}$$

DETERMINATION OF ω_0 WITH BOX. The value of ω_0 with the closed-box enclosure is, from Table 8.2 and C_A given above,

$$\omega_0 = \frac{1}{\sqrt{M_A C_A}} = \frac{10^4}{\sqrt{(10.27)(0.984)}} = 318 \text{ radians/sec}$$

or

$$f_0 \doteq 50 \text{ cps}$$

DETERMINATION OF Q_T.

$$Q_T = \frac{\omega_0 M_A}{R_A} = \frac{318 M_A}{R_A}$$

Values of Q_T, M_A/R_A (for $f > 150$ cps), and ω/ω_0 are tabulated in Table 8.3. Also shown in Table 8.3 are the values of the directivity index as determined from Fig. 4.20 for a piston in a long tube.

TABLE 8.3.　Determination of M_A, Q_T, and M_A/R_A

f.........	20–150 cps	200	250	300	400	500
ka........	0.0582–0.436	0.582	0.727	0.873	1.164	1.455
M_{AD}......	6.98	6.98	6.98	6.98	6.98	6.98
M_{AB}......	1.85	1.85	1.85	1.85	1.29	1.03
M_{A1}......	1.44	1.44	1.44	1.44	1.27	1.02
M_A........	10.27	10.27	10.27	10.27	9.54	9.03

R_g	Q_T	M_A/R_A				
3	0.27	8.2×10^{-4}	8×10^{-4}	7.6×10^{-4}	7×10^{-4}	6×10^{-4}
130	2.5	5.7×10^{-3}	4.6×10^{-3}	3.6×10^{-3}	2.5×10^{-3}	1.6×10^{-3}
ω/ω_0......	0.4–3.0	4.0	5.0	6.0	8.0	10.0
DI, db....	0–0.9	1.4	1.9	2.4	3.4	4.6

CALCULATION OF FREQUENCY-RESPONSE CURVE.　The data of Table 8.3 permit the calculation of the relative sound-pressure-level response curve for a microphone position on the principal axis of the loudspeaker.　Figure 8.12 is used for the frequency range below 150 cps and Fig. 8.13 for 200 cps and above.　The results of the calculations are shown in Fig. 8.15.　Calculations for intermediate resistances are also shown.　The value of $R_g = 3$ ohms corresponds to constant voltage across the terminals.　$R_g = 14$ ohms is equal to the nominal impedance of the loudspeaker.　$R_g = 30$ ohms is equivalent to changing Bl by a factor of 0.5.　$R_g = 130$ ohms is equivalent to changing Bl by a factor of 5.

Each curve has its own reference pressure.　However, to show the effect on the relative level of the response curve (for constant e_g) as the amplifier impedance R_g was varied, the reference pressures (in decibels) are shifted relative to each other by

$$(\text{Reference 1 minus reference 2}) = 20 \log_{10} \frac{R_{g2} + R_E}{R_{g1} + R_E} \quad \text{db}$$

If the power-available-efficiency response (PAE) is desired, subtract the directivity index (DI) of Table 8.2 from the calculated response curves of Fig. 8.13.　This gives the *shape* of the PAE_db curve.

MEASURED FREQUENCY-RESPONSE CURVE. The frequency-response curve was measured in an anechoic chamber using a 3-ohm 50-watt amplifier as the source. The power available was about 1 watt. The data were taken at a distance of 7 ft on the principal axis of the loudspeaker. To be in the far-field of the loudspeaker a distance

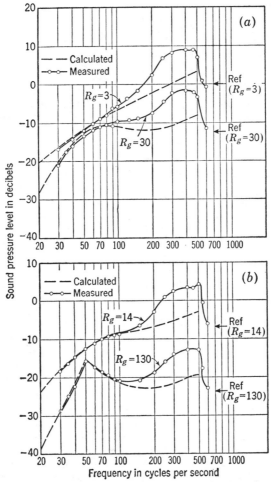

FIG. 8.15a and b. Graphs of the on-axis sound pressure level produced by a low-frequency direct-radiator loudspeaker below the frequency of rim resonance. The dashed curves are calculated. Solid curves with points are measured. Graphs are plotted relative to the reference pressure for $R_g = 3$ ohms. Differences between calculations and data at frequencies above 100 cps are due to cone resonances. The rim resonance occurs at about 500 cps.

greater than 7 ft would have been desirable, but the small size of the available anechoic chamber prevented this. The data for $R_g = 3$, 14, 30, and 130 ohms with a constant value of the open-circuit amplifier voltage e_g are also shown by the circles and solid line in Fig. 8.15.

The agreement between theory and experiment at frequencies below 100 cps is excellent, which indicates that the diaphragm was vibrating as a rigid piston.

Between 100 and 500 cps the loudspeaker radiates more sound along the principal axis than the theory predicts, assuming the directivity index for a 15-in. piston in a long tube.

Polar plots of the sound pressure level were also made with the microphone held in one position and the loudspeaker (and baffle) rotated. The measured directivity index is compared with the theoretical in Table 8.4. The differences between the

TABLE 8.4. Comparison of Measured DI for a 15-in. Loudspeaker in a Closed-box Baffle to that for a 15-in. Piston in a Long Tube. Comparison of the DI Differences is Also Made with the Differences between the Curves of Fig. 8.15

f	Directivity index, db		Differences in DI db	Differences from Fig. 8.15, db
	Loudspeaker	Piston		
100	0.8	0.4	0.4	0.6
200	3.0	1.4	1.6	4.0
300	5.5	2.4	3.1	8.0
400	8.0	3.4	4.6	8.0
500		4.6		

directivity indexes are also compared with the differences between the theoretical and measured curves of Fig. 8.15. Comparison of these curves shows that not only was the loudspeaker more directive than expected but it also radiated more power by the amount indicated in the fifth column. The increased directivity is assumed to be due to the conical shape of the diaphragm, whereas the increased power output is attributable to the breakup of the cone into higher-order resonances. A similar difference between measured and calculated curves was also found for the 8-in. loudspeaker of Fig. 7.8.

At about 500 cps, the major rim resonance occurs, with the expected sharp dip in response. As the loudspeaker was not designed to operate above 500 cps, no data were taken beyond that point.

PART XX　*Bass-reflex Enclosures*

8.9. General Description. The bass-reflex enclosure is a closed box in which an opening, usually called the *port*, has been made.[5-8] The area of the port is commonly made equal to or smaller than the effective area of

[5] A. L. Thuras, "Sound Translating Device," U.S. Patent No. 1,869,178, July, 1932 (filed 1930).

[6] H. F. Olson, "Elements of Acoustical Engineering," 2d ed., pp. 154–156, D. Van Nostrand Company, Inc., New York, 1947.

[7] D. J. Plach and P. B. Williams, Loudspeaker Enclosures, *Audio Engineering*, **35**: 12ff. (July, 1951).

[8] J. J. Baruch and H. C. Lang, "Some Vented Enclosures for Loudspeakers" unpublished report of Acoustics Laboratory, Massachusetts Institute of Technology, Cambridge, Mass., January, 1953.

the diaphragm of the driving unit. A common construction of this type of loudspeaker is shown in Fig. 8.16. When the diaphragm vibrates, part of its displacement compresses the air inside the box and the remainder of its displacement moves air outward through the port. Thus the port is a second "diaphragm," driven by the back side of the loudspeaker diaphragm. The port is, at low frequencies, equivalent to a short length of

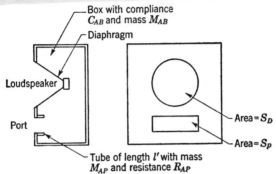

FIG. 8.16. Bass-reflex baffle. The port has an area S_p, and the diaphragm has an area S_D. The inner end correction for the tube is included in the magnitude of M_{AP}.

tube with an acoustic reactance and a series acoustic resistance. This tube has an end correction on the inner end and a radiation impedance on the outer, or radiating, end.

We shall assume for the remainder of this analysis that $ka < 0.5$. In other words, we are restricting ourselves to the very low frequency region where the radiation from both the port and the loudspeaker is nondirectional.

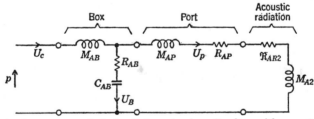

FIG. 8.17. Analogous acoustical circuit for a loudspeaker box with a port. The volume velocity of the diaphragm is U_c, that of the port is U_p, and that of the box is U_B.

8.10. Acoustical Circuit. The acoustical circuit for the box and the port is given in Fig. 8.17. The compliance and resistance of the box are C_{AB} and R_{AB}. The mass loading on the back side of the diaphragm is M_{AB}. The mass and resistance of the air in the port that penetrates the side of the box, including the inner end correction, are M_{AP} and R_{AP}, respectively. Finally, the series radiation mass and resistance from the front side of the port are, respectively, M_{A2} and $\Re_{AR2}$. The values of these quantities are M_{AB} as in Eq. (8.8); R_{AB} as in Eq. (8.10); C_{AB} as in Eq. (8.7); M_{A2} as in Eq. (8.5), but with a_2 instead of a, that is, $M_{A2} = 0.23/a_2$; $\Re_{AR2}$ as in Eq. (8.4); and

$M_{AP} = (t + 0.6a_2)\rho_0/\pi a_2^2 =$ acoustic mass of the air in the port in kilograms per meter^4. This quantity includes the inner end correction.

$R_{AP} =$ acoustic resistance of the air in the port in mks acoustic ohms. [See Eq. (5.54). Use the number (1) in the parentheses.]

$\rho_0 =$ Density of air in kilograms per cubic meter (normally about 1.18 kg/m^3).

$a_2 =$ effective radius in meters of the port in the vented enclosure. If the port is not circular, then let $a_2 = \sqrt{S_p/\pi}$, where S_p is the effective area of the opening in square meters.

$S_p = \pi a_2^2 =$ effective area of the port in square meters.

$t =$ length of the tube or the thickness of the wall of the enclosure. in which the port is cut in meters.

In case the port is comprised of a number of identical *small* openings or tubes, the following procedure is followed:

Let N equal the number of such openings in the enclosure. For each opening the acoustic mass and resistance including M_{A2} and $\Re_{AR2}$ are

$M_A = (t + 1.7a_3)\rho_0/(\pi a_3^2)$ kg/m^4 [see Eq. (5.57)]

$R_A =$ acoustic resistance of each opening in mks acoustic ohms [see Eq. (5.56)]

$a_3 =$ effective radius of each opening in meters

The total acoustic mass and resistance for the N identical openings are

$M_{A2} + M_{AP} = M_A/N$ kg/m^4

$\Re_{AR2} + R_{AP} = R_A/N$ mks acoustic ohms

The directivity factor for a group of holes is about equal to that for a piston with an area equal to the area within a line circumscribing the entire group of holes.

8.11. Electro-mechano-acoustical Circuit. The complete circuit for a loudspeaker in a bass-reflex enclosure is obtained by combining Figs. 8.4

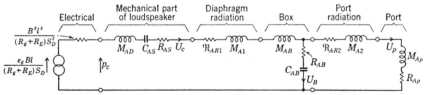

FIG. 8.18. Complete electro-mechano-acoustical circuit for a bass-reflex loudspeaker. The total force produced at the voice coil by the electric current is $p_c S_D$, where S_D is the area of the diaphragm. The volume velocity of the diaphragm is U_c, and that of the port is U_p. The box is assumed to be unlined. If the box is lined with a sound-absorbing material, a resistance must be inserted in series with M_{AB}. Note that M_{AP} includes the inner mass loading for the port.

and 8.17. To do this, the portion of the circuit labeled "box" in Fig. 8.4 is removed, and the circuit of Fig. 8.17 is substituted in its place. The resulting circuit is shown in Fig. 8.18.

SUMMARY OF BASS-REFLEX DESIGN

1. *To determine the volume of the box and the type of port to go in it:*

   a. Find the values of M_{MD}, C_{MS}, and S_D from Par. 8.7 (pages 228 to 231). Approximate values may be obtained from Fig. 8.5.

   b. Determine $M_{AD} = M_{MD}/S_D^2$ and $C_{AS} = C_{MS}S_D^2$.

   c. Select the volume of baffle box you desire, making it reasonably large, if possible.   Calculate C_{AB} from Eq. (8.7).

   d. Determine M_{AB} from Eq. (8.8), and compute $M_{A1} = 0.23/a$, where a is the effective radius of the diaphragm in meters.

   e. Determine M_A from Eq. (8.51), and calculate $M_{AT} = M_A C_{AS}/C_{AB}$.   The quantity M_{AT} is the acoustic mass of the port.   It is composed of two parts M_{A2} and M_{AP}.

   f. Decide on the area of the port.   Usually the area is between 0.5 and 1.0 times S_D.   Calculate $M_{A2} = 0.23/a_2$, where a_2 is the effective radius of the port in meters.   If the port is not round, $a_2 = \sqrt{S_p/\pi}$, where S_p is the port area in square meters.   Subtract M_{A2} from M_{AT} to get M_{AP}.

   g. Determine the length t of the tube that goes into the port from

$$t = 0.85 M_{AP} S_P - 0.6 a_2$$

   h. Study Pars. 8.16 and 8.17 (pages 252 to 254) for construction, adjustment, and performance.

2. *To determine the response of the loudspeaker at frequencies below the rim resonance* (about 500 cps):

   a. Find the element sizes for Figs. (8.18) and (8.19) from the text.

   b. Determine the reference sound pressure at distance r from Eq. (8.64).

   c. Determine the values of Q_1, Q_M, and Q_2 from Eqs. (8.56) to (8.59) and the ratio C_{AS}/C_{AB}.

   d. Determine the sound-pressure response in decibels relative to the reference sound pressure at the three critical frequencies from Figs. 8.22 to 8.24.   The values of the critical frequencies are found from Figs. 8.20 and 8.21.

The quantities not listed in the previous paragraph are

e_g = open-circuit voltage in volts of the audio amplifier.

B = flux density in the air gap in webers per square meter (1 weber/m² = 10^4 gauss).

l = length in meters of voice-coil wire.

R_g = output electrical resistance in ohms of the audio amplifier.

R_E = electrical resistance in ohms of the voice coil.

a = effective radius of the diaphragm in meters.

$M_{AD} = M_{MD}/S_D^2$ = acoustic mass of the diaphragm and the voice coil in kilograms per meter⁴.

$C_{AS} = C_{MS}S_D^2$ = acoustic compliance of the diaphragm suspension in meters⁵ per newton.

$R_{AS} = R_{MS}/S_D^2$ = acoustic resistance of the diaphragm suspension in mks acoustic ohms.

M_{A1} = acoustic-radiation mass for the front side of the loudspeaker diaphragm = $0.195\rho_0/a$ kg/m⁴.   Note that we assume the loudspeaker unit is equivalent to a piston in the end of a tube.

$\mathfrak{R}_{AR1} = 0.01076f^2 = $ acoustic-radiation resistance for the front side of the loudspeaker diaphragm in mks acoustic ohms (see Fig. 5.7 for $ka > 1.0$).

If the port is closed off so that U_p, the volume velocity of the air in the port, equals zero, then Fig. 8.18 reduces to Fig. 8.4. At very low frequencies the mass of air moving *out* of the lower opening is nearly equal to that moving *into* the upper opening at all instants. In other words, at very low frequencies, the volume velocities at the two openings are nearly equal in magnitude and opposite in phase.

8.12. Radiated Sound. The port in the box of a bass-reflex baffle is generally effective only at fairly low frequencies. At those frequencies its dimensions are generally so small it can be treated as though it were a simple source. The loudspeaker diaphragm can also be treated as a simple source because its area is often nearly the same as that of the opening.

Referring to Eq. (4.3), we find that the sound pressure a distance r away from the bass-reflex loudspeaker is

$$p = p_1 + p_2 \doteq \frac{jf\rho_0}{2r} (U_c e^{-jkr_1} - U_p e^{-jkr_2}) \tag{8.47}$$

where p_1 and p_2 = complex rms sound pressures, respectively, from the diaphragm and the port at distance r.

$\quad\quad\quad r$ = average distance of the point of observation from the diaphragm and the port. Note that r is large compared with the diaphragm and port radii.

r_1 and r_2 = actual distances, respectively, of the point of observation from the diaphragm and the port.

$\quad\quad U_c$ = complex rms volume velocity of the diaphragm.

$\quad\quad U_p$ = complex rms volume velocity of the port. Note that the negative sign is used for U_p because, except for phase shift introduced by R_{AB} and C_{AB}, the air from the port moves outward when the air from the diaphragm moves inward.

Also, the complex rms volume velocity necessary to compress and expand the air in the box is

$$U_B = U_c - U_p \tag{8.48}$$

If we now let $r_1 = r_2 = r$ by confining our attention to a particular point in space in front of the loudspeaker where this is true, we get

$$p \doteq \frac{f\rho_0}{2r} (U_c - U_p)e^{-j\psi} \tag{8.49}$$

where ψ is a phase angle equal to $kr - \pi/2$ radians.

Since $U_c - U_p = U_B$, we have simply that

$$|p| \doteq \frac{f\rho_0 |U_B|}{2r} \tag{8.50}$$

Amazing as it seems, the sound pressure produced at faraway points equidistant from cone and port of a bass-reflex loudspeaker is directly proportional to the volume velocity necessary to compress and expand the air inside the box!

At very low frequencies, where the reactance of C_{AB} is very high, U_c becomes nearly equal to U_p and the pressure, measured at points $r = r_1 = r_2$ approaches zero. In fact, the two sources behave like a dipole so that the radiated sound pressure decreases by a factor of 4 for each halving of frequency. In addition, if we are below the lowest resonance frequency of the circuit of Fig. 8.18, the diaphragm velocity U_c

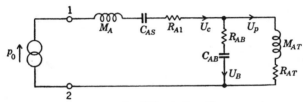

FIG. 8.19. Simplification of the circuit of Fig. 8.18. The generator pressure p_0 equals $e_g Bl/(R_g + R_E)S_D$.

halves for each halving of frequency. Hence, in this *very low* frequency region, the sound pressure decreases by a factor of 8, which is 18 db, for each halving of frequency. Note that this decrease is greater than that for a loudspeaker in a closed box or in an infinite baffle.

8.13. Resonance Frequencies. In the interest of simplifying our analysis, let us redraw Fig. 8.18 to be as shown in Fig. 8.19. The new quantities shown on that circuit are defined as follows:

$$M_A = M_{AD} + M_{A1} + M_{AB} \tag{8.51}$$

$$R_{A1} = \frac{B^2 l^2}{(R_g + R_E)S_D{}^2} + R_{AS} + \Re_{AR1} \tag{8.52}$$

$$M_{AT} = M_{A2} + M_{AP} \tag{8.53}$$

$$R_{AT} = R_{AP} + \Re_{AR2} \tag{8.54}$$

The resonance frequency of the vented enclosure ($\omega_0 = 1/\sqrt{C_{AB}M_{AT}}$) may or may not be set equal to that for the loudspeaker itself. As is common practice, let the resonance frequencies of the enclosure and the loudspeaker equal ω_0, that is, let

$$M_A C_{AS} = M_{AT} C_{AB} \equiv \frac{1}{\omega_0{}^2} \tag{8.55}$$

At the angular frequency ω_0, the impedance of the enclosure becomes very large, and nearly all the radiation takes place from the port. In other words, U_c approaches zero, and U_p becomes large.

As further steps in handling the circuit of Fig. 8.19, let us define Q's for the various parts of the circuit as follows:

$$Q_1 \equiv \frac{\omega_0 M_A}{R_{A1}} = \frac{1}{\omega_0 C_{AS} R_{A1}} \tag{8.56}$$

$$Q_c \equiv \frac{1}{\omega_0 C_{AB} R_{AB}} \tag{8.57}$$

$$Q_M \equiv \frac{\omega_0 M_{AT}}{R_{AT}} \tag{8.58}$$

$$Q_2 \equiv \frac{\omega_0 M_{AT}}{R_{AT} + R_{AB}} = \frac{1}{\omega_0 C_{AB}(R_{AT} + R_{AB})} \tag{8.59}$$

The total impedance Z_A looking to the right of the terminals 1 and 2 in Fig. 8.19 equals

$$Z_A = \left[\frac{R_{AT}\left(\frac{\omega_0^2}{\omega^2} + \frac{1}{Q_c^2}\right) + R_{AB}\left(\frac{\omega^2}{\omega_0^2} + \frac{1}{Q_M^2}\right)}{\frac{1}{Q_2^2} + \left(\frac{\omega}{\omega_0} - \frac{\omega_0}{\omega}\right)^2} + R_{A1} \right]$$

$$+ j\left[\frac{1}{\omega_0 C_{AB}} \frac{\frac{\omega}{\omega_0 Q_c^2} - \frac{\omega_0}{\omega Q_M^2} - \left(\frac{\omega}{\omega_0} - \frac{\omega_0}{\omega}\right)}{\frac{1}{Q_2^2} + \left(\frac{\omega}{\omega_0} - \frac{\omega_0}{\omega}\right)^2} + \omega_0 M_A\left(\frac{\omega}{\omega_0} - \frac{\omega_0}{\omega}\right) \right] \tag{8.60}$$

For zero reactance we find that

$$\frac{C_{AS}}{C_{AB}} = \frac{M_{AT}}{M_A} = \frac{\frac{1}{Q_2^2} + \left(\frac{\omega}{\omega_0} - \frac{\omega_0}{\omega}\right)^2}{1 + \frac{1}{\frac{\omega}{\omega_0} - \frac{\omega_0}{\omega}}\left(\frac{1}{Q_M^2}\frac{\omega}{\omega_0} - \frac{1}{Q_c^2}\frac{\omega_0}{\omega}\right)} \tag{8.61}$$

Also, we shall need U_B in terms of U_c.

$$U_B = \frac{\left(\frac{1}{Q_M} + j\frac{\omega}{\omega_0}\right) U_c}{(1/Q_2) + j[(\omega/\omega_0) - (\omega_0/\omega)]} \tag{8.62}$$

It is difficult to handle these equations if Q_2 is small. There is no advantage, furthermore, to a bass-reflex box over a closed box if Q_2 is small. If R_{AT} is large, the box becomes the same as though it were tightly closed. If R_{AB} is large, then U_B is small and it would be better to use a closed-box baffle. We shall limit our studies to Q_2 greater than 2.5.

We shall also limit our studies to cases where the resistance is principally in either the port or in the box but not in both.   Then, Q_2 equals either Q_M or Q_c.

Under these conditions there are generally three frequencies ω_L, ω_M, and ω_H at which the reactance to the right of terminals 1 and 2 becomes zero for given values of Q_2 and C_{AS}/C_{AB}.   One of these frequencies, ω_M, is near the resonance frequency ω_0.   Another frequency, ω_H, occurs at a value greater than ω_0.   The third frequency, ω_L, occurs at a value less than ω_0.

The upper and lower resonance frequencies are found from Eq. (8.61) as follows: Set $Q_c = \infty$ and $Q_2 = Q_M$.   Then, ω_0/ω_L and ω_H/ω_0 are determined.   The results are plotted in Figs. 8.20 and 8.21, respectively. Next, set $Q_M = \infty$ and $Q_2 = Q_c$.   Then, ω_0/ω_L and ω_H/ω_0 are determined. The results are plotted in Figs. 8.21 and 8.20, respectively.   In other words, Fig. 8.20 gives the lower resonance frequency for $Q_c = \infty$ and the upper resonance frequency for $Q_M = \infty$.   Figure 8.21 gives the upper resonance frequency for $Q_c = \infty$ and the lower resonance frequency for $Q_M = \infty$.

If the value of Q_2 is greater than about 10, the two resonance frequencies, ω_L and ω_H, are symmetrically located on either side of ω_0 for a given value of C_{AS}/C_{AB}.   That is to say, $\omega_0/\omega_L \doteq \omega_H/\omega_0$.

The resonance frequencies for loudspeaker systems where the series and parallel branches of the circuit of Fig. 8.19 do not have the same angular resonance frequency are more difficult to find.   If we set $Q_2 = \infty$ and if the angular resonance frequencies of the two parts of the circuit are ω_1 and ω_2, we find that, for zero reactance to the right of terminals 1 and 2 of Fig. 8.18,

$$\frac{C_{AS}}{C_{AB}} = \frac{M_{AT}}{M_A} = G^2 \frac{K^4 + 1}{K^2} - (G^4 + 1) \tag{8.63}$$

where ω_0 is the geometric-mean angular frequency equal to $\sqrt{\omega_1\omega_2}$ and $\omega_1 = G\omega_0$, $\omega_2 = \omega_0/G$, and $\omega_H/\omega_0 = K = \omega_0/\omega_L$.   Note that this equation is symmetrical about ω_0 just as Eq. (8.61) is for $Q_2 = \infty$.

8.14. Reference Sound Pressure.   The concept of reference sound pressure has been discussed in Part XIX and here is expressed as follows:

$$|p_c| = \frac{e_g B l \rho_0}{(R_g + R_E) M_A 4\pi r S_D} \tag{8.64}$$

By way of review, p_c is the sound pressure that would be produced at r if the frequency were above the critical frequency ω_H and if the loudspeaker were nondirectional.

8.15. Radiated Sound Pressure at the Critical Frequencies.   We expect that there are important fluctuations in the response curves at the three critical frequencies ω_H, ω_M, and ω_L.   At $\omega_M \doteq \omega_0$, the radiation occurs principally from the port, and the radiated pressure there depends

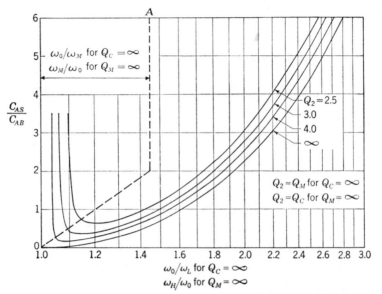

FIG. 8.20. Value of two of the zero-reactance frequencies of a bass-reflex loudspeaker system as a function of $C_{AS}/C_{AB} = M_{AT}/M_{AS}$ with Q_2 as the parameter. ω_0 is the resonance frequency of either the series or parallel branch of Fig. 8.19. The region to the right of the dashed line A gives ω_0/ω_2 for $Q_C = \infty$ or ω_H/ω_0 for $Q_M = \infty$. The near-vertical lines at the left of the dashed line A give ω_M/ω_0 for $Q_M = \infty$ or ω_0/ω_M for $Q_C = \infty$.

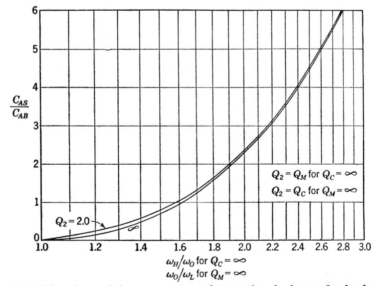

FIG. 8.21. Value of one of the zero-reactance frequencies of a bass-reflex loudspeaker system as a function of $C_{AS}/C_{AB} = M_{AT}/M_{AS}$ with Q_2 as the parameter. ω_0 is the resonance frequency of either the series or the parallel branch of Fig. 8.19.

both on Q_1 and Q_2. Regardless of the values of Q_2 and C_{AS}/C_{AB}, the upper frequency for zero reactance, ω_H, is about the same as though the port were closed off.   This can be seen by comparing the curve of Fig. 8.21 marked $Q_2 = 2$ with the curve for $Q_2 = \infty$.   If $\omega_H/\omega_0 \doteq 1$ is small, the

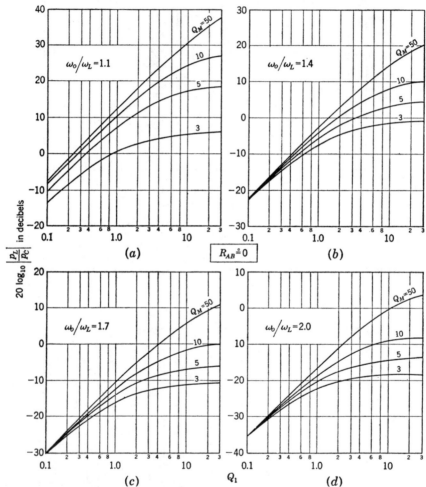

Fig. 8.22. Relative response at the lower critical frequency, ω_L for $R_{AB} \doteq 0$ as a function of Q_M and Q_1.   Box unlined.

response of the loudspeaker with vented enclosure will be nearly the same as that for the same loudspeaker in an infinite baffle.   This condition requires a large box (large C_{AB}) which obviously approaches being an infinite baffle.

For $R_{AB} \doteq 0$ $(Q_c \doteq \infty)$, No Acoustical Material in Box.   We find the magnitude of the ratio of the rms sound pressure to the reference pressure at the three frequencies, ω_L, ω_M, and ω_H, from Eqs. (8.50), (8.62), and (8.64) and by solving for U_c from Fig. 8.19 with the aid of Eq. (8.60).

FOR $\omega = \omega_L$.   The pressure ratio at the lowest critical frequency is

$$\left|\frac{p_L}{p_c}\right| = \frac{\sqrt{1 + Q_M{}^2(\omega_L/\omega_0)^2}}{\sqrt{\left(1 - \dfrac{\omega_0{}^2}{\omega_L{}^2}\right)^2 + \dfrac{\omega_0{}^2}{Q_M{}^2\omega_L{}^2}\left[\dfrac{Q_M}{Q_1} + \dfrac{\omega_0{}^2/\omega_L{}^2}{1 + \dfrac{1}{Q_M{}^2(\omega_L{}^2/\omega_0{}^2 - 1)}}\right]}} \tag{8.65}$$

Plots of this equation, in decibels, are given in Fig. 8.22 for four values of ω_0/ω_L.

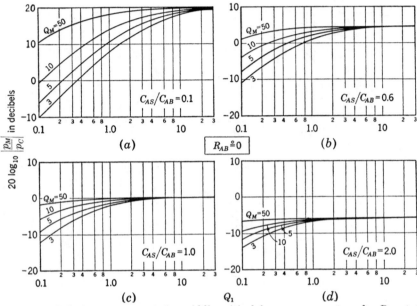

FIG. 8.23. Relative response at the middle critical frequency, $\omega_M \doteq \omega_0$ for $R_{AB} \doteq 0$ as a function of Q_M and Q_1.   Box unlined.

FOR $\omega = \omega_M \doteq \omega_0$.   The pressure ratio at the middle critical frequency is

$$\left|\frac{p_M}{p_c}\right| = \frac{\sqrt{1 + 1/Q_M{}^2}}{C_{AS}/C_{AB} + 1/Q_1 Q_M} \tag{8.66}$$

Plots of this equation, in decibels, are given in Fig. 8.23 for four values of C_{AS}/C_{AB}.

FOR $\omega = \omega_H$.   The ratio at the highest critical frequency is,

$$\left|\frac{p_H}{p_c}\right| = \frac{\sqrt{1 + Q_M{}^2(\omega_H/\omega_0)^2}}{\sqrt{\left(1 - \dfrac{\omega_0{}^2}{\omega_H{}^2}\right)^2 + \dfrac{\omega_0{}^2}{Q_M{}^2\omega_H{}^2}}\right)\left[\dfrac{Q_M}{Q_1} + \dfrac{\omega_0{}^2/\omega_H{}^2}{1 + \dfrac{1}{Q_M{}^2(\omega_H{}^2/\omega_0{}^2 - 1)}}\right]} \tag{8.67}$$

This equation, in decibels, is plotted in Fig. 8.24 for four values of ω_H/ω_0.

For $R_{AT} \doteq 0$ $(Q_M \doteq \infty)$, *Acoustical Material in Box.* The magnitude of the ratio of the rms sound pressure to the reference sound pressure is found by using the same equations and figure as for $Q_c = \infty$.

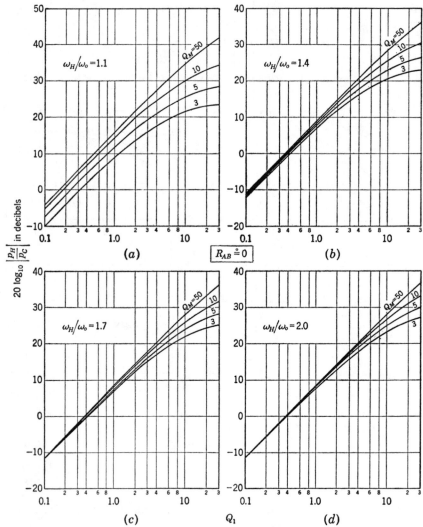

FIG. 8.24. Relative response at the upper critical frequency, ω_H for $R_{AB} \doteq 0$ as a function of Q_M and Q_1. Box unlined.

FOR $\omega = \omega_L$. The pressure ratio at the lowest critical frequency is

$$\left| \frac{p_L}{p_c} \right| = \frac{(\omega_L/\omega_0)Q_c}{\sqrt{\left(1 - \frac{\omega_0^2}{\omega_L^2}\right)^2 + \frac{\omega_0^2}{Q_c^2\omega_L^2} \left[\frac{\omega_0^2/\omega_L^2}{1 + \frac{1}{Q_c^2(\omega_0^2/\omega_L^2 - 1)}} + \frac{Q_c}{Q_1}\right]}} \tag{8.68}$$

This equation, in decibels, is plotted in Fig. 8.25 for four values of ω_0/ω_L.

FOR $\omega = \omega_M \doteq \omega_0$. The pressure ratio at the middle critical frequency is

$$\left| \frac{p_M}{p_C} \right| = \frac{1}{C_{AS}/C_{AB} + 1/Q_1 Q_c} \tag{8.69}$$

This equation, in decibels, is plotted in Fig. 8.26 for four values of C_{AS}/C_{AB}.

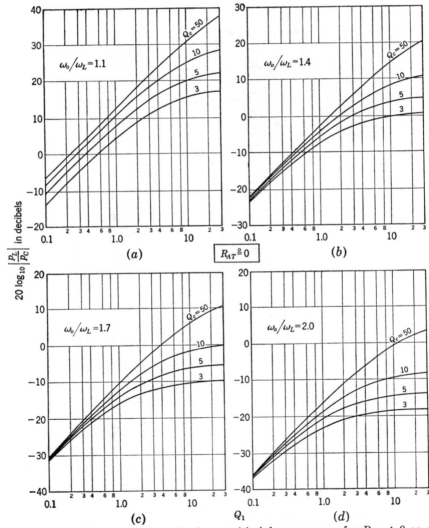

FIG. 8.25. Relative response at the lower critical frequency, ω_L for $R_{AT} \doteq 0$ as a function of Q_c and Q_1. Box lined.

FOR $\omega = \omega_H$. The pressure ratio at the highest critical frequency is

$$\left|\frac{p_H}{p_C}\right| = \frac{(\omega_H/\omega_0)Q_c}{\sqrt{\left(1 - \frac{\omega_0^2}{\omega_H^2}\right)^2 + \frac{\omega_0^2}{Q_c^2\omega_H^2}\left[\frac{\omega_0^2/\omega_H^2}{1 + \frac{1}{Q_c^2(\omega_0^2/\omega_H^2 - 1)}} + \frac{Q_c}{Q_1}\right]}} \qquad (8.70)$$

This equation, in decibels, is plotted in Fig. 8.27 for four values of ω_H/ω_0.

8.16. Performance. With the formulas and charts just given, it is possible to calculate the response of the loudspeaker in a bass-reflex

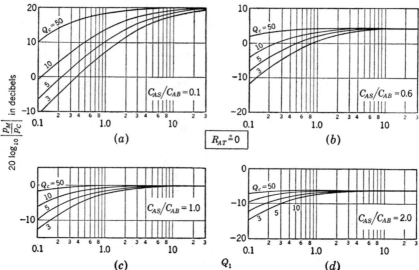

FIG. 8.26. Relative response at the middle critical frequency, $\omega_M \doteq \omega_0$ for $R_{AT} \doteq 0$ as a function of Q_c and Q_1. Box lined.

enclosure at the three critical frequencies. A complete example is given in Par. 8.18.

From Fig. 8.19, we see that for frequencies below ω_0, radiation from the port (proportional to $-U_p$) is out of phase with the radiation from the diaphragm (proportional to U_c). As a result, the response at very low frequencies is usually not enhanced by the addition of the port. Above the resonance frequency ω_0, radiation from the port is in phase with that from the diaphragm, with a resulting enhancement over the closed-baffle response. The amount of the increase in response generally averages about 5 db over a frequency range of one to two octaves.

An important reason for using a bass-reflex enclosure is that the loudspeaker produces less distortion at frequencies between ω_0 and $2\omega_0$ for a given acoustic power radiated than would be the case if the box were closed. The assumption on which this statement is made is that the

motion of the air in the port is distortionless even though the amplitude
of vibration is large.   This is true generally because there is no suspension
or magnetic circuit in the port in which nonlinear effects can occur.   A

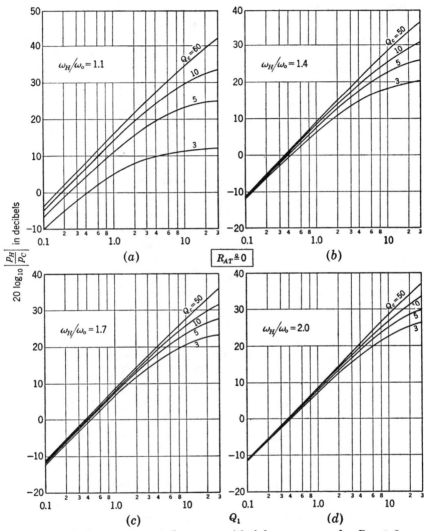

Fig. 8.27. Relative response at the upper critical frequency, ω_H for $R_{AT} \doteq 0$ as a
function of Q_c and Q_1.   Box lined.

large loudspeaker diaphragm usually is superior to a small one because
the amplitude of its motion is less, thereby reducing nonlinear distortion.
    The port should be placed close to the loudspeaker to avoid an irregular
directivity pattern up to as high a frequency as possible and to achieve
the widest possible frequency range of enhancement.

An advantage of a bass-reflex enclosed loudspeaker is that, where space is a factor, a properly tuned bass-reflex system helps to offset the effect of the small enclosure volume. For a small enclosure volume, it is best to tune the port exactly to the loudspeaker resonance.

For a very large enclosure, it is permissible to tune the port to a frequency below the loudspeaker resonance. This permits the use of a port that is not over about 1.5 to 2.0 times the effective speaker area.

8.17. Construction and Adjustment Notes. The box should be very rigid in order to resist vibration. The joints should be tight—glued or caulked—and the larger panels should be braced by gluing reinforcing strips to them. The access side should be screwed on securely.

In constructing the box to achieve a desired volume, account should be taken of the volume of air displaced by the loudspeaker and by the inward extension of the port if such occurs. As we said earlier, the volume of air displaced by the loudspeaker (in cubic meters) equals about 0.4 times the fourth power of the advertised diameter (in meters).

When the cabinet has been completed and the loudspeaker has been installed, the correctness of the tuning may be determined by connecting an audio oscillator with an output impedance about 100 times that of the loudspeaker to the electrical terminals. Next, connect a voltmeter across the loudspeaker terminals. Then vary the frequency of the oscillator until the two critical frequencies ω_L and ω_H are observed as peaks in the voltmeter reading. These should occur at the calculated frequencies if the design is correct.

The resonance frequency of the enclosure can be increased by decreasing the enclosed volume V_B. This can be accomplished by putting blocks of wood inside through the port opening. The port area may be decreased to lower the resonance frequency of the enclosure.

8.18. Example of Bass-reflex Enclosure Design. In the previous part we discussed in detail the design of a closed-box baffle for a low-frequency (woofer) loudspeaker. We presented methods for the determination of its physical constants, and we showed a comparison between measurements and calculations.

In this part we shall use the same loudspeaker unit and box. A port will be introduced into the box that resonates with the box compliance to the same frequency as the series branch of the circuit of Fig. 8.19, that is, $\omega_0 = 1/\sqrt{M_A C_{AS}}$.

The element sizes are as given in Example 8.3 (page 235), except that the resonance frequency ω_0 is different for the analysis here, viz.,

$$\omega_0 = 1/\sqrt{M_A C_{AS}} = 10^3/\sqrt{(10.27)(1.82)} = 231 \text{ radians/sec}$$
$$f_0 = \omega_0/2\pi = 36.8 \text{ cps}$$
$$C_{AS}/C_{AB} = 1.82/2.14 = 0.85$$
$$M_{AT} = C_{AS}M_A/C_{AB} = (0.85)(10.27) = 8.74 \text{ kg/m}^4$$

The area of the port must be chosen. If the circumference of the port is less than a half wavelength, the port behaves as a simple source and it will radiate the same amount of power regardless of its size for a given volume velocity. Arbitrarily, let us select the area S_p to be a little more than one-half of the area of the diaphragm, say,

$$S_p = 0.055 \text{ m}^2$$

Calculate the acoustic mass at the outer end of the port.

$$M_{A2} = \frac{0.6a_2\rho_0}{S_p} = \frac{0.6\rho_0}{\sqrt{S_p\pi}}$$
$$= \frac{(0.6)(1.18)}{\sqrt{(0.055)(3.1416)}} = 1.70 \text{ kg/m}^4$$

The remaining mass of the port is

$$M_{AP} = M_{AT} - M_{A2} = 8.74 - 1.7 = 7.04 \text{ kg/m}^4$$
$$= 7.04 = \frac{(t + 0.6a_2)\rho_0}{S_p} = \frac{t(1.18)}{0.055} + 1.7$$

So t, the length of the tube behind the port, is

$$t = \frac{7.04 - 1.7}{21.4} = 0.25 \text{ m}$$

This says that the length of the tube behind the port should be about 9.8 in.

The acoustic resistance of the port is found from Eq. (5.54).

$$R_{AP} = \frac{1.18}{0.055} \sqrt{2\omega(15.6)} \times 10^{-3} \left(\frac{0.25}{0.133} + 1\right)$$
$$= \sqrt{\omega} \,(0.12) \text{ mks acoustic ohms}$$

To get an average value of R_{AP}, let us take $\omega = \omega_0$.

$$R_{AP} = \sqrt{231} \,(0.12) \doteq 1.82 \text{ mks acoustic ohms}$$

The resistance from the front side of the port is

$$\mathfrak{R}_{AR2} = \frac{\pi f^2 \rho_0}{c} = \mathfrak{R}_{AR1} = 0.0107 f^2$$

Let $\mathfrak{R}_{AR1}$ and $\mathfrak{R}_{AR2}$ equal their value at $\omega = \omega_0$, that is, $f_0 = 36.8$. Then,

$$\mathfrak{R}_{AR1} = \mathfrak{R}_{AR2} = (0.0107)(1355) = 14.5 \text{ mks acoustic ohms}$$

Finally,

$$R_{AT} = R_{AP} + \mathfrak{R}_{AR2} \doteq 16.3 \text{ mks acoustic ohms}$$

and, for later use,

$$R_{A1} = \frac{9.7 \times 10^4}{R_g + 5.5} + 356 + 15 \text{ mks acoustic ohms}$$

$$Q_M = \frac{(231)(8.74)}{16.3} \doteq 114$$

This value of Q is very high. Let us see whether or not Q_c is smaller. The box contained one 18- by 35- by 3-in. piece of PF Fiberglas board with a flow resistance of about 2000 mks rayls/in., or 6000 mks rayls for 3 in. Hence, from Par. 8.9,

$$R_{AM} = \frac{6000}{(3)(0.406)} = 4920$$

$$R_{AB} \doteq \frac{4920}{(231)^2(2.14)^2 \times 10^{-12}(4920)^2 + (7)^2}$$

$$= \frac{4920}{5.9 + 49} = 90 \text{ mks acoustic ohms}$$

$$Q_c = \frac{10^6}{(90)(231)(2.14)} \doteq 22$$

and

$$Q_2 = \frac{(231)(8.74)}{16.3 + 90} \doteq 19$$

Hence, we may assume that $Q_2 \doteq Q_c$.

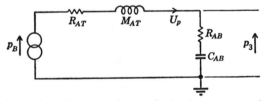

FIG. 8.28. Circuit showing the port volume velocity for an external source of sound when the loudspeaker cone is blocked so that it cannot move. The sound pressure p_B is that sound pressure which would exist at the entrance to the port if the port were blocked off so that U_p were zero. The magnitude of the volume velocity U_p equals $|p_3|C_{AB}\omega$. It is assumed that $R_{AB} \ll 1/\omega C_{AB}$.

The Q_c of the box may be measured by replacing the loudspeaker unit by a rigid board and placing the box in free space, along with a source of sound of variable frequency. Reference to Figs. 8.19 and 8.28 shows that if $U_c = 0$, then $-U_p = U_B$. Placing a microphone inside the box measures the sound-pressure drop p_3 across the compliance C_{AB}. A plot is then made as a function of frequency of the product of this sound pressure and the frequency of the oscillator. This plot is proportional to the volume velocity U_p. Then, Q_2 is found by dividing the frequency (cycles per second) for the peak of the curve by the width of the curve (cycles per second) at the points where the volume velocity is 0.707 of its peak value.

This measurement made on our box yielded a value of $Q_2 = 7$, which is much lower than that calculated. A plausible explanation for this is that the plywood absorbs energy internally when it is flexed. According to information in the literature, a 3/4-in. plywood box absorbs *at* 40 *cps* about half as much sound energy per square foot as does the 3-in.-thick layer of Fiberglass at the same frequency. This is, of course, a function of panel size, type of glue, and bracing, so that at best only a crude guess can be made. At frequencies above 100 cps, the 3-in.-thick layer of Fiberglass will absorb much more energy than the wood box.

Since the area of the walls of the box is about six times that of the Fiberglas, the box absorption is equivalent to adding an additional three times as much material, making the total to be four times as much area and volume of acoustical material as that now assumed. An approximate revised R_{AB} would be

$$R_{AB} \doteq 330 \text{ mks acoustic ohms}$$

Hence,

$$Q_c \doteq \frac{10^6}{(330)(231)(2.14)} = 6$$

This number is very close to the value of 7 measured; so we shall assume

$$Q_c = Q_2 = 7$$

From Figs. 8.20 and 8.21 we find the three critical frequencies

$$\omega_M \doteq \omega_0 = 231 \text{ radians/sec}$$
$$f_0 = 36.8 \text{ cps}$$
$$\omega_H = 1.55\omega_0 = 358 \text{ radians/sec}$$
$$f_H = 57.0 \text{ cps}$$
$$\omega_L = \frac{\omega_0}{1.56} = 148 \text{ radians/sec}$$
$$f_L = 23.6 \text{ cps}$$

To determine the relative sound-pressure response at the three critical frequencies, we use Figs. 8.25 to 8.27. First, we shall determine Q_1 at these three frequencies for the four values of amplifier resistance R_g, namely, 3, 14, 30, and 130 ohms.

The values of R_{A1} and Q_1 are given in Table 8.5.

TABLE 8.5.   Values of R_{A1} and Q_1 vs. Amplifier Resistance

R_g, ohms...........	3	14	30	130
R_{A1}	1.18×10^4	5.3×10^3	3.1×10^3	1.09×10^3
Q_1	0.20	0.45	0.77	2.2

We obtain the values of the sound-pressure-level response relative to the reference sound-pressure-level response at the three critical frequencies given in Table 8.6.

TABLE 8.6.   Sound-pressure-level Response at the Critical Frequencies Relative to the Reference Sound-pressure-level Response

R_g / f	3	14	30	130
23.4 cps	−22 db	−15.6	−11.8	− 5.6
36.8 cps	− 3.9	− 1.4	− 0.3	+ 0.8
57.7 cps	− 5.7	+ 1.2	+ 5.8	+14.2

Comparison of the numbers given in Table 8.6 with the calculated curves of Fig. 8.15 for a closed box shows that the average enhancement of the response at the upper two critical frequencies was about 6 db.   By comparison, the measurement on this particular loudspeaker showed an average increase in response of approximately 4 db.   No explanation is offered for this 2-db difference.   Data reported by other observers usually show an enhancement of the order of 6 db.   It is possible that transmission through the side walls of the box, which would tend to produce an out-of-phase wave, contributed to the difference in the results.   Time did not permit enclosing the box in a sand or a concrete jacket to determine the magnitude of this effect.

CHAPTER 9

HORN LOUDSPEAKERS

PART XXI *Horn Driving Units*

9.1. Introduction. Horn loudspeakers usually consist of a moving-coil driving unit coupled to a horn. When well-designed, the large end of the horn, called the "mouth," has an area sufficiently large to radiate sound efficiently at the lowest frequency desired. The small end of the horn, called the "throat," has an area selected to match the acoustic impedance of the driving unit and to produce as little nonlinear distortion of the acoustic signal as possible.

Horn loudspeakers are in widespread use in cinemas, theaters, concert halls, stadiums, and arenas where large acoustic powers must be radiated and where control of the direction of sound radiation is desired. The efficiency of radiation of sound from one side of a well-designed direct-radiator loudspeaker was shown in Chaps. 7 and 8 to be a few per cent. By comparison, the efficiency of radiation from a horn loudspeaker usually lies between 10 and 50 per cent.

The principal disadvantages of horn loudspeakers compared with the direct-radiator loudspeakers are higher cost and larger size.

Before proceeding with an analysis of the horn loudspeaker, it should be mentioned again that the radiating efficiency of a direct-radiator loudspeaker can be increased at low frequencies by mounting several units side by side in a single baffle. The mutual interaction among the radiating units serves to increase the radiation resistance of each unit substantially. For example, two identical direct-radiator loudspeakers very near each other in an infinitely large plane baffle, and vibrating in phase, will produce four times the intensity on the principal axis as will one of them alone.

Direct-radiator loudspeakers used in multiple often are not as satisfactory at high frequencies as one horn loudspeaker because of the difficulty of obtaining uniform phase conditions from different direct-radiator diaphragms. That is to say, the conditions of vibration of a loudspeaker cone are complex, so that normal variations in the uniformity of cones result in substantial differences in the phases of the radiated signals of different cones at high frequencies. A very irregular and unpredictable response curve and directivity pattern result.

This problem does not arise with a horn where only a single driving unit is employed. When two or more driving units are used to drive a single horn, the frequency range in which the response curve is not adversely affected by the multiplicity of driving units is that where the diaphragms vibrate in one phase.

9.2. Electro-mechano-acoustical Circuit.[1] The driving unit for a horn loudspeaker is essentially a small direct-radiator loudspeaker that couples to the throat of a flaring horn as shown in Fig. 9.1. In the next part we shall discuss the characteristics of the horn itself. In this paragraph

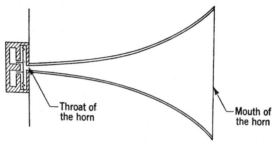

Throat of
the horn

Mouth of
the horn

FIG. 9.1. Cross section of a typical horn loudspeaker with an exponential cross section. For this design, the radius of the throat is 0.2, the radius of the mouth 3.4, and the length 5.0 (arbitrary units).

we restrict ourselves to that part of the frequency range where the complex mechanical impedance Z_{MT} looking into the *throat* of a horn is a pure resistance,

$$Z_{MT} = \frac{1}{z_{MT}} = \rho_0 c S_T \qquad \text{mks mechanical ohms} \qquad (9.1)$$

where ρ_0 = density of air in kilograms per cubic meter

c = velocity of sound in meters per second

$\rho_0 c$ = 406 mks ohms at 22°C and 10^5 newtons/m² ambient pressure

S_T = area of the throat in square meters

z_{MT} = mechanical mobility at the throat of the horn in mks mechanical mohms (mobility ohms)

Cross-sectional sketches of two typical driving units for horn loudspeakers are shown in Fig. 9.2. Each has a diaphragm and voice coil with

[1] An authoritative discussion of horn loudspeakers is found in H. F. Olson, "Elements of Acoustical Engineering," 2d ed., Chap. VII, D. Van Nostrand Company, Inc., New York, 1947.

a total mass M_{MD}, a mechanical compliance C_{MS}, and a mechanical resistance $R_{MS} = 1/r_{MS}$. The quantity r_{MS} is the mechanical responsiveness of the diaphragm in mohms (mobility ohms).

Behind the diaphragm is a space that is usually filled with a soft acoustical material. At low frequencies this space acts as a compliance C_{MB} which can be lumped in with the compliance of the diaphragm. At high frequencies the reactance of this space becomes small so that the space behind the diaphragm becomes a mechanical radiation resistance $R_{MB} = 1/r_{MB}$ with a magnitude also equal to that given in Eq. (9.1). This resistance combines with the mechanical radiation resistance of the throat, and the diaphragm must develop power both to its front and its back. Obviously, any power developed behind the diaphragm is wasted, and at high frequencies this sometimes becomes as much as one-half of the total generated acoustic power.

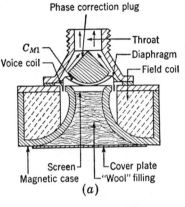

In front of the diaphragm there is an air space with compliance C_{M1}. At low frequencies the air in this space behaves like an incompressible fluid, that is, ωC_{M1} is small, and all the air displaced by the diaphragm passes into the throat of the horn. At high frequencies the mechanical reactance of this air space becomes sufficiently low (*i.e.*, the air becomes compressible) so that all the air displaced by the diaphragm does not pass into the throat of the horn.

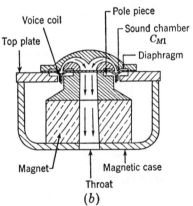

Fig. 9.2. Cross section of two typical horn driving units. The diaphragm couples to the throat of the horn through a small cavity with a mechanical compliance C_{M1}.

The voice coil has an electrical resistance R_E and inductance L. As stated above, z_{MT} is the mechanical mobility at the throat of the horn.

By inspection, we draw the mobility-type analogous circuit shown in Fig. 9.3. In this circuit forces "flow" *through* the elements, and the velocity "drops" *across* them. The generator open-circuit voltage and resistance are e_g and R_g. The electric current is i; the linear velocity of the voice coil and diaphragm is u_c; the linear velocity of the air at the throat of the horn is u_T; and the force at the throat of the horn is f_T. As before, the area of the diaphragm is S_D, and that of the throat is S_T.

9.3. Reference Power Available Efficiency.   In the middle-frequency range many approximations usually can be made to simplify the analogous circuit of Fig. 9.3.   Because the driving unit is very small, the mass of the diaphragm and the voice coil M_{MD} is very small.   This in turn usually means that the compliance of the suspension C_{MS} is large in order to keep the resonance frequency low.   Also, the responsiveness of the suspension r_{MS} usually is large, and the reactance ωC_{M1} is small.   Hence, in this fre₂

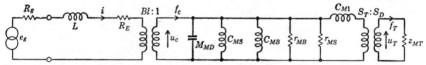

Fig. 9.3. Electro-mechano-acoustical analogous circuit of the mobility type for the driving unit, assuming that the mechanical impedance at the horn throat is $\rho_0 c S_T$, that is, the mechanical mobility is $Z_{MT} = 1/\rho_0 c S_T$.

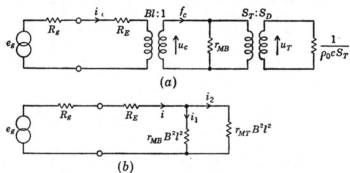

Fig. 9.4. Simplified analogous circuits of the mobility type for the driving unit in the region where the motion of the diaphragm is resistance-controlled by the horn.

quency range, the circuit reduces essentially to that of Fig. 9.4a, where the responsiveness behind the diaphragm is

$$r_{MB} \equiv \frac{1}{\rho_0 c S_D} \qquad \text{mks mechanical mohms} \qquad (9.2)$$

With the area-changing and electromechanical transformers removed, we get Fig. 9.4b, where the radiation responsiveness at the throat is

$$r_{MT} \equiv \frac{S_T}{\rho_0 c S_D{}^2} \qquad \text{mks mechanical mohms} \qquad (9.3)$$

As before, S_T is the area of the throat and S_D is the area of the diaphragm in square meters.   We have assumed here that the cavity behind the diaphragm in this frequency range is nearly perfectly absorbing, which may not always be true.   Usually, however, this circuit is valid over a considerable frequency range because of the heavy damping provided by the responsiveness of the horn r_{MT}.   Also, r_{MT} usually is smaller than

r_{MB} so that most of the power supplied by the diaphragm goes into the horn.

Solution of Fig. 9.4b gives us the *reference current* $i = i_{ref}$,

$$i = i_{ref} = \frac{e_g}{(R_g + R_E) + \dfrac{B^2l^2S_T}{S_D{}^2\rho_0 c(1 + S_T/S_D)}} \tag{9.4}$$

The maximum electrical power available to the loudspeaker from the generator, assuming R_g fixed, is

$$\text{Max power available} = \frac{e_g{}^2}{4R_g} \tag{9.5}$$

The *reference power available efficiency* (PAE$_{ref}$) is equal to the reference power delivered to the horn, $|i_2|^2B^2l^2r_{MT}$, times 100 divided by the maximum electrical power available.

$$\begin{aligned}
\text{PAE}_{ref} &= \frac{|i|^2_{ref}[r_{MB}/(r_{MT} + r_{MB})]^2 r_{MT}B^2l^2}{e_g{}^2/4R_g} \times 100 \\
&= \frac{400|i|^2_{ref}R_g r_{MT}B^2l^2}{(1 + S_T/S_D)^2 e_g{}^2}
\end{aligned} \tag{9.6}$$

From Eqs. (9.3), (9.4), and (9.6) we get

$$\text{PAE}_{ref} = \frac{400R_g r_{MT}B^2l^2}{[B^2l^2 r_{MT} + (R_g + R_E)(1 + S_T/S_D)]^2} \tag{9.7}$$

9.4. Frequency Response. The frequency response of a complete horn loudspeaker, in the range where the throat impedance of the horn is a resistance as given by Eq. (9.1), is determined by solution of the circuit of Fig. 9.3. For purposes of analysis, we shall divide the frequency range into three parts, A, B, and C, as shown in Fig. 9.5.

Mid-frequency Range.[2] In the mid-frequency range, designated as B in Fig. 9.5, the response is equal to the reference PAE given by Eq. (9.7). Here, the response is "flat" with frequency, and, for the usual high-frequency units used in cinemas with 300 cps cutoff frequencies, the flat region extends from a little above 500 to a little below 3000 cps. In this region the velocity of the diaphragm is constant with frequency, rather than decreasing in inverse proportion to frequency as was the case for a direct-radiator loudspeaker.

Resonance Frequency. It is apparent from Fig. 9.3 that since ωC_{M1} is small, zero reactance will occur at the frequency where

$$f_0 = \frac{1}{2\pi \sqrt{M_{MD}[C_{MS}C_{MB}/(C_{MS} + C_{MB})]}} \tag{9.8}$$

---

[2] See also Fig. 7.2, p. 188, of reference 1. Note that Olson's efficiency is the ratio $100r_{MT}B^2l^2/(R_E + r_{MT}B^2l^2)$, assuming that $r_{MB} \gg r_{MT}$.

In practice, this resonance usually is located in the middle of region B of Fig. 9.5 and is heavily damped by the responsiveness r_{MT}, so that the velocity of the diaphragm is resistance-controlled.

Low Frequencies. At frequencies well below the resonance frequency the response will drop off 6 db for each octave decrease in frequency if the

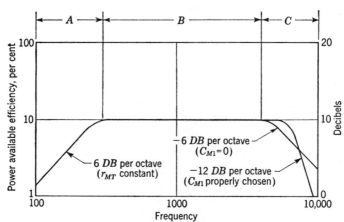

FIG. 9.5. Power available efficiency of a horn driver unit, in the frequency region where the mechanical impedance of the horn at the throat is a pure resistance $\rho_0 c S_T$. The ordinate is a logarithmic scale, proportional to decibels.

throat impedance is a resistance as given by Eq. (9.1). This case is shown as region A in Fig. 9.5.

In practice, however, the throat impedance Z_{MT} of the horn near the lowest frequency at which one wishes to radiate sound is *not* a pure resistance. Hence, region A needs more careful study.

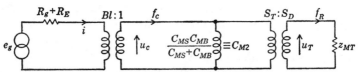

FIG. 9.6. Analogous circuit for a horn driver unit in the region where the diaphragm would be stiffness-controlled if the horn mobility were infinite. The actual value of the mechanical mobility of the horn at the throat is z_{MT}.

Let us simplify Fig. 9.3 so that it is valid only for the low-frequency region, well below the resonance of the diaphragm. Then the inductance L, the mass M_{MD}, the compliance C_{M1}, and the responsivenesses r_{MS} and r_{MB} may all be dropped from the circuit, giving us Fig. 9.6.

Solving for the mechanical mobility at the diaphragm of the driving unit yields

$$z_{Mc} = \frac{u_c}{f_c} = \frac{j\omega C_{M2}(S_T/S_D)^2 z_{MT}}{j\omega C_{M2} + (S_T/S_D)^2 z_{MT}} \tag{9.9}$$

where

$$C_{M2} = \frac{C_{MS}C_{MB}}{C_{MS} + C_{MB}} \tag{9.10}$$

z_{MT} = mechanical mobility at the throat of the horn with area S_T

The mechanical impedance at the diaphragm of the driving unit is the reciprocal of z_{Mc},

$$Z_{Mc} = \frac{f_c}{u_c} = \left(\frac{S_D}{S_T}\right)^2 Z_{MT} - j\frac{1}{\omega C_{M2}} \tag{9.11}$$

where $Z_{MT} = 1/z_{MT}$ = mechanical impedance at the throat of the horn with area S_T.

As we shall show in the next part, the mechanical impedance at the throat of ordinary types of horn at the lower end of the useful frequency range is equal to a mechanical resistance in series with a negative compliance. That is to say,

$$Z_{MT} \equiv \Re_{MT} + j\frac{1}{\omega C_{MT}} \tag{9.12}$$

The German $\Re$ in $\Re_{MT}$ indicates that this resistance varies with frequency. Usually, its variation is between zero at very low frequencies and $\rho_0 c S_T$ [as given by Eq. (9.1)] at some frequency in region A of Fig. 9.5. Hence, the mobility $z_{MT} = 1/Z_{MT}$ is a resistance in series with a negative mass reactance. In the frequency range where this is true, therefore, the reactive part of the impedance Z_{Mc} can be canceled out by letting [see Eqs. (9.11) and (9.12)]

$$\frac{S_D{}^2}{S_T{}^2}\frac{1}{C_{MT}} = \frac{1}{C_{M2}} = \left(\frac{1}{C_{MB}} + \frac{1}{C_{MS}}\right) \tag{9.13}$$

Then,

$$Z_{Mc} = \Re_{MT}\left(\frac{S_D}{S_T}\right)^2 \equiv \frac{1}{\mathfrak{r}_{Mc}} \tag{9.14}$$

where $\mathfrak{r}_{Mc}$ is the acoustic responsiveness of the throat of the horn at low frequencies transformed to the diaphragm.

The power available efficiency for frequencies where the approximate circuit of Fig. 9.6 holds, and where the conditions of Eq. (9.13) are met, is

$$\text{PAE} = \frac{400R_g B^2 l^2 \mathfrak{r}_{Mc}}{[(R_g + R_E) + B^2 l^2 \mathfrak{r}_{Mc}]^2} \tag{9.15}$$

The responsiveness $\mathfrak{r}_{Mc}$ usually varies from "infinity" at very low frequencies down to $S_T/(S_D{}^2 \rho_0 c)$ at some frequency in region A of Fig. 9.5.

High Frequencies. At very high frequencies, the response is limited principally by the combined mass of the diaphragm and the voice coil M_{MD}. If C_{M1} were zero, the response would drop off at the rate of 6 db per octave (see region C of Fig. 9.5). It is possible to choose C_{M1} to resonate with M_{MD} at a frequency that extends the response upward beyond where it would extend if it were limited by M_{MD} alone. We can understand this situation by deriving a circuit valid for the higher frequencies as shown in Fig. 9.7. It is seen that a damped antiresonance occurs at a selected high frequency. Above this resonance frequency, the response drops off 12 db for each octave increase in frequency (see region C of Fig. 9.5).

Because the principal diaphragm resonance [Eq. (9.8)] is highly damped by the throat resistance of the horn, it is possible to extend the region of flat response of a driver unit over a range of four octaves by proper choice

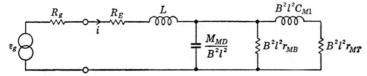

Fig. 9.7. Analogous circuit for a horn driver unit at high frequencies where the diaphragm mass reactance is much larger than its compliance reactance.

of C_{M1} at higher frequencies and by meeting the conditions of Eq. (9.13) at lower frequencies.

9.5. Examples of Horn Calculations

Example 9.1. As an example of the reference power available efficiency of a theater horn driver unit designed to operate in the frequency range above 500 cps, let us insert typical values of the loudspeaker constants into Eq. (9.7). We have

$$R_g = R_E = 24 \text{ ohms}$$
$$B = 19,000 \text{ gauss} = 1.9 \text{ webers/m}^2$$
$$l = 3.49 \text{ m}$$
$$B^2l^2 = 44 \text{ webers}^2/\text{m}^2$$
$$S_T = 3.14 \text{ cm}^2 = 3.14 \times 10^{-4} \text{ m}^2$$
$$S_D = 28.3 \text{ cm}^2 = 28.3 \times 10^{-4} \text{ m}^2$$
$$S_T/S_D = 0.111$$
$$S_T/S_D^2 = 3.14 \times 10^{-4}/(28.3 \times 10^{-4})^2 = 39.2 \text{ m}^{-2}$$
$$r_{MT} = 39.2/406 = 0.0965 \text{ mks mechanical mohms [see Eq. (9.3)]}$$
$$\text{PAE}_{\text{ref}} = \frac{(400)(24)(0.0965)(44)}{[4.25 + (48)(1.111)]^2}$$
$$= \frac{40,750}{3310} = 12.3\%$$

Solution. To increase the PAE_{ref}, it is seen from this example that the electrical resistance R_E should be reduced as far as possible without decreasing the length l of the wire on the voice coil. Within given space limitations, this can be done by winding the voice coil from wire with a rectangular cross section rather than with a circular cross section. This means that the voice-coil mass will be increased. Reduction of R_E further will demand wire of larger cross section which will require a larger air gap, with a corresponding reduction in B. Also, the voice coil must not become too large as its mass will limit the high-frequency response. We note further that the value of r_{MT}, and hence the ratio S_T/S_D^2, would seem to need to be large for high efficiency. However, if S_T/S_D^2 becomes too large, reference to Fig. 9.4b shows that too much power will be dissipated in r_{MB} and the efficiency will be low. A compromise is needed to effect the desired frequency response at high efficiency and minimum cost. The optimum value of S_T/S_D is calculated as shown in the next example.

Example 9.2. Determine the optimum value of the area of the throat of the horn for use with the driving unit of the preceding example. Calculate the reference PAE for this optimum condition.

Solution. Let us express the ratio of the throat area to the diaphragm area as

$$a = \frac{S_T}{S_D}$$

Then Eq. (9.7) becomes

$$\text{PAE}_{\text{ref}} = \frac{400 R_g a B^2 l^2 / \rho_0 c S_D}{\left[(R_g + R_E) + a \left(R_g + R_E + \frac{B^2 l^2}{\rho_0 c S_D} \right) \right]^2}$$

To determine the maximum value of PAE, we differentiate the equation with respect to a and equate the result to zero. This operation yields

$$a_{\text{opt}} = \frac{R_g + R_E}{R_g + R_E + B^2 l^2 / \rho_0 c S_D}$$

Inverting,

$$\left(\frac{S_D}{S_T} \right)_{\text{opt}} = \frac{1}{a_{\text{opt}}} = 1 + \frac{B^2 l^2}{(R_g + R_E) \rho_0 c S_D}$$

Substitution of the constants from the previous example gives us

$$\left(\frac{S_D}{S_T} \right)_{\text{opt}} = 1 + \frac{44}{(48)(406)(28.3) \times 10^{-4}} \doteq 1.8$$

or $a_{\text{opt}} = 0.555$. Hence,

$$(S_T)_{\text{opt}} = \frac{28.3 \times 10^{-4}}{1.8} = 15.7 \times 10^{-4} \, \text{m}^2$$

Substitution of $a = 0.555$ into the equation for PAE above gives

$$\text{PAE}_{\text{ref}} = \frac{(400)(24)(0.555)(44) \times 10^4}{(406)(28.3) \left[48 + 0.555 \left(48 + \frac{44 \times 10^4}{(406)(28.3)} \right) \right]^2}$$

$$= \frac{20.4 \times 10^4}{[48 + 48]^2} = 22.1\%$$

By comparison with the previous example, we see that, by proper choice of the throat area, we are able nearly to double the reference (mid-frequency) efficiency. In addition, let us see what would happen if the output impedance of the power amplifier were made very low (say, $R_g = 1$ ohm),

$$\frac{S_D}{S_T} \doteq 2.54$$

or

$$S_T = \frac{28.3 \times 10^{-4}}{2.54} = 11.1 \times 10^{-4} \, \text{m}^2$$

and

$$\text{PAE}_{\text{ref}} = \frac{6.04 \times 10^3}{[25 + 24.9]^2} = 2.4\%$$

This result shows that for maximum efficiency the amplifier impedance should approximately equal the loudspeaker impedance.

PART XXII *Horns*

9.6. General Description. A horn is in effect an acoustic transformer. It transforms a small-area diaphragm into a large-area diaphragm without the difficulties of cone resonances discussed in Part XVIII. A large-area diaphragm has a radiation impedance that is more nearly resistive over the desired frequency range than is the radiation impedance for a small-area diaphragm (see Fig. 5.3). As a result, more power is radiated at low frequencies for a given volume velocity of air.

In designing a horn for a particular application we usually wish to select the parameters so as to radiate the maximum amount of acoustic power over the desired frequency range with suitably low nonlinear distortion. Once we have stated the frequency range, tolerable distortion, and desired radiated power, we can choose the driving unit and then proceed to calculate the throat and the mouth diameters and the length and shape of the horn.

9.7. Mouth Size. The large end (mouth) of the horn should have a circumference large enough so that the radiation impedance is nearly resistive over the desired frequency range. Reference to Fig. 5.3 shows that this will be true for $ka > 1$: that is, $C/\lambda > 1$, where C is the circumference of the mouth of the horn and λ is the wavelength of the lowest tone that it is desired to radiate. If the mouth of the horn is not circular but square, it will behave in nearly the same way, as far as radiated power is concerned, for equal mouth areas. Hence, for good design, the mouth circumference C or mouth area S_M,

$$C = 2\sqrt{\pi S_M} > \lambda \quad (9.16)$$

where λ is the longest wavelength of sound that is to be radiated efficiently.

9.8. Exponential Horns. Many types of longitudinal section are possible for a horn. It may be parabolic, conical, hyperbolic, or one of a number of others in shape. For a horn to be a satisfactory transformer, its cross-sectional area near the throat end should increase gradually with x (see Fig. 9.8). If

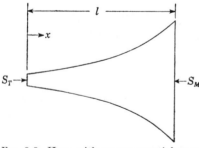

FIG. 9.8. Horn with an exponential cross section. The mouth area S_M equals S_T exp (ml), where m is the flare constant in units of inverse meters, l is the length of the horn in meters, and S_T is the throat area in square meters.

it does, the transformation ratio remains reasonably constant with frequency over a wide range.

The most common shape of longitudinal section meeting the requirement is *exponential*, shown by Fig. 9.8. For this shape, the cross-sectional area at any point x along the axis is given by the formula

$$S = S_T e^{mx} \qquad (9.17)$$

where S = cross-sectional area at x in square meters
$\quad S_T$ = cross-sectional area of the throat in square meters
$\quad m$ = flare constant in inverse meters
$\quad x$ = distance from throat in meters

If the horn is a number of wavelengths long and if the mouth circumference is larger than the wavelength, we may call the horn "infinite" in length. This simplification leads to equations that are easy to understand and are generally useful in design.

Theoretical Considerations. The general differential equation for sound propagation in an exponential horn whose length l is "infinite" is

$$\frac{\partial^2 p}{\partial t^2} - c^2 m \frac{\partial p}{\partial x} - c^2 \frac{\partial^2 p}{\partial x^2} = 0 \qquad (9.18)$$

where p = sound pressure at a point along the length of the horn. (It is assumed that the pressure is uniform across the cross section of the horn.)

$\quad c$ = speed of sound.

$\quad m$ = flare constant. Obviously, m determines the magnitude of the second term of the equation above, which expresses the rate at which the sound pressure changes with distance down the horn. If $m = 0$, Eq. (9.18) becomes the equation for propagation in a cylindrical tube, *i.e.*, a horn with zero flare.

A solution to Eq. (9.18) for the steady state is

$$p(t) = P_+ e^{-mx/2} e^{-jx\sqrt{4k^2 - m^2}/2} e^{j\omega t} \qquad (9.19)$$

where

$$k = \frac{2\pi}{\lambda} = \frac{\omega}{c} \qquad (9.20)$$

The volume velocity U at any point x is [see Eq. (2.58)]

$$U = \frac{S}{j\omega\rho_0}\left[\frac{m}{2} + j\frac{\sqrt{4k^2 - m^2}}{2}\right] p \qquad (9.21)$$

Impedances. The acoustic impedance Z_A, at a point x along the horn where the cross-sectional area is S, is

$$Z_A = \frac{p}{U} = \frac{2j\omega\rho_0}{S(m + j\sqrt{4k^2 - m^2})}$$

$$= \frac{\rho_0 c}{S}\sqrt{1 - \frac{m^2}{4k^2}} + j\frac{\rho_0 c m}{2kS} \qquad \text{mks acoustic ohms} \qquad (9.22)$$

At the throat, where $S = S_T$, the mechanical impedance $Z_{MT} = Z_{AT}S_T^2$ is

$$Z_{MT} = \frac{f}{u} = \rho_0 c S_T \sqrt{1 - \frac{m^2}{4k^2}} + j\frac{\rho_0 c m S_T}{2k}$$

$$\equiv R_{MT} + j\frac{1}{\omega C_{MT}} \qquad \text{mks mechanical ohms} \qquad (9.23)$$

The acoustic mobility $z_A = 1/Z_A$ at the throat is

$$z_{AT} = \frac{U}{p} = \frac{S_T}{\rho_0 c} \sqrt{1 - \frac{m^2}{4k^2}} - j\frac{S_T m}{2\rho_0 c k} \qquad (9.24)$$

The real and imaginary parts of Z_A and z_A behave alike with frequency and differ only by the magnitude $(S/\rho_0 c)^2$ and the sign of the imaginary part. Notice that this derivation is not restricted to a circular cross section. Limitations on cross-sectional shapes will be discussed later. Let us see next how varying the flare constant m affects the acoustic impedance Z_A at any point along the horn where the area is S.

Flare Constant and Throat Impedance. When the flare constant m is *greater* than 4π divided by the wavelength ($m > 2k$, low frequencies), the acoustic resistance R_{AT} and the acoustic reactance X_{AT}, at the throat of the horn where the area is S_T, are

$$R_{AT} = 0$$
$$X_{AT} = \frac{\rho_0 c}{S_T}\left(\frac{m}{2k} - \sqrt{\frac{m^2}{4k^2} - 1}\right) \qquad (9.25)$$

When the flare constant m *equals* 4π divided by the wavelength, the acoustic resistance and reactance are

$$R_{AT} = 0$$
$$X_{AT} = \frac{\rho_0 c m}{2k S_T} = \frac{\rho_0 c}{S_T} \qquad (9.26)$$

For all cases where m is *less* than 4π divided by the wavelength ($m < 2k$, high frequencies), the acoustic resistance and reactance at any point x along the horn where the cross-sectional area is S are

$$R_{AT} = \frac{\rho_0 c}{S_T}\sqrt{1 - \frac{m^2}{4k^2}}$$
$$X_{AT} = \frac{\rho_0 c m}{2k S_T} = \frac{\rho_0 c^2 m}{2\omega S_T} \equiv \frac{1}{\omega C_{AT}} \qquad (9.27)$$

where $C_{AT} = 2S_T/\rho_0 c^2 m$.

For very high frequencies, the reactance approaches zero and the resistance approaches $\rho_0 c/S_T$ or $\rho_0 c/S$ in general. This is also the impedance for a plane progressive sound wave in a tube of uniform cross section S.

Cutoff Frequency. The special case of $m = 4\pi/\lambda$ occurs at a frequency which we shall designate f_c, where

$$f_c = \frac{mc}{4\pi} \qquad (9.28)$$

This frequency f_c is called the *cutoff frequency* because for frequencies lower than this no power will be transmitted down the horn, *i.e.*, the impedance at all positions along the horn is purely reactive [see Eq. (9.25)].

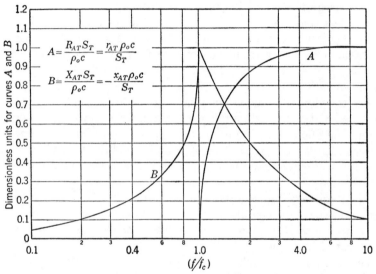

FIG. 9.9. Plot of the quantities A and B, which are defined by the relations given on the graph.

To obtain the acoustic impedance at the throat of the horn in terms of the cutoff frequency, we observe that $f_c/f = m/2k$. Substituting in Eq. (9.22) yields

$$Z_{AT} = \frac{\rho_0 c}{S_T}\left(\sqrt{1 - \left(\frac{f_c}{f}\right)^2} + j\frac{f_c}{f}\right) = R_{AT} + jX_{AT} \qquad (9.29)$$

where S_T = throat area in square meters

$\rho_0 c$ = characteristic impedance of air in mks rayls

f_c = cutoff frequency

f = driving frequency

Graphs of two quantities A and B that are directly proportional to the resistive and reactive parts of the acoustic impedance at the throat of an infinitely long exponential horn are shown in Fig. 9.9. The quantities A and B also are directly proportional to the real and imaginary parts of the acoustic mobility at the throat. The relations among A, B, R_{AT}, X_{AT}, r_{AT}, and x_{AT} are given on the graph.

When the frequency is greater than approximately double the cutoff frequency f_c, the throat impedance is substantially resistive and very near its maximum value in magnitude.

Finite-length Horn. The equation for the acoustic impedance at the throat of an exponential horn of finite length was given by Eq. (5.80). For exact calculations, that rather complicated equation must be used whenever the bell diameter is not large or when the horn length is short. To illustrate what the words "large bell diameter" and "long length" mean, let us refer to Fig. 9.10.

If the circumference of the mouth of the horn divided by the wavelength is less than about 0.5 (*i.e.*, the diameter of the mouth divided by the wavelength is less than about 0.16), the horn will resonate like a cylindrical tube, *i.e.*, at multiples of that frequency where the length is equal to a half wavelength. This condition is shown clearly by the two lower-frequency resonances in *a* of Fig. 9.10.

When the circumference of the mouth of the horn divided by the wavelength is greater than about 3 (*i.e.*, diameter divided by wavelength greater than about 1.0), the horn acts nearly like an infinite horn. This is shown clearly by comparison of *d* and *e* of Fig. 9.10, for the region where f/f_c is greater than about 2, which is the case of mouth diameter/$\lambda > 0.5$.

In the frequency region where the circumference of the mouth to wavelength ratio lies between about 1 and 3, the exact equation for a finite exponential horn [Eq. (5.80)] must be used, or the results may be estimated from *b* and *c* of Fig. 9.10.

When the length of the horn becomes less than one-fourth wavelength, it may be treated as a simple discontinuity of area such as was discussed in Par. 5.11 (pp. 139 to 141).

Obviously, if one chooses a certain mouth area and a throat area to obtain maximum efficiency, the length of the horn is automatically set by the flare constant m, which is in turn directly dependent upon the desired cutoff frequency.

Nonlinear Distortion. A sound wave produces an expansion and a compression of the air in which it is traveling. We find from Eq. (2.6) that the relation between the pressure and the volume of a small "box" of the air at 20°C through which a sound wave is passing is

$$P = \frac{0.726}{V^{1.4}} \tag{9.30}$$

where V = specific volume of air in $m^3/kg = 1/\rho_0$

P = absolute pressure in bars, where 1 bar = 10^5 newtons/m^2

This equation is plotted as curve AB in Fig. 9.11

Assuming that the displacement of the diaphragm of the driver unit is sinusoidal, it acts to change the volume of air near it sinusoidally. For

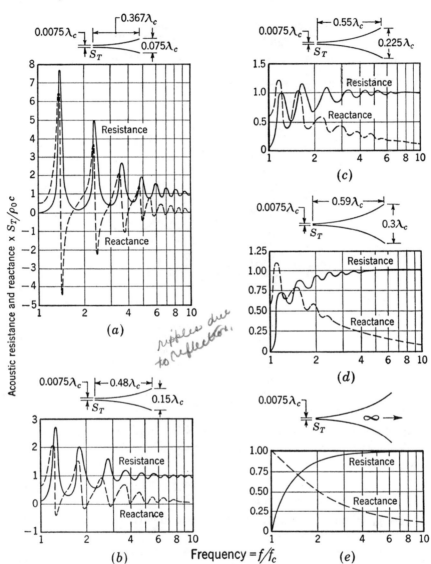

FIG. 9.10. Graphs showing the variation in specific acoustic impedance at the throat of four exponential horns as a function of frequency with bell diameter as the parameter. The cutoff frequency $f_c = mc/4\pi$ and the throat diameter $= 0.0075\ c/f_c$; both are held constant. Bell circumferences are (a) $C = 0.236\lambda_c$, (b) $C = 0.47\lambda_c$, (c) $C = 0.71\lambda_c$, (d) $C = 0.94\lambda_c$, and (e) $C = \infty$. The mouth of the horn is assumed to be terminated in an infinite baffle. (*After Olson.*).

large changes in volume, the pressure built up in the throat of the horn is no longer sinusoidal, as can be seen from Fig. 9.11.   The pressure wave so generated travels away from the throat toward the mouth.

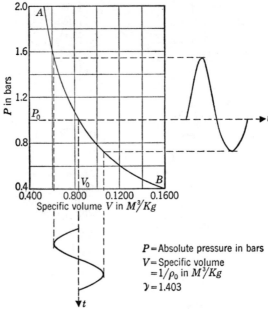

FIG. 9.11. Plot of the gas equation $PV^\gamma = 1.26 \times 10^4$, valid at 20°C.   Normal atmospheric pressure (0.76 m Hg) is shown as $P_0 = 1$ bar.

If the horn were simply a long cylindrical pipe, the distortion would increase the farther the wave progressed according to the formula (air assumed)[3,4]

$$\frac{p_2}{p_1} = \frac{\gamma + 1}{\sqrt{2}\gamma} k \frac{p_1}{P_0} x = 1.21k \frac{p_1}{P_0} x \qquad (9.31)$$

where p_1 = rms sound pressure of the fundamental frequency in newtons per square meter

p_2 = rms sound pressure of the second harmonic in newtons per square meter

P_0 = atmospheric pressure in newtons per square meter

$k = \omega/c = 2\pi/\lambda$ = wave number in meters^{-1}

$\gamma = 1.4$ for air

x = distance the wave has traveled along the cylindrical tube in meters

[3] A. L. Thuras, R. T. Jenkins, and H. T. O'Neil, Extraneous Frequencies Generated in Air Carrying Intense Sound Waves, *J. Acoust. Soc. Amer.*, **6**: 173–180 (1935).

[4] L. H. Black, A Physical Analysis of the Distortion Produced by the Non-linearity of the Medium, *J. Acoust. Soc. Amer.*, **12**: 266–267 (1940).

Equation (9.31) breaks down when the second-harmonic distortion becomes large, and a more complicated expression, not given here, must be used.

In the case of an exponential horn, the amplitude of the fundamental decreases as the wave travels away from the throat, so that the second-harmonic distortion does not increase linearly with distance. Near the throat it increases about as given by Eq. (9.31), but near the mouth the pressure amplitude of the fundamental is usually so low that very little additional distortion occurs.

The distortion introduced into a sound wave after it has traveled a distance x down an exponential horn for the case of a constant power supplied to unit area of the throat is found as follows:

1. Differentiate both sides of Eq. (9.31) with respect to x, so as to obtain the rate of change in p_2 with x for a constant p_1. Call this equation (9.31a).

2. In Eq. (9.31a), substitute for p_1 the pressure $p_T e^{-mx/2}$, where p_T is the rms pressure of the fundamental at the throat of the horn in newtons per square meter and m is the flare constant.

3. Then let $p_T = \sqrt{I_T \rho_0 c}$, where I_T is the intensity of the sound at the throat in watts per square meter and $\rho_0 c$ is the characteristic acoustic impedance of air in mks rayls.

4. Integrate both sides of the resulting equation with respect to x. This yields

Per cent second-harmonic distortion

$$= \frac{50(\gamma + 1)}{\gamma} \frac{\sqrt{I_T \rho_0 c}}{P_0} \frac{f}{f_c} [1 - e^{-mx/2}] \quad (9.32)$$

For an infinitely long exponential horn, at normal atmospheric pressure and temperature, the equation for the total distortion introduced into a wave that starts off sinusoidally at the throat is

Per cent second-harmonic distortion $= 1.73 \dfrac{f}{f_c} \sqrt{I_T} \times 10^{-2} \quad (9.33)$

where f = driving frequency in cycles per second

f_c = cutoff frequency in cycles per second

I_T = intensity in watts per square meter at the throat of the horn

Equation (9.33) is shown plotted in Fig. 9.12. Actually, this equation is nearly correct for finite horns because most of the distortion occurs near the throat.

Equation (9.33) reveals that for minimum distortion the cutoff frequency f_c should be as large as possible, which in turn means as large a flare constant as possible. In other words, the horn should flare out rapidly in order to reduce the intensity rapidly as one travels along the horn toward the mouth.

Unfortunately, a high cutoff frequency is not a feasible solution for horns that are designed to operate over a wide frequency range. In this case, it is necessary to operate the horn at low power at the higher frequencies if the distortion is to be low at these frequencies. This goal is achieved automatically to some extent in reproducing speech and music, because above 1000 cps the intensity for these sounds decreases by about a factor of 10 for each doubling of frequency.

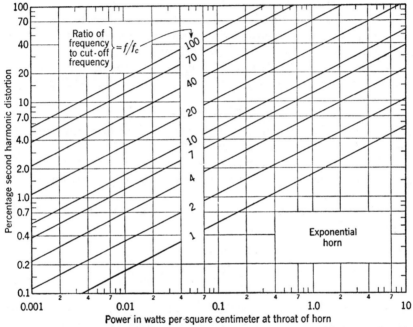

FIG. 9.12. Percentage second-harmonic distortion in an exponential horn as a function of the intensity at the horn throat with the ratio of the frequency to the cutoff frequency as parameter.

9.9. Other Types of Horn. As we stated before, other than exponential types of horn shapes may be used. Representative types include conical horns, parabolic horns, and hyperbolic horns. The equations for these types are

Parabolic Horn (see Fig. 9.13).

$$S = S_T x \tag{9.34}$$

where S = cross-sectional area at x in square meters

S_T = cross-sectional area of throat in square meters

Conical Horn (see Fig. 9.13).

$$S = S_T x^2 \tag{9.35}$$

Hyperbolic Horn (see Fig. 9.14).

$$S = S_T \left(\cosh \frac{x}{x_0} + M \sinh \frac{x}{x_0} \right)^2 \qquad (9.36)$$

where x = axial distance from the throat in meters

x_0 = reference axial distance from the throat in meters

　M = parameter which never exceeds unity

　f_c = cutoff frequency = $c/2\pi x_0$ cps

　c = speed of sound in meters per second

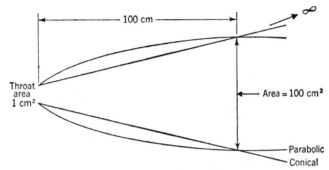

FIG. 9.13. Longitudinal sections of conical and parabolic horns.

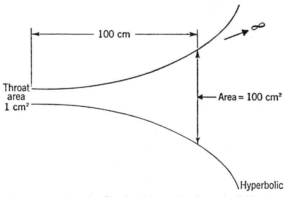

FIG. 9.14. Longitudinal sections of a hyperbolic horn.

At very high frequencies, all these types behave about alike. At low frequencies, however, there are considerable differences. These differences can be shown by comparison of the throat impedances for the conical and hyperbolic horns with that for the exponential horn. In Fig. 9.15, the throat impedances for the first two types are shown. The throat impedance for the exponential type was shown in Fig. 9.9.

For all horns, the throat resistance is very low, or zero, below the cutoff frequency. Above the cutoff frequency, the specific throat resistance rises rapidly to its ultimate value of $\rho_0 c$ for those cases where the rate of taper is *small* near the throat of the horn. For example, the specific throat resistance for the hyperbolic horn reaches $\rho_0 c$ at about one-twentieth the frequency at which the specific throat resistance for the conical horn reaches this value. Similarly for the hyperbolic horn, the specific throat resistance approaches unity at about one-third the frequency for the exponential horn.

It would seem that for best loading conditions on the horn-driver unit over the frequency range above the cutoff frequency one should use the hyperbolic horn. However, it should also be remembered that the non-linear distortion will be higher for the hyperbolic horn because the wave travels a longer distance in the horn before the pressure drops off owing to area increase than is the case for the other horns. For minimum distortion at given power per unit area, the conical horn is obviously the best

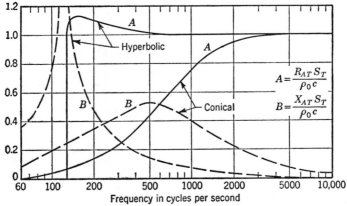

FIG. 9.15. Quantities A and B proportional to the acoustic resistance R_{AT} and acoustic reactance X_{AT}, respectively, for the conical and hyperbolic horns of Figs. 9.13 and 9.14.

of the three. The exponential horn is usually a satisfactory compromise in design because it falls between these two extremes.

9.10. Bends in Horns. A horn loudspeaker for use at low frequencies is very large and long, because m must be small for a low cutoff frequency and the area of the mouth must be large to radiate sound properly. As a consequence, it has become popular to "fold" the horn so that it will fit conveniently into the home.

Many types of folded horns have been devised that are more or less successful in reproducing music and speech with satisfactory frequency response. In order to be successful, the bends in folded horns must not be sharp when their lateral dimensions approach a half wavelength, or they will change the spectrum of the radiated sound.

Good data on the comparative performance of folded horns are not available. This is partly because it is difficult to measure the response of large folded horns in an anechoic chamber and partly because commercial companies guard their data. Some data are available, however, on the effect of bends of various types in rectangular ducts such as are used in ventilating systems. These data are shown in Fig. 9.16.

It is seen that when the duct width is near one-half or three-halves wavelength the attenuation of the sound by a 90° bend is quite large. At high frequencies, the losses for even 19° bends are large. If possible, the wavelength should be long compared with the width of the duct at the bend. Then the attenuation is very small.

9.11. Cross-sectional Shapes. Earlier it was stated that the cross-sectional shape of a horn is not too important. This is true provided the lateral dimensions of the horn are not comparable with a wavelength. When the lateral dimensions are large enough, standing waves exist across the duct, similar to the standing waves in a closed end tube.

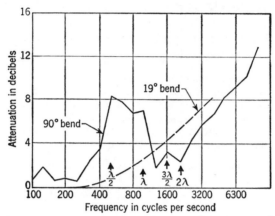

FIG. 9.16. Attenuation of sound due to the introduction of 19° and 90° bends as a function of frequency. For a sketch of the duct see Fig. 11.20, p. 351.

These waves are usually not important in an exponential horn that is circular or square in cross section because, generally, only that section of the horn near the mouth is greater than a half wavelength.

In a rectangular horn that is constructed with two sides parallel and the other two sides varying according to the exponential or hyperbolic law, standing waves may exist between the two parallel walls. These resonances occur at wavelengths that are submultiples of the width of the duct, *i.e.*, at frequencies equal to

$$f = \frac{nc}{2l_x} \tag{9.37}$$

or wavelengths equal to

$$\lambda = \frac{2l_x}{n} \tag{9.38}$$

where n is an integer, that is, 1, 2, 3, 4, ...

For example, suppose that the width of the horn was 1.5 ft. Then resonances (standing waves) would occur at 377, 754, 1508, etc., cps.

At these frequencies, reduced power output generally occurs.   In general, the upper frequency limit for operation of a horn should be chosen sufficiently low so that troubles from transverse standing waves are avoided.

9.12. Materials.   The material from which a horn is constructed is very important.   If the side walls of the horn resonate mechanically at one or more frequencies in the range of operation, "dips" in the power-output curve will occur.   Undamped thin metal is the least desirable material because the horn from which it is made will resonate violently at fairly low frequencies.   Heavy metals, covered on the outside with thick mastic material so that mechanical resonances are damped, are much better.   A concrete or plaster horn 1 or 2 in. in thickness is best because of its weight and internal damping.

Plywood is commonly used in the construction of large horns.   Although it is not as satisfactory as concrete, it gives satisfactory results if its thickness exceeds ¾ in. and if it is braced with wooden pieces glued on at frequent, irregular intervals.

Example 9.3.   A horn combination consisting of two horns, one for radiating low frequencies and the other for radiating high frequencies, is required.   It is desired that the frequency response be flat between 70 and 6000 cps and that the horn combination be designed to fit into an average-sized living room.   The system will be used for reproducing high-fidelity music in the home.   The maximum acoustic powers to be radiated in six frequency regions are estimated as follows:[5]

Frequency range	Power level, db re 10^{-13} watt
70–250	102
250–400	102
400–1000	102
1000–2200	99
2200–3000	96
3000–8000	96

Solution for Low-frequency Horn.   We shall select the exponential horn as the best compromise shape of horn for our use.   Because the lowest frequency at which good radiation is desired is 70 cps, we choose the mouth area from Eq. (9.16).

$$\text{Mouth area } S_M = \frac{\lambda^2}{4\pi} = \frac{c^2}{4\pi f^2} = 1.93 \text{ m}^2 = 20.8 \text{ ft}^2$$

Twenty square feet is probably too large a mouth area for the living room of most homes, so that a compromise in design is necessary.

Let us choose arbitrarily a mouth area of about 10 ft², that is, 0.93 m².   This corresponds to the bell opening shown in Fig. 9.10c.   We see from this chart that below $f = 2f_c$ there will be two resonances that are not desirable, but they are fairly well damped.

Let us design for a cutoff frequency of

$$f_c = 60 \text{ cps}$$

[5] H. F. Hopkins and N. R. Stryker, A Proposed Loudness-efficiency Rating for Loudspeakers and the Determination of System Power Requirements for Enclosures, *Proc. IRE*, **36**: 315–335 (1948).

The flare constant m equals [see Eq. (9.28)]

$$m = \frac{4\pi f_c}{c} = \frac{4\pi 60}{344.8} = 2.18 \text{ m}^{-1}$$

Let us choose a 12-in. direct-radiator unit with an effective diameter of 0.25 m as the driver.   The effective area of this driver unit is

$$S_D = \pi(0.125)^2 = 0.049 \text{ m}^2$$

Assume the other constants to be as follows:

$$R_g = R_E = 6 \text{ ohms}$$
$$Bl = 15 \text{ webers/m}$$

From Example 9.2, it appears that for maximum efficiency S_D/S_T should equal 2. However, to keep the length down, let us make

$$\frac{S_D}{S_T} = 1$$

Then,

$$r_{MT} = \frac{S_T}{\rho_0 c S_D^2} = \frac{1}{(406)(0.049)} = 0.05 \text{ mks mechanical mohms}$$

Let us calculate the reference PAE.   Assume in Fig. 9.4b that $r_{MB} \gg r_{MT}$.   From Eq. (9.7),

$$\text{PAE} = \frac{(400)(6)(0.05)(15)^2}{[(15^2)(0.05) + (12)(2)]^2} = 22\%$$

As a trial, let us make $S_D/S_T = 2.0$.   Then $r_{MT} = 0.025$, and PAE $= 24\%$.   Finally, let $S_T/S_D = 2$.   Then, $r_{MT} = 0.1$, and PAE $= 15.8\%$.

It is seen that the ratio of the throat and diaphragm areas may be made equal with little loss of efficiency, thereby making our horn of reasonably short length.   The length of our horn is found, from Eq. (9.17),

$$e^{mx} = \frac{0.93}{0.049} = 19$$

or

$$mx = 2.94$$
$$x = \frac{2.94}{2.18} = 1.35 \text{ m} = 4.4 \text{ ft}$$

The intensities for a horn with a throat area of 0.049 m² are as follows, assuming uniform pressure distribution:

Frequency	Power, watts	Watts/cm² at throat
70–250	1.58×10^{-3}	3.22×10^{-6}
250–400	1.58×10^{-3}	3.22×10^{-6}
400–1000	1.58×10^{-3}	3.22×10^{-6}
1000–2200	7.94×10^{-4}	1.62×10^{-6}
2200–3000	4×10^{-4}	8.16×10^{-7}
3000–8000	4×10^{-4}	8.16×10^{-7}

Let us set the upper limit of operation at 600 cps. Then $f/f_c = 10$. Extrapolation of the line for 10 in Fig. 9.12 to 3.22×10^{-6} shows that the per cent second-harmonic distortion in the horn will be about 0.02 per cent, which is negligible. In fact, the power level could be increased 30 db before the distortion would become as large as 1 per cent.

This calculation would seem to indicate that the low-frequency unit could be operated successfully above 600 cps. However, it seems from experience that for psychological reasons the changeover from the low-frequency to the high-frequency horn should occur at a frequency below 600 cps for best auditory results.

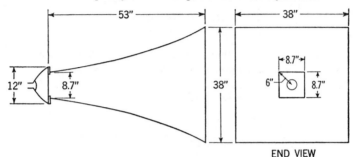

END VIEW

FIG. 9.17. Plans for a simple straight exponential horn with a cutoff frequency of 60 cps, a throat area of 0.049 m², and a mouth area of 0.93 m².

Let us see what value the total compliance in the driving-unit circuit ought to have if it is to balance out the mass reactance of the horn at frequencies below the diaphragm resonance frequency. From Eqs. (9.12), (9.13), and (9.23),

$$C_{M2} = \frac{S_T{}^2}{S_D{}^2} C_{MT} = \frac{2S_T}{S_D{}^2 \rho_0 c^2 m} = \frac{2S_T}{S_D{}^2 \gamma P_0 m}$$

$$C_{M2} = \frac{2}{(1.4)(0.049) \times 10^5 (2.18)} = 1.34 \times 10^{-4} \text{ m/newton}$$

The quantity C_{M2} includes the combined compliance of the loudspeaker and the enclosure behind it. Reference to Fig. 8.5d shows that this is a reasonable value of compliance to expect from a loudspeaker of this diameter. In case the compliance is not correct, we can vary the size of the throat, or even m somewhat, in order to achieve the desired value for C_{M2}.

Two possible horns for our design are the straight square horn shown in Fig. 9.17 or the folded horn of the Klipsch type[6] shown in Fig. 9.18. If the straight horn is used, it will probably be necessary to put it partially above the ceiling or below the floor in order to make its presence nonobjectionable in the room.

Solution for High-frequency Horn. As a cutoff frequency, let us choose

$$f_c = 300 \text{ cps}$$

We shall use an electrical crossover network of 500 cps, which will make effective use of both horns and is a good choice of frequency from the standpoint of the psychology of listening.

The flare constant is [see Eq. (9.28)]

$$m = \frac{4\pi f_c}{c} = \frac{4\pi 300}{344.8} = 10.9 \text{ m}^{-1}$$

---

[6] P. W. Klipsch, A Low-frequency Horn of Small Dimensions, *J. Acoust. Soc. Amer.*, **13**: 137–144 (1941).

Let us assume that the driver unit is the one discussed in Example 9-1.   For this, $S_T = 3.14 \times 10^{-4}$ m².

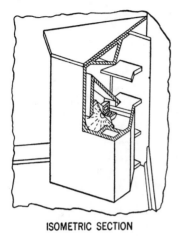

ISOMETRIC SECTION

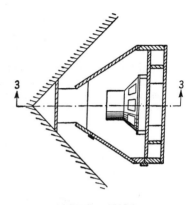

HORIZONTAL SECTION
(Section 2·2)

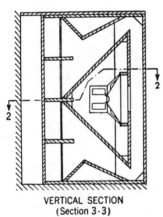

VERTICAL SECTION
(Section 3·3)

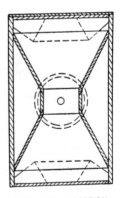

TRANSVERSE SECTION

FIG. 9.18. Sketches for a Klipsch type of folded exponential horn.   This particular horn is about 40 in. high and has smooth response below 200 cps.

The horn should radiate sound well at 400 cps, so that the mouth-opening area should be, if possible, greater than that given by Eq. (9.16),

$$S_M = \frac{c^2}{4\pi f^2} = \frac{(344.8)^2}{4\pi(400)^2} = 0.0591 \text{ m}^2$$
$$= 91.5 \text{ in.}^2$$

As we learned in Chap. 4, in order to get a wide directivity pattern, say $\pm30°$ over a wide range of frequencies, the horn should have a curved mouth.   Let us select a design that is about 6 in. in height and has a circular curved mouth with an arc length of 30 in.   The mouth area for these dimensions is 180 in.², or 0.1163 m², which is double that called for above.

The length of the horn is found from Eq. (9.17),

$$e^{mx} = \frac{(11.63) \times 10^{-2}}{(3.14) \times 10^{-4}} = 370$$

$$mx = 5.91$$

$$x = \frac{5.91}{10.9} = 0.541 \text{ m} = 21.3 \text{ in.}$$

The horn will have the shape and cross section shown in Fig. 9.19.

The cutoff frequency is at 300 cps and is far enough below the 500 cps crossover frequency so that the throat impedance will be resistive over the entire useful range of the loudspeaker; hence there is no real need to balance $(\omega C_{M2})^{-1}$ against the mass

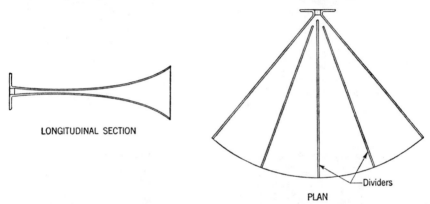

LONGITUDINAL SECTION

PLAN

Fig. 9.19. Plan for a simple straight exponential horn with a cutoff frequency of 300 cps, a throat area of 3.14×10^{-4} m², and a mouth area of 11.6×10^{-2} m². The dividers guide the wave along the horn and tend to produce a plane wave front of uniform intensity at the mouth.

reactance of the horn. Nevertheless, let us calculate the value of C_{M2} from Eqs. (9.12), (9.13), and (9.23),

$$C_{M2} = \frac{2S_T}{S_D{}^2 \gamma P_0 m}$$

$$= \frac{(2)(39)}{(1.4) \times 10^5 (10.9)} = 5.1 \times 10^{-5} \text{ m/newton}$$

The magnitude of this compliance, also, is not an unreasonable value to achieve in a driving unit of the type given in Example 9.1.

Finally, let us determine the power-handling capacity of this horn. The intensity at the throat of the horn I_T in the 3000- to 8000-cps band will equal

$$I_T = \frac{4 \times 10^{-4}}{3.14} = 1.27 \times 10^{-4} \text{ watt/cm}^2$$

At our upper design frequency of 6000 cps, which gives us

$$\frac{f}{f_c} = \frac{6000}{300} = 20$$

we see from Fig. 9.12 that the second-harmonic distortion will equal about 0.4 per cent. This is low distortion, and we conclude that our design is satisfactory.

CHAPTER 10

SOUND IN ENCLOSURES

PART XXIII *Sound Fields in Small Regularly Shaped Enclosures*

10.1. Introduction. The study of sound in enclosures involves not only a search into how sounds are reflected backward and forward in an enclosure but also investigations into how to measure sound under such conditions and the effect various materials have in absorbing and controlling this sound. Also, of great importance in applying one's engineering knowledge of the behavior of sound in such enclosed spaces is an understanding of the personal preferences of listeners, whether listening in the room where the music is produced or listening at a remote point to a microphone pickup. Psychological criteria for acoustic design have occupied the attention of many investigators and must always be borne in mind. This chapter is confined to physical acoustics. Psychological factors will be discussed in Chap. 13, which deals with psychoacoustical phenomena and criteria for acoustic design.

Two extremes to the study of sound in enclosures can be analyzed and understood easily. At the one extreme we have small enclosures of simple shape, such as rectangular boxes, cylindrical tubes, or spherical shells. In these cases the interior sound field is describable in precise mathematical terms, although the analysis becomes complicated if the walls of the enclosures are covered in whole or in part with acoustical absorbing materials.

At the other extreme we have very large irregularly shaped enclosures where no precise description can be made of the sound field but where a statistically reliable statement can be made of the average conditions in the room. This is analogous to a study that a physician might make of a particular man to determine the number of years he will live, as opposed to a study of the entire population on a statistical basis to determine how long a man, on the average, will live. As might be expected, the statistical study leads to simpler formulas than the detailed study of a particular case.

10.2. Stationary and Standing Waves. One type of small regularly shaped enclosure, the rigidly closed tube, has been discussed already in Part IV. This case provides an excellent example of the acoustical situation that exists in large enclosures.

First, we noted that along the x axis of the tube the sound field could be described as the combination of an outward-traveling wave and a backward-traveling wave. Actually, the outward-traveling wave is the sum of the original free-field wave that started out from the source plus the outward-going waves that are making their second, third, fourth, and so on, round trips. Similarly, the backward-traveling wave is a combination of the first reflected wave and of waves that are making the return leg of their second, third, fourth, and so on, round trips. These outward- and backward-traveling waves add in magnitude to produce what is called a *stationary wave*[1] if the intensity along the tube is zero or what is called a *standing wave* if there is absorption at the terminating end of the tube so that power flows along the tube away from the source (intensity not equal to zero).

10.3. Normal Modes and Normal Frequencies. We saw from Eq. (2.48) that whenever the driving frequency is such that $\sin kl \to 0$, the pressure in the tube reaches a very large value. That is to say, the pressure is very large whenever

$$kl = n\pi \tag{10.1}$$

Then, because

$$k \equiv \frac{2\pi f}{c} = \frac{2\pi}{\lambda}$$

we have

$$f_n = \frac{nc}{2l} \tag{10.2}$$

or

$$\frac{l}{\lambda_n} = \frac{n}{2} \tag{10.3}$$

where

$$n = 1, 2, 3, 4, \ldots, \infty \tag{10.4}$$

f_n = nth resonance (normal) frequency of the tube

λ_n = c/f_n = nth resonance (normal) wavelength of the tube

Equation (10.3) tells us that the pressure is very large whenever the length of the tube equals some integral multiple of a half wavelength ($\lambda/2$).

The condition where the frequency equals $nc/2l$ so that a very large sound pressure builds up in the tube is called a *resonance* condition or a

[1] The definitions for standing and stationary waves are found in "American Standard Acoustical Terminology," Z24.1—1951, American Standards Association, Inc., New York N.Y.

normal mode of vibration of the air space in the tube.   The frequency f_n of a normal mode of vibration is called a *normal frequency*.   There are an infinite number of normal modes of vibration for a tube because n can take on all integral values between 0 and infinity.   We may look on the tube, or in fact on any enclosure, as a large assemblage of acoustic resonators, each with its own normal frequency.

In the closed-tube discussion of Part IV, we made no mention of the effect on the results of the cross-sectional shape or size of the tube.   It was assumed that the transverse dimensions were less than about 0.1 wavelength so that no transverse resonances would occur in the frequency region of interest.

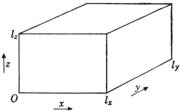

If the transverse dimensions are greater than one-half wavelength, we have a small room which, if rectangular, can be described by the dimensions shown in Fig. 10.1.   Waves can travel in the room backward and forward be-

FIG. 10.1. Dimensions and coordinate system for a rectangular enclosure.

tween any two opposing walls.   They can travel also around the room involving the walls at various angles of incidence.   If these angles are chosen properly, the waves will return on themselves and set up stationary or standing waves.   Each standing wave is a normal mode of vibration for the enclosure.

It would be interesting in such a rectangular enclosure to solve mathematically and to describe exactly the distribution of sound as determined by the strength and type of source.   That study is beyond the scope of this text.   We shall describe, however, the simplest cases and suggest extra reading for those interested.[2,3]

The number of modes of vibration in a rectangular enclosure is much greater than that for the rigidly closed tube whose diameter is small compared with a wavelength.   In fact, the normal frequencies of such an enclosure are given by the equation,

$$f_n = \frac{\omega_n}{2\pi} = \frac{c}{2} \sqrt{\left(\frac{n_x}{l_x}\right)^2 + \left(\frac{n_y}{l_y}\right)^2 + \left(\frac{n_z}{l_z}\right)^2} \qquad (10.5)$$

where f_n = nth normal frequency in cycles per second.

n_x, n_y, n_z = integers that can be chosen separately.   They may take on all integral values between 0 and ∞.

l_x, l_y, l_z = dimensions of the room in meters.

c = speed of sound in meters per second.

[2] P. M. Morse, "Vibration and Sound," 2d ed., Chap. VIII, McGraw-Hill Book Company, Inc., New York, 1948.

[3] P. M. Morse and R. H. Bolt, Sound Waves in Rooms, *Rev. Mod. Phys.*, **16**: 69–150 (1944).

As an example, let us assume that the z dimension, l_z, is less than 0.1 of all wavelengths being considered. This corresponds to n_z being zero at all times. Hence,

$$f_{n_x,n_y,0} = \frac{c}{2}\sqrt{\left(\frac{n_x}{l_x}\right)^2 + \left(\frac{n_y}{l_y}\right)^2}$$ (10.6)

Let $l_x = 12$ ft and $l_y = 10$ ft. Find the normal frequencies of the $n_x = 1$, $n_y = 1$ and the $n_x = 3$, $n_y = 2$ normal modes of vibration. We have

$$f_{1,1,0} = {}^{1131\!\!/_2}\sqrt{{}^1\!/_{144} + {}^1\!/_{100}}$$
$$= 73.5 \text{ cps}$$

and

$$f_{3,2,0} = {}^{1131\!\!/_2}\sqrt{{}^9\!/_{144} + {}^4\!/_{100}}$$
$$= 181 \text{ cps}$$

Morse[2] shows that the sound-pressure distribution in a rectangular box for each normal mode of vibration with a normal frequency ω_n is proportional to the product of three cosines,

$$p_{n_x,n_y,n_z} \propto \cos\frac{\pi n_x x}{l_x}\cos\frac{\pi n_y y}{l_y}$$
$$\cos\frac{\pi n_z z}{l_z}e^{j\omega_n t}$$ (10.7)

where the origin of coordinates is at the corner of the box. It is assumed in writing Eq. (10.7) that the walls have very low absorption. If the absorption is high, the sound pressure cannot be represented by a simple product of cosines.

If we inspect Eq. (10.7) in detail, we see that n_x, n_y, and n_z indicate the number of planes of zero pressure occurring along the x, y, and z coordinates, respectively. Such a distribution of sound pressure

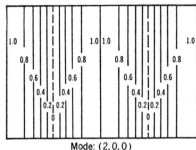

Mode: (2, 0, 0)

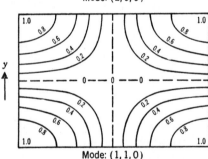

Mode: (1, 1, 0)

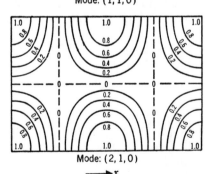

Mode: (2, 1, 0)

FIG. 10.2. Sound-pressure contour plots on a section through a rectangular room. The numbers on the plots indicate the relative sound pressure.

levels can be represented by forward- and backward-traveling waves in the room. This situation is analogous to that for the closed tube (one-dimensional case). Examples of pressure distributions for three modes of vibration in a rectangular room are shown in Fig. 10.2. The lines indicate planes of constant pressure extending from floor to ceiling along the z dimension. Note that n_x and n_y indicate the number of planes of zero pressure occurring along the x and y coordinates, respectively.

The angles θ_x, θ_y, and θ_z at which the forward- and backward-traveling waves are incident upon and reflect from the walls are given by the relations

$$\theta_x = \tan^{-1} \frac{\sqrt{(n_y/l_y)^2 + (n_z/l_z)^2}}{n_x/l_x} = \cos^{-1} \frac{n_x c}{2l_x f_n} \qquad (10.8)$$

$$\theta_y = \tan^{-1} \frac{\sqrt{(n_x/l_x)^2 + (n_z/l_z)^2}}{n_y/l_y} = \cos^{-1} \frac{n_y c}{2l_y f_n} \qquad (10.9)$$

$$\theta_z = \text{similarly} \qquad (10.10)$$

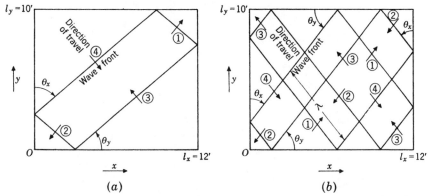

FIG. 10.3. Wave fronts and direction of travel for (a) $n_x = 1$, $n_y = 1$ normal mode of vibration; and (b) $n_x = 3$ and $n_y = 2$ normal mode of vibration. These represent two-dimensional cases where $n_z = 0$. The numbers one and three indicate forward-traveling waves, and the numbers two and four indicate backward-traveling waves.

For the examples where $n_x = 1$, $n_y = 1$ and $n_x = 3$, $n_y = 2$, the traveling waves reflect from the $x = 0$ and $x = l_x$ walls at

$$(\theta_x)_{1,1,0} = \tan^{-1} \frac{l_x}{l_y} = \tan^{-1} \frac{12}{10} = 50.2°$$

$$(\theta_x)_{3,2,0} = \tan^{-1} \frac{2l_x}{3l_y} = \tan^{-1} 0.8 = 38.65°$$

The angles of reflection at the $y = 0$ and $y = l_y$ walls are

$$(\theta_y)_{1,1,0} = \tan^{-1} \frac{l_y}{l_x} = 39.8°$$

$$(\theta_y)_{3,2,0} = \tan^{-1} \frac{3l_y}{2l_x} = \tan^{-1} \frac{30}{24} = 51.35°$$

The wave fronts travel as shown in (a) and (b) of Fig. 10.3. It is seen that there are two forward-traveling waves (1 and 3) and two backward-

traveling waves (2 and 4).   In the three-dimensional case, there will be
four forward- and four backward-traveling waves.

When the acoustical absorbing materials are placed on some or all
surfaces in an enclosure, energy will be absorbed from the sound field at
these surfaces and the sound-pressure distribution will be changed from
that for the hard-wall case.   For example, if an absorbing material were
put on one of the $l_x l_z$ walls, the sound pressure at that wall would be

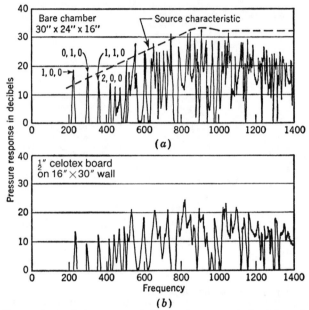

FIG. 10.4. Comparison of two transmission curves recorded with and without an
absorbing sample on a 24- by 30-in. wall of a model chamber with dimensions 16 by
24 by 30 in.   The microphone was in one corner, and the source was diagonally
opposite.   The dashed line shows the relative response of the small loudspeaker
($\frac{3}{8}$ in. diameter) measured at 2 in. in free space.   Zero decibel for the source curve
is about 50 db *re* 0.0002 microbar and for the transmission curve is about 78 db r₀
0.0002 microbar. [*From Hunt, Beranek, and Maa, Analysis of Sound Decay i:
Rectangular Rooms, J. Acoust. Soc. Amer.,* **11**: 80–94 (1939)].

lower than at the other $l_x l_z$ wall and the traveling wave would undergo a
phase shift as it reflected from the absorbing surface.

All normal modes of vibration cannot be excited to their fullest extent
by a sound source placed at other than a maximum pressure point in the
room.   In Fig. 10.2, for example, the source of sound can excite only a
normal mode to its fullest extent if it is at a 1.0 contour.   Obviously,
since the peak value of sound pressure occurs on a 1.0 contour, the
microphone also must be located on a 1.0 contour to measure the maximum pressure.

If the source is at a corner of a rectangular room, it will be possible for
it to excite every mode of vibration to its fullest extent provided it radi-

ates sound energy at every normal frequency.   Similarly, if a microphone is at the corner of the room, it will measure the peak sound pressure for every normal mode of vibration provided the mode is excited.

If either the source or the microphone is at the center of a rectangular room, only one-eighth of the normal modes of vibration will be excited or detected, because at the center of the room seven-eighths of the modes have contours of zero pressure.   In Fig. 10.2, as an illustration, two out of the three normal modes portrayed have contours of zero pressure at the center of the room.   In fact, only those modes of vibration having even

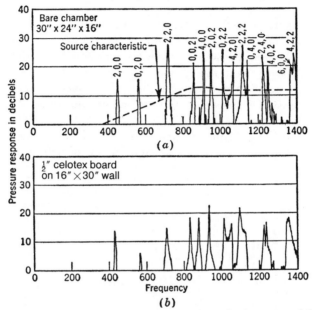

FIG. 10.5. Same as Fig. 10.4, except that the source was in the center of the room and the zero decibel reference for the source characteristic is about 71 db re 0.0002 microbar.

numbers simultaneously for n_x, n_y, and n_z will not have zero sound pressure at the center.

Examples of the transmission of sound from a small loudspeaker to a miniature microphone in a model sound chamber are shown in Figs. 10.4 and 10.5.   The curves were obtained by slowly varying the frequency (a pure tone) of the loudspeaker and simultaneously recording the output of the microphone.   The eightfold increase in the number of modes of vibration that were excited with the source at the corner over that with the source at the center is apparent.   It is apparent also that the addition of sound-absorbing material decreases the height of resonance peaks and smooths the transmission curve, particularly at the higher frequencies, where the sound-absorbing material is most effective.

10.4. Steady-state and Transient Sound Pressures.   When a source of sound is turned on in a small enclosure, such as that of Fig. 10.1, it will excite one or more of the stationary-wave possibilities, *i.e.*, normal modes of vibration in the room.   Let us assume that the source is constant in strength and is of a single frequency and that its frequency coincides with one of the normal frequencies of the enclosure.   The sound pressure for that normal mode of vibration will build up until the magnitude of its rms value (averaged in time and also in space by moving the microphone backward and forward over a wavelength) equals[4]

$$|p_n| = \frac{K}{k_n} \tag{10.11}$$

where K = source constant determined principally by the strength and location of the source and by the volume of the room.

  k_n = damping constant determined principally by the amount of absorption in the room and by the volume of the room.   The more absorbing material that is introduced into the room, the greater k_n becomes, and the smaller the value of the average pressure.

When the driving frequency does not coincide with the normal frequency, the pressure for that particular mode of vibration builds up

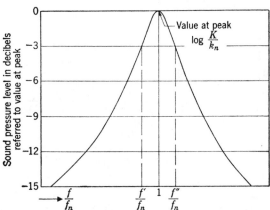

FIG. 10.6. Resonance curve for a normal mode of vibration.   Sound pressure level vs. the ratio of frequency to f_n.

according to a standard resonance curve as shown in Fig. 10.6.   The width of the resonance curve at the half-power (3 db down) points is equal to[5]

$$f'' - f' = \frac{k_n}{\pi} \tag{10.12}$$

[4] F. V. Hunt, L. L. Beranek, and D. Y. Maa, Analysis of Sound Decay in Rectangular Rooms, *J. Acoust. Soc. Amer.*, **11**: 80–94 (1939).

[5] L. L. Beranek, "Acoustic Measurements," pp. 329–336, John Wiley & Sons, Inc., New York, 1949.

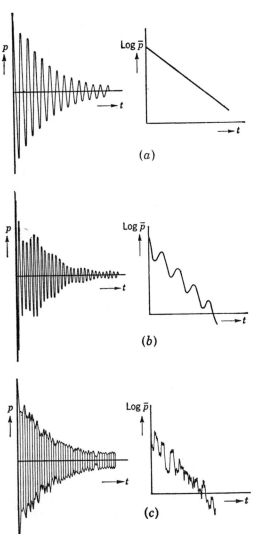

FIG. 10.7. (a) Sound-pressure decay curve for a single mode of vibration. (b) Sound-pressure decay curves for two closely spaced modes of vibration with the same decay constant. (c) Sound-pressure decay curves for a number of closely spaced modes of vibration with the same decay constant. The graph on the left shows the course of the instantaneous sound pressure, and that on the right shows the curve of the envelope of the left graph, plotted in a log $\bar{p}$ vs. t coordinate system.

The magnitude of the sound pressure p_n is given by

$$|p_n| = \frac{2K\omega}{\sqrt{4\omega_n^2 k_n^2 + (\omega^2 - \omega_n^2)^2}} \tag{10.13}$$

where ω is the angular driving frequency and ω_n is the angular normal frequency given approximately by Eq. (10.5).

Obviously, if the driving frequency lies between two normal frequencies, or if k_n is large so that the resonance curve is broad, more than one normal mode of vibration will be excited significantly, each to the extent shown by Eq. (10.13).

When the source of sound is turned off, each normal mode of vibration behaves like an electrical parallel-resonance circuit in which energy has been stored initially. The pressure for each normal mode of vibration will decay exponentially at its own normal frequency as shown in Fig. 10.7.

If only one mode of vibration is excited, the decay is as shown in Fig. 10.7a and as given by

$$p_n = \frac{K}{k_n} e^{-k_n t} \cos \omega_n t \tag{10.14}$$

Stated differently, on a log p_n scale vs. time, the magnitude of the rms sound pressure level decays linearly with time.

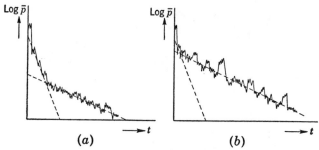

Fig. 10.8. Decay curves with double slopes produced by normal modes of vibration with different decay constants.

If two or more modes of vibration are decaying simultaneously, beats will occur because each has its own normal frequency (Fig. 10.7b and c). Also, it is very possible that each will have its own decay constant, dependent upon the position of the absorbing materials in the room. In that case the magnitude of the sound-pressure-level curves will decay with two or more slopes as shown in Fig. 10.8.

In summary, we see that when a sound source of a given frequency is placed in a rectangular enclosure, it will excite one or more of the infinity of resonance conditions, called normal modes of vibration. Each of those normal modes of vibration has a different distribution of sound pressures in the enclosure, its own normal frequency, and its own damping constant. The damping constant determines the maximum height and the width of the steady-state sound-pressure resonance curve.

In addition, when the source of sound is turned off, the sound pressure associated with each mode of vibration decays exponentially with its own normal frequency and at a rate determined by its damping constant. The room is thus an assemblage of resonators that act independently of each other when the sound source is turned off. The larger the room and the higher the frequency, the nearer together will be the normal frequencies and the larger will be the number of modes of vibration excited by a single-frequency source or by a source with a narrow band of frequencies.

10.5. Examples of Rectangular Enclosures

Example 10.1. Determine the normal frequencies and directional cosines for the lowest six normal modes of vibration in a room with dimensions 20 by 14 by 8 ft.

Solution. From Eq. (10.5) we see that

$$f_{1,0,0} = 113\tfrac{2}{2} \times \tfrac{1}{20} = 28.3 \text{ cps}$$
$$f_{0,1,0} = 113\tfrac{2}{2} \times \tfrac{1}{14} = 40.4 \text{ cps}$$
$$f_{1,1,0} = 113\tfrac{2}{2} \sqrt{\tfrac{1}{400} + \tfrac{1}{196}} = 49.3 \text{ cps}$$
$$f_{2,0,0} = 113\tfrac{2}{2} \times \tfrac{2}{20} = 56.6 \text{ cps}$$
$$f_{2,1,0} = 113\tfrac{2}{2} \sqrt{\tfrac{4}{400} + \tfrac{1}{196}} = 69.6 \text{ cps}$$
$$f_{0,0,1} = 113\tfrac{2}{2} \times \tfrac{1}{8} = 70.7 \text{ cps}$$

From Eqs. (10.8) to (10.10) we find the direction cosines as follows:

For (1,0,0) mode: $\theta_x = 0$; $\theta_y = 90°$; $\theta_z = 90°$

$\quad$ (0,1,0) mode: $\theta_x = 90°$; $\theta_y = 0°$; $\theta_z = 90°$

$\quad$ (1,1,0) mode: $\theta_x = \cos^{-1} \dfrac{566}{20 \times 49.3} = 55°$

$$\theta_y = \cos^{-1} \dfrac{566}{14 \times 49.3} = 35°$$
$$\theta_z = 90°$$

$\quad$ (2,0,0) mode: $\theta_x = 0$; $\theta_y = 90°$; $\theta_z = 90°$

$\quad$ (2,1,0) mode: $\theta_x = \cos^{-1} \dfrac{1132}{20 \times 69.6} = 35.5°$

$$\theta_y = \cos^{-1} \dfrac{566}{14 \times 69.6} = 54.5°$$
$$\theta_z = 90°$$

$\quad$ (0,0,1) mode: $\theta_x = 90°$; $\theta_y = 90°$; $\theta_z = 0°$

Example 10.2. If a source with a frequency of 50 cps is used to excite the room of the previous example, what will be the relative pressure amplitudes of the six lowest modes of vibration? Assume that when the frequency of the source equals any one of the normal frequencies, the peak amplitude for that normal mode of vibration will be the same as for any other normal mode of vibration. Also, assume that k_n equals 2.0 sec^{-1}.

Solution. We find from Eq. (10.13) that

$$|p_n|_{(1,0,0)} = \frac{2K \cdot 314}{\sqrt{(4)(178)^2(4) + (314^2 - 178^2)^2}} = 0.0094K$$

$$|p_n|_{(0,1,0)} = \frac{2K \cdot 314}{\sqrt{(4)(254)^2(4) + (314^2 - 254^2)^2}} = 0.01835K$$

$$|p_n|_{(1,1,0)} = \frac{2K \cdot 314}{\sqrt{(4)(310)^2(4) + (314^2 - 310^2)^2}} = 0.225K$$

$$|p_n|_{(2,0,0)} = \frac{2K \cdot 314}{\sqrt{(4)(356)^2(4) + (314^2 - 356^2)^2}} = 0.0224K$$

$$|p_n|_{(2,1,0)} = \frac{2K \cdot 314}{\sqrt{(4)(437)^2(4) + (314^2 - 437^2)^2}} = 0.00678K$$

$$|p_n|_{(0,0,1)} = \frac{2K \cdot 314}{\sqrt{(4)(444)^2(4) + (314^2 - 444^2)^2}} = 0.00637K$$

Inspection of Eq. (10.13) shows that if $\omega_n = \omega$ and $k_n = 2$, then $|p_n| = 0.5K$. Hence, expressed in decibels, the magnitude in decibels of the six normal modes of vibration given above, relative to $0.5K$, is

$$(1,0,0): -34.5 \text{ db}$$
$$(0,1,0): -28.7 \text{ db}$$
$$(1,1,0): -6.9 \text{ db}$$
$$(2,0,0): -27.0 \text{ db}$$
$$(2,1,0): -37.4 \text{ db}$$
$$(0,0,1): -37.9 \text{ db}$$

It is obvious that for a k_n as small as 2 and for the lowest normal frequencies only the normal mode of vibration located nearest the driving frequency receives appreciable excitation.

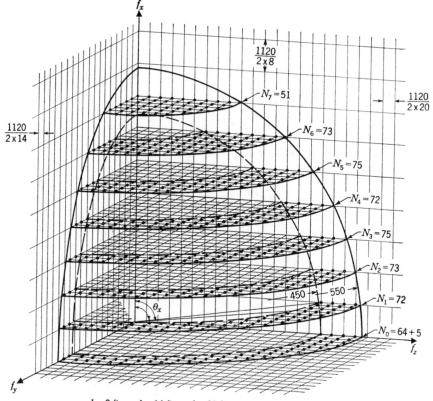

$$l_x = 8 \text{ ft.} \qquad l_y = 14 \text{ ft.} \qquad l_z = 20 \text{ ft.}$$
Frequency band 450-550 CPS

FIG. 10.9. Normal frequency diagram, drawn to scale for a 20 by 14 by 8 ft rectangular room with hard walls. Most of the vertical lines are omitted to avoid confusion. [*After Hunt, Beranek, and Maa*,[4] *Analysis of Sound Decay in Rectangular Rooms, J. Acoust. Soc. Amer.*, **11**: 80–94 (1939).]

Example 10.3. A rectangular room with dimensions $l_x = 8$ ft, $l_y = 14$ ft, and $l_z = 20$ ft is excited by a sound source located in one corner of the room. The sound pressure level developed is measured at another corner of the room. The sound source produces a continuous band of frequencies between 450 and 550 cps, with a uniform spectrum level, and a total acoustic-power output of 1 watt. When the sound source is turned off, a linear decay curve (log p vs. t) is obtained which has a slope of 30 db/sec. (a) Determine graphically the number of normal modes of vibration excited by the source; (b) determine the approximate angle of incidence of the traveling-wave field involving the walls at $x = 0$ and $x = l_x$ in each of the principal groupings of normal frequencies shown in the graphical construction; (c) determine the reverberation time T; (d) determine the decay constant k; (e) determine the width of the resonance curve at the one-half power (3 db down) points; (f) determine the sound pressure level at the resonance peak for each mode of vibration, assuming equal division of energy among the modes of vibration and that K in Eq. (10.11) equals $U_0 P_0/V$, where U_0 is the rms strength of the source defined on page 93 of Part X.

Solution. a. A graphical solution to Eq. (10.5) is given in Fig. 10.9. The frequency of any given normal mode of vibration is the distance from the origin of coordinates to one of the black spheres shown. That frequency will be made up of three components given by $cn_x/2l_x$, $cn_y/2l_y$, and $cn_z/2l_z$. Notice that along the vertical coordinate the normal frequencies occur in increments of $1120/16$; along the right-hand axis in increments of $1120/40$; and along the remaining axis in increments of $1120/28$. On the layer labeled N_0, there are 69 normal frequencies. The total number of normal frequencies between 450 and 550 cps for this room is 560. The average frequency is 500 cps.

b. The θ_x angles of incidence can be divided into eight principal groups as shown in Fig. 10.9. The angles are as follows:

$$\theta_x(0,n_y,n_z) \doteq 90°$$
$$\theta_x(1,n_y,n_z) \doteq \cos^{-1}(1120/16 \cdot 1/500) \doteq 82°$$
$$\theta_x(2,n_y,n_z) \doteq \cos^{-1}\left(\frac{2 \cdot 1120}{16 \cdot 500}\right) \doteq 74°$$
$$\theta_x(3,n_y,n_z) \doteq \cos^{-1}(0.42) \doteq 65°$$
$$\theta_x(4,n_y,n_z) \doteq \cos^{-1}(0.56) \doteq 56°$$
$$\theta_x(5,n_y,n_z) \doteq \cos^{-1}(0.70) \doteq 45°$$
$$\theta_x(6,n_y,n_z) \doteq \cos^{-1}(0.84) \doteq 33°$$
$$\theta_x(7,n_y,n_z) \doteq \cos^{-1}(0.98) \doteq 11°$$

c. Decay rate = 30 db/sec.

$$T = \text{time for 60 db of decay (see Part XXIV)} = 2 \text{ sec.}$$

d. $k = 6.91/T$ (see Eq. 10.32) = 3.46 sec^{-1}

e. $f'' - f' = k_n/\pi = 1.1$ cps.

f. The power supplied to each mode of vibration $W_n = 1/560 = 0.00179$ watt. From Eq. (4.4) we see that, for a simple source, U_0 is related to $W_n = 4\pi r^2 I$, by

$$U_0^2 = \frac{W_n c}{\pi f^2 \rho_0} = \frac{(0.00179)(344.8)}{\pi(500)^2(1.18)}$$

or

$$U_0 = 0.000816 \text{ m/sec}$$
$$p_n = \frac{U_0 \gamma P_0}{k_n V} = \frac{8.16 \times 10^{-4} \times 1.4 \times 10^5}{3.46 \times (8 \times 14 \times 20) \times 0.0283}$$
$$= 0.521 \text{ newton/m}^2$$
$$\text{SPL} = 20 \log_{10} \frac{0.521}{2 \times 10^{-5}} = 88.3 \text{ db}$$

PART **XXIV** *Sound Fields in Large Irregularly Shaped Enclosures*

10.6. Introduction. Large irregularly shaped enclosures also have normal modes of vibration that respond when driven at their normal frequencies. However, their number is so large and their pressure distribution so complex that when a source of sound excites the room it sets up standing waves that involve each wall at nearly *all* angles of incidence, even if the frequencies in the source are bunched in a narrow frequency band. Also, at any point in the room, sound waves are traveling in all directions, so that we can speak of the sound field in the room as being a *diffuse sound field*.

If a source with a narrow band of many frequencies excites a large irregular room, we observe fluctuations in the sound pressure as a microphone is moved about the room, just as we would for the simple case of a small regular room. However, for the large irregular room, the maxima and minima in sound pressure lie very much nearer to each other in position in the room than would be the case for a regular room, and it is a relatively simple matter to move the microphone backward and forward and thereby to obtain a satisfactory space average of the sound pressure.

10.7. Energy-density Sound-pressure Relation. If the microphone reads the effective (rms) sound pressure level, averaged in space by moving the microphone backward and forward over a wavelength distance, we may use Eq. (2.83) to give us the *sound energy density*,

$$D_{avg} = \frac{|p_{avg}|^2}{\rho_0 c^2} \qquad \text{watt-sec/m}^3 \qquad (10.15)$$

where $|p_{avg}|$ = rms magnitude of the sound pressure averaged in space and in time in newtons per square meter.

$\rho_0 c$ = characteristic impedance of air in mks rayls. At normal temperature and pressure, $\rho_0 c = 407$ mks rayls.

10.8. Mean Free Path. For a large irregular room, we can visualize the acoustical conditions by imagining a wave traveling around inside the room. This wave travels in a straight line until it strikes a surface. It is then reflected off the surface at an angle equal to the angle of incidence and travels in this new direction until it strikes another surface. Because sound travels about 1130 ft/sec, many reflections will occur during each second.

From a statistical standpoint, we define a *mean free path* as the average distance a sound wave travels in a room between reflections from the bounding surfaces.

Knudsen[6] determined experimentally the mean free path in 11 very differently shaped large enclosures and found it to be very nearly

$$\text{Mean free path} = d = \frac{4V}{S} \quad \text{m (or ft)} \quad (10.16)$$

where V is the volume of the room in cubic meters (or cubic feet) and S is the total area of the surfaces of the room in square meters (or square feet) including the floor but not the area of loose items of furniture. This equation can also be derived theoretically.

10.9. Sound (Energy) Absorption Coefficients, α_n and $\bar{\alpha}$. The sound energy density of a wave traversing a room is given by Eq. (10.15) in terms of the sound pressure. After this wave has undergone a reflection from a wall that is absorbing, the energy density will be less during its next traverse of the room.

Let us give to each reflecting surface a sound absorption coefficient α_n defined as the ratio of the energy absorbed by the surface to the energy incident upon the surface. As the wave travels around the room, it involves the surfaces at various angles of incidence. In a large irregular room the number of waves traveling are so numerous that at each surface all directions of incident flow are equally probable. The sound absorption coefficient is, therefore, taken to be averaged for all angles of incidence. The sound absorption coefficients given in published tables are measured generally under as nearly these conditions as possible. Hence, our α_n is obtainable directly from official publications.[7,8]

In addition to averaging the absorption coefficients for all angles of incidence on one surface, we shall average the absorption coefficients for different surfaces in the room, weighted according to the area of each, as follows:

$$\bar{\alpha} = \frac{S_1\alpha_1 + S_2\alpha_2 + S_3\alpha_3 + \cdots + S_n\alpha_n}{S} \quad (10.17)$$

$$S = S_1 + S_2 + S_3 + \cdots + S_n \quad (10.18)$$

where S_1, S_2, S_3, . . . are the areas of particular absorbing surfaces in square meters or in square feet; α_1, α_2, α_3, . . . are the absorption coefficients associated respectively with those areas; and $\bar{\alpha}$ *is the average sound absorption coefficient for the room as a whole.*

If the conditions of this analysis, *viz.*, that sound waves of nearly equal energy density are traveling equally probably in all directions, are to

[6] V. O. Knudsen, "Architectural Acoustics," pp. 132–141, John Wiley & Sons, Inc., New York, 1932.

[7] "Sound Absorption Coefficients of Architectural Acoustical Materials," *Acoust. Materials Assoc. Bull.* 14, 1953, New York, N.Y.

[8] "Sound Absorption Coefficients of the More Common Acoustical Materials," *Natl. Bur. Standards Bull., Letter Circ.* LC870,

TABLE 10.1.   Coefficients of General Building Materials†

Building material	Thick-ness, in.	Coefficients					
		125	250	500	1000	2000	4000
Brick wall, unpainted...............	18	0.02	0.02	0.03	0.04	0.05	0.05
Brick wall, painted.................	18	0.01	0.01	0.02	0.02	0.02	0.02
Plaster, gypsum, on hollow tile, plain or painted.......................	.....	0.02	0.02	0.02	0.03	0.04	0.04
Plaster, gypsum, scratch and brown coats on metal lath, on wood studs...	.....	0.04	0.04	0.04	0.06	0.06	0.03
Plaster, lime, sand finish on metal lath.	¾	0.04	0.05	0.06	0.08	0.04	0.06
Plaster, on wood wool..............	.....	0.40	0.30	0.20	0.15	0.10	0.10
Plaster, fibrous....................	2	0.35	0.30	0.20	0.55	0.10	0.04
Poured concrete, unpainted..........	.....	0.01	0.01	0.02	0.02	0.02	0.03
Poured concrete, painted............	.....	0.01	0.01	0.01	0.02	0.02	0.02
Wood, solid and polished............	2	0.1	....	0.05	....	0.04	0.04
Wood, paneling, 2 to 4 in. air space behind.........................	⅜–½	0.30	0.25	0.20	0.17	0.15	0.10
Wood platform with large space beneath............................	.....	0.4	0.3	0.2	0.17	0.15	0.1
Glass..............................	.....	0.04	0.04	0.03	0.03	0.02	0.02
Floors:							
Slate on solid....................	.....	0.01	0.01	0.01	0.02	0.02	0.02
Wood on solid....................	.....	0.04	0.04	0.03	0.03	0.03	0.02
Cork, linoleum, gypsum, or rubber tile on solid....................	3⁄16	0.04	0.03	0.04	0.04	0.03	0.02
Wood block, pitch pine............	.....	0.05	0.03	0.06	0.09	0.10	0.22
Carpets:							
Wool pile, with underpad..........	⅝	0.20	0.25	0.35	0.40	0.50	0.75
Wool pile, on concrete.............	⅜	0.09	0.08	0.21	0.26	0.27	0.37
Draperies and fabrics:							
Velour, hung straight							
10 oz/yd².....................	.....	0.04	0.05	0.11	0.18	0.30	0.35
14 oz/yd².....................	.....	0.05	0.07	0.13	0.22	0.32	0.35
18 oz/yd².....................	.....	0.05	0.12	0.35	0.48	0.38	0.36
Velour, draped to half area							
14 oz/yd².....................	.....	0.07	0.31	0.49	0.75	0.70	0.60
18 oz/yd².....................	.....	0.14	0.35	0.55	0.75	0.70	0.60
Seats and people ($\alpha_n S$ in square feet per person or per seat):							
Seats							
Chair: upholstered back, leather seats.......................	.....	2.0	2.5	3.0	3.0	3.0	2.5
Chair: theater, heavily upholstered.......................	.....	3.5	3.5	3.5	3.5	3.5	3.5
Orchestra chairs, wood..........	.....	0.1	0.15	0.2	0.35	0.5	0.6
Cushions for pews, per person....	1.5	1.0	1.5	1.7	1.7	1.6	1.4

† Collected and averaged by author from published data. Numerous inconsistencies in the published literature make accurate numbers impossible.

TABLE 10.1. Coefficients of General Building Materials.—(*Continued*)

Building material	Thick-ness, in.	Coefficients					
		125	250	500	1000	2000	4000
People							
In upholstered seats (add to leather-seat chair absorption)...	.....	0.7	0.6	0.5	1.3	1.6	2.0
In heavily upholstered seats......	.....	0.7	0.6	0.6	1.0	1.0	1.0
In orchestra seats with instruments (add to wood-seat absorption).....................	.....	4.0	7.5	11.0	13.0	13.5	11.0
Child in high school, seated, including seat.................	.....	2.2	3.0	3.3	4.0	4.4	4.5
Child in elementary school, seated, including seat...............	.....	1.8	2.3	2.8	3.2	3.5	4.0
Standing......................	.....	2.0	3.5	4.7	4.5	5.0	4.0
In church pew (no seat cushion)..	.....	2.5	2.7	3.3	3.8	4.0	3.8

hold, it is necessary that no one part of the room has a preponderance of the absorbing material. Equation (10.17) is written on the assumption either that $\bar{\alpha}$ is very small or that all surfaces have nearly the same absorption coefficients. It is common design practice for satisfactory listening conditions that the absorbing materials should not be concentrated on surfaces in one part of the room. Furthermore, it is unlikely that a diffuse sound field exists in a room if $\bar{\alpha}$ becomes much greater than 0.3, because the traveling waves die out very rapidly.

Note that the absorption coefficient α_n is actually a measure of the absorbing power of unit area (square meter or square foot) of a bounding surface in the room. If there is an open window in the room, all the energy incident on its open area will pass outdoors and none will be reflected. Hence, for an open window, $\alpha_n \doteq 1.0$.[†] The absorption of an area of acoustical material or another surface of a room is, therefore, expressed in terms of the equivalent area of open window. For example, 10 ft² of wallboard may absorb the same percentage of incident sound energy as 4 ft² of open window. Hence, α_n for the wallboard equals 0.4.

All materials have absorption coefficients that are different at different frequencies. For a complete specification of the absorbing properties of a material, a curve of α as a function of frequency would need to be drawn. It is standard practice to specify sound absorption coefficients in manu-

[†] This statement is strictly true only if the window is several wavelengths or more wide and high. When its dimensions are comparable with or smaller than a wavelength, its absorption coefficient will appear to be greater or less than 1.0 because of the effects of diffraction.

facturers' literature at 125, 250, 500, 1000, 2000, and 4000 cps.   In discussing the performance of a material in a cursory way, it is usual to specify its performance at 500 cps.

The acoustical effect of people, chairs, seats, tables, desks, and so forth, which are objects that are not part of the bounding surfaces of the room, must be considered also.   It is customary to assign an $S\alpha_n$ number to each person or thing and to add it into the numerator of Eq. (10.17). For example, if 20 seated persons are in a room, and if the $S\alpha_n$ number for each seated person at a particular frequency is 4, then 80 units must be added into the numerator of Eq. (10.17).   No modification is made of the total area S.

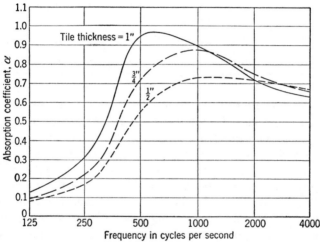

FIG. 10.10. Average absorption characteristics of perforated tiles cemented directly to hard backing.   The number of holes in the perforated face is between 400 and 500 per square foot and their diameter is between $\frac{5}{32}$ and $\frac{3}{16}$ in.

For use in solving problems, typical values of absorption coefficients for common materials and for seats and people as measured in reverberation chambers[9] are given in Table 10.1.   Graphs of the absorption coefficients for typical acoustical tiles with faces that are perforated with approximately 450 holes per square foot, each hole being $\frac{3}{16}$ in. in diameter, are given in Figs. 10.10 and 10.11.   In Fig. 10.12 we show the effect of varying the hole size and the spacing of holes in a metal sheet used to cover 3 in. of rock wool.   Typical data for porous, homogeneous tiles of various thicknesses are shown in Fig. 10.13.   Detailed data on commercially available materials may be obtained from the sources given in references 7 and 8.

[9] Beranek, *op. cit.*, pp. 860–869.

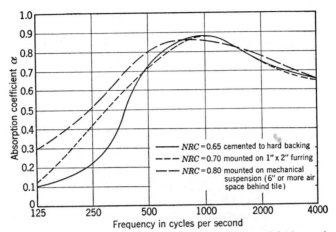

FIG. 10.11. Effect of mounting on noise-reduction coefficient of ¾-in. perforated tile. The quantity NRC is the noise-reduction coefficient obtained by averaging the coefficients at 250, 500, 1000, and 2000 cps, inclusive, to the nearest 0.05.

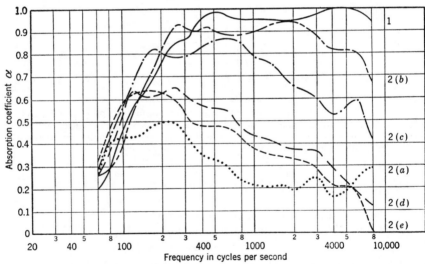

FIG. 10.12. Variation of absorption with spacing of holes in a paintable sheet metal cover.

1. 3-in. rock wool, uncovered.

2. 3-in. rock wool covered with 22 Bwg steel sheet (*a*) unperforated; (*b*) ⅛-in. holes at ⁵⁄₁₆-in. centers; (*c*) ⅛-in. holes at ⅝-in. centers; (*d*) ⅛-in. holes at 1¼-in. centers; (*e*) ⅛-in. holes at 1⅞-in. centers. (*From National Physical Laboratory, England.*)

10.10. Decay Rate. The decay rate of the sound in a large irregular room may be found easily.[10,11] Assume that the sound source has just stopped and that the sound is decaying in the room.   Let

$$t' = \frac{d}{c} = \frac{4V}{cS} \quad \text{sec} \tag{10.19}$$

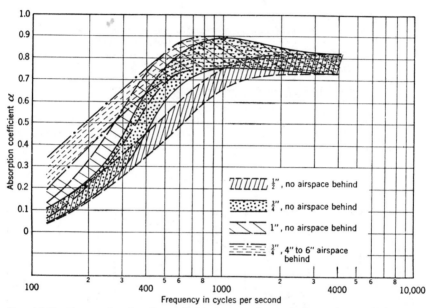

Fɪɢ. 10.13. Curves showing the probable ranges of sound-absorption coefficients for plain or fissured porous acoustic tiles of various thicknesses mounted in relation to a rigid wall as indicated. Measured published values will usually lie within the spreads indicated, although important exceptions can be found.

where t' = time in seconds for the sound wave to travel one mean free
       path
    c = speed of sound in meters per second
    $d = 4V/S$ = mean free path in meters [see Eq. (10.16)]

Assume that the initial sound energy density was D' watt-sec/m³. Then after nt' sec, the sound wave will have undergone n reflections.   On each reflection the energy density at that instant will be reduced by $\bar{\alpha}$ (see

[10] W. C. Sabine, "Collected Papers on Acoustics," pp. 43–45, Harvard University Press, Cambridge, Mass., 1927 (out of print).

[11] R. F. Norris, "A Derivation of the Reverberation Formula," published in Appendix II of V. O. Knudsen, "Architectural Acoustics," John Wiley & Sons, Inc., New York, 1932.

Fig. 10.14). Hence, the energy density D after each successive reflection is

$$D(t') = D'(1 - \bar{\alpha})$$
$$D(2t') = D'(1 - \bar{\alpha})^2$$
$$D(nt') = D'(1 - \bar{\alpha})^n \quad (10.20)$$

But

$$n = \frac{t}{t'} \quad (10.21)$$

so that, from Eq. (10.19),

$$D(t) = D'(1 - \bar{\alpha})^{(cS/4V)t} \quad (10.22)$$

Now,

$$1 - \bar{\alpha} \equiv e^{\log_e (1-\bar{\alpha})} \quad (10.23)$$

Then,

$$D(t) = D'e^{-(cS/4V)[-\log_e (1-\bar{\alpha})]t} \quad (10.24)$$

where $D(t)$ is the energy density after time t.

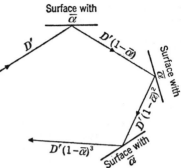

FIG. 10.14. Path of a sound wave with energy density D' as it reflects from surfaces with average absorption coefficient $\bar{\alpha}$. The surfaces are spaced a mean-free-path distance $d = 4V/S$ apart.

If we convert this to sound pressure level with the aid of Eqs. (10.15) and (1.18) we get,†

$$10 \log_{10} \left[\frac{|p_{\text{avg}}|^2}{(0.00002)^2} \right]_{t=t} = 10 \log_{10} \left[\frac{|p_{\text{avg}}|^2}{(0.00002)^2} \right]_{t=0}$$
$$- 4.34 \frac{cS}{4V} [-\log_e (1 - \bar{\alpha})]t \quad \text{db} \quad (10.25)$$

$$\text{SPL}_{t=t} - \text{SPL}_{t=0} = 1.085 \frac{cS}{V} [-2.30 \log_{10} (1 - \bar{\alpha})]t \quad \text{db} \quad (10.26)$$

Hence, the sound pressure level decays at the rate of

$$1.085 \frac{cS}{V} [-2.30 \log_{10} (1 - \bar{\alpha})] \quad \text{db/sec} \quad (10.27)$$

Metric Absorption Units. We shall define arbitrarily a constant a' as

$$a' \equiv S[-2.30 \log_{10} (1 - \bar{\alpha})]$$
$$\equiv \text{metric absorption units in m}^2 \quad (10.28)$$

where S is given by Eq. (10.18).

Absorption Units in Sabins. For the special case of English units, the quantity a' of Eq. (10.28) is called a, the number of *absorption units in sabins.* The quantity a has the dimensions of square feet.

In practice, $\bar{\alpha}$ is found from Eq. (10.17), and $[-2.30 \log_{10} (1 - \bar{\alpha})]$ is found from $\bar{\alpha}$ using the chart given in Fig. 10.15. Then a is obtained by multiplying S of Eq. (10.18) in square feet by this result.

† Note that $\log_{10} e^x = 0.434x$; $\log_e 10^x = 2.30x$; and $\log_e y = 2.30 \log_{10} y$.

10.11 Boundary Absorption Only. *Reverberation Time T.* The reverberation time T is defined as the time required for the sound energy density to decay 60 db, that is, to 10^{-6} of its original value. From Eqs. (10.27) and (10.28), we see that the reverberation time is

$$T = \frac{60V}{1.085ca'} \quad \text{sec} \tag{10.29}$$

where V = volume of room in cubic meters
       c = speed of sound in meters per second
       a' = metric absorption units in square meters

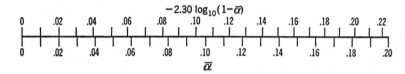

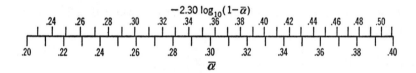

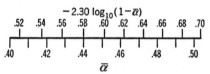

Fig. 10.15. Chart giving $-2.30 \log_{10}(1 - \bar{\alpha})$ as a function of $\bar{\alpha}$.

Metric Units. For metric units, at normal temperature (22°C or 71.6°F), the reverberation time is

$$T = 0.161 \frac{V}{a'} \quad \text{sec} \tag{10.30}$$

English Units. For English units, at normal temperature (22°C or 71.6°F), the reverberation time is

$$T = 0.049 \frac{V}{a} \quad \text{sec} \tag{10.31}$$

where V = volume of room in cubic feet
       a = absorption units in sabins (square feet)

The reverberation time T, as a function of the volume of the room and of the $S\bar{\alpha}$ product for the room [see Eq. (10.17)], can be found directly from Fig. 10.16, for rooms whose maximum dimension is less than five times the minimum dimension. Use of this chart obviates the use of Fig. 10.15, as the relation between the $S\bar{\alpha}$ product and a is incorporated automatically in the formula to within 5 per cent accuracy.

Decay Constant k_n. As an added item of interest, we should note that if the exponent of Eq. (10.24), which gives the rate of decay of the sound

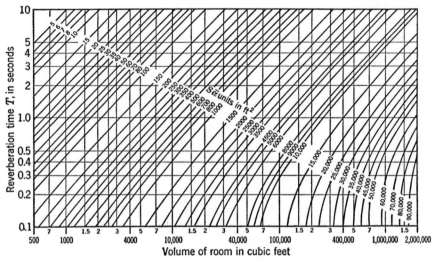

Fig. 10.16. Relations among reverberation time, volume, and $S\bar{\alpha}$ units in square feet. The $-2.30 \log_{10}(1 - \bar{\alpha})$ correction is incorporated; maximum error in reverberation time is less than 5 per cent for rooms with maximum dimension less than five times minimum dimension.

energy density, is equated to twice the exponent of Eq. (10.14) (that is, $2k_n$), which gives the same quantity, we get

$$2k_n = \frac{cS}{4V}[-\log_e(1 - \bar{\alpha})] = \frac{ca'}{4V}$$

or, from Eq. (10.29),

$$k_n = \frac{15}{(2)(1.085T)} = \frac{6.91}{T} \qquad \text{sec}^{-1} \qquad (10.32)$$

In other words, if the sound pressure level decays linearly with time over a range of 60 db, then all significant modes of vibration have the same decay constant k_n and this constant is equal to 6.91 divided by the reverberation time T.

10.12. Air and Boundary Absorption. Sound is absorbed not only at the boundaries of a room but also in the air itself. If the room is small, the number of reflections from the boundaries in each second is large and the time the wave spends in the air between reflections is correspondingly small. In this case, the air absorption is generally not important. In a very large auditorium or church, the time a wave spends in the air between reflections becomes large so that the absorption of energy in the air itself can be neglected no longer.

The reverberation equations (10.29) to (10.31) must be modified to take into account the air absorption if accurate results are required. This is necessary particularly at the higher frequencies (above 1000 cps), as we shall see shortly.

Referring back to Eq. (10.20), we see that each time a wave traverses the room, it experiences a reflection and the energy density is reduced by the absorption coefficient $\bar{\alpha}$. After the time t, the sound has experienced t/t' reflections, where t' is the time for the wave to travel a mean free path.

In the period that the wave is traversing a mean free path of distance, the energy density will decrease owing to air absorption alone by the relation

$$D(t') = D'e^{-md} \tag{10.33}$$

where $D(t')$ = energy density after the wave has traveled one mean free path in watt-seconds per cubic meter

D' = initial energy density at $t = 0$

m = energy attenuation constant in meters^{-1}

d = mean-free-path length in meters

After the time t, the wave will have experienced t/t' reflections and will have traveled a distance of td/t'. Hence, the energy density is given, with the aid of Eqs. (10.16), (10.19), and (10.22), by

$$D(t) = D'(1 - \bar{\alpha})^{(cS/4V)t}e^{-(cS/4V)(4V/S)mt} \tag{10.34}$$

or

$$D(t) = D'e^{-(cS/4V)[-\log_e (1-\bar{\alpha})-4Vm/S]t} \tag{10.35}$$

Reverberation Time. When converted to the time required for 60 db of decay, we obtain the reverberation equations.

METRIC UNITS.

$$T = \frac{60V}{1.085c(a' + 4mV)} \tag{10.36}$$

where T = reverberation time in seconds

V = volume of room in cubic meters

c = speed of sound in meters per second

a' = number of metric absorption units given by Eq. (10.28)

m = energy attenuation constant in meters^{-1} as given from Eqs. (10.38) and (10.39a), following.

ENGLISH UNITS. For English units, at normal temperature (22°C or 71.6°F), the reverberation time including the effects of air absorption is

$$T = 0.049 \frac{V}{a + 4mV} \tag{10.37}$$

where V = volume of room in cubic feet

a = absorption units in sabins (square feet)

m = energy attenuation constant in feet^{-1} as given from Eqs. (10.38) and (10.39b) or from Fig. (10.18)

Determination of Energy Attenuation Constant m. The quantity m, the energy attenuation constant, is made up of two parts. One part arises from the effects of viscosity and heat conduction in the gas and the other part from the effects of molecular absorption and dispersion in polyatomic gases involving an exchange of translational and vibrational energy between colliding molecules.

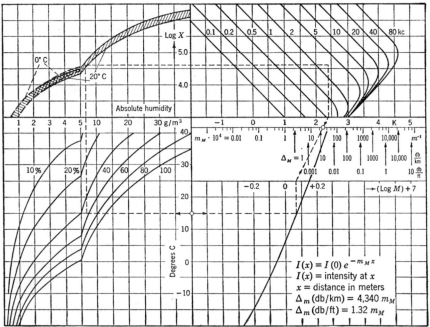

FIG. 10.17. Nomogram for determining the attenuation constant in air caused by molecular absorption, m_M.

For air, the first part is given by the equation

$$m_1 = 4.24 f^2 \times 10^{-11} \text{ m}^{-1} \tag{10.38}$$

The second part, m_m, is found from Fig. 10.17. To determine m_m from Fig. 10.17, the following procedure is used:

1. Begin with the proper temperature in degrees centigrade and move *left* from the lower center line of the nomogram (15°C in the example shown).

2. Stop the line at the proper relative humidity (50% R.H. in the example shown).

3. Move vertically to the lower edge of the shaded region.   Then move right to the proper frequency contour (3000 cps in the example shown).

4. Then move downward to the K scale ($K = 2.4$ in the example shown).

5. Start with the temperature again, and move to the *right* to the lower right curve and thence upward to the log $M + 7$ scale (in the example shown, log $M + 7 = +0.07$).

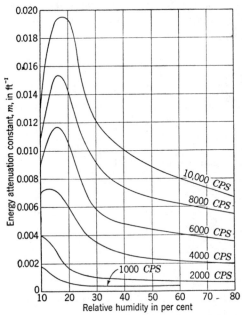

FIG. 10.18. Measured values of the energy attenuation constant m as a function of relative humidity for different frequencies, $I(x) = I_0 \exp(-mx)$. The temperature is assumed to be about 68°F. (*After Knudsen and Harris,*[12] "*Acoustical Designing in Architecture,*" p. 160, Fig. 8.10, John Wiley & Sons, Inc., New York, 1950.)

6. Then join the end points of the two tracings with a straight line (the value of m_m for the example shown is 0.003 m^{-1}).

The value of m is equal to the sum of the two parts,

$$m = m_1 + m_m \qquad \text{m}^{-1} \qquad (10.39a)$$

or,

$$m = 0.305(m_1 + m_m) \qquad \text{ft}^{-1} \qquad (10.39b)$$

where m_1 is given by Eq. (10.38) and m_m is given by Fig. 10.17, both in metric units.

At normal room temperatures ($F \doteq 68°F$), *measured* quantities equivalent to Eq. (10.39b) are plotted in Fig. 10.18 as a function of relative humidity in per cent.[12]   These quantities are higher than those calculated

[12] V. O. Knudsen and C. M. Harris, "Acoustical Designing in Architecture," p. 160, Fig. 8.10, John Wiley & Sons, Inc., New York, 1950.

from Eq. (10.39b) for reasons that are not fully understood. It is common practice to use the measured values in solutions of practical problems.

As incidental information, the number of decibels that the wave is attenuated in traveling through the air for each unit of distance is

$$\Delta_m = 4.34m \quad \text{db/unit distance} \quad (10.40)$$

The reverberation time T is of prime importance in the design of auditoriums, concert halls, churches, broadcast studios, and rooms. We shall tell in Chap. 13 how T should be chosen for any enclosure.

10.13. Reverberant Steady-state Sound Energy Density. After the sound source is turned on, energy is supplied to the room by the source at a rate faster than it is absorbed at the boundary surfaces until a condition of equilibrium is achieved. In the equilibrium condition, the energy absorbed and the energy supplied by the source are equal.

It is convenient to divide the sound field into two distinct parts, the *direct sound field* and the *reverberant sound field*. By definition, the sound field is "direct" until just after the wave from the source has undergone its first reflection. The reverberant sound field comprises all sound waves after they have experienced their initial reflection.

The sound source radiates a power W watts into the room. Of this power, W_D will be absorbed by the wall at the time of the first reflection, and W_R will be absorbed during all subsequent reflections. Obviously $W = W_D + W_R$, neglecting absorption in the air for the moment.

Boundary Absorption Only. If the sound source is nondirectional and is centrally located in the room, the spherical sound wave emitted by it will undergo a reflection after it has traveled a mean-free-path distance. The walls from which it reflects have an average absorption coefficient $\bar{\alpha}$. Hence, the power supplied by the source associated with the direct wave equals the power absorbed on first reflection. This power is, by definition,

$$W_D = W\bar{\alpha} \quad (10.41)$$

and the power supplied by the source to the reverberant sound field is

$$W_R = W(1 - \bar{\alpha}) \quad (10.42)$$

Let us discuss the reverberant sound field first. The steady-state value of the reverberant energy density is D'. The number of reflections in 1 sec from surfaces of average absorption $\bar{\alpha}$ is equal to $1/t' = cS/4V$. Then the total reverberant energy per second removed from the room is

$$D'V \frac{\bar{\alpha}}{t'} = W_R$$

$$\frac{D'cS\bar{\alpha}}{4V} V = W(1 - \bar{\alpha})$$

or

$$D' = \frac{4W(1 - \bar{\alpha})}{cS\bar{\alpha}} \tag{10.43}$$

where D' = steady-state reverberant sound energy density in watt-seconds per cubic meter

W = total power supplied by the source in watts

V = volume of room in cubic meters

c = speed of sound in meters per second

S = area of surfaces of room in square meters

$\bar{\alpha}$ = average sound absorption coefficient

Combining Eqs. (10.15) and (10.43), we get

$$|p_{\text{avg}}|_R^2 = \frac{4\rho_0 c W}{R'} \tag{10.44}$$

where, by definition,

$$R' \equiv \frac{S\bar{\alpha}}{1 - \bar{\alpha}} = \text{room constant, m}^2 \tag{10.45}$$

where S is in square meters, or

$$R = \frac{S\bar{\alpha}}{1 - \bar{\alpha}} = \text{room constant, ft}^2 \tag{10.46}$$

where S is in square feet.

The mean square sound pressure, averaged in time and in space, is seen from Eq. (10.44) to be directly proportional to the power emitted by the source and inversely proportional to the room constant R.

Air and Boundary Absorption. The reduction in the energy density of the sound wave due to air absorption while it traverses one mean free path of distance equals $D' \exp(-4mV/S)$. In t sec the sound wave will make t/t' mean free path traversals, where $t' = 4V/cS$ is the time for the wave to make one traversal of the room. The total power absorbed from the reverberant waves in the room will equal that absorbed at the boundaries plus that absorbed in the air.

$$\text{Power absorbed at boundaries} = D'V\left(\frac{cS}{4V}\right)\bar{\alpha} \tag{10.47}$$

$$\text{Power absorbed in air} = D'V\frac{cS}{4V}(1 - e^{-4mV/S}) \tag{10.48}$$

Total power absorbed from reverberant sound field
$$= We^{-4mV/S}(1 - \bar{\alpha}) \tag{10.49}$$

Equations (10.47) plus (10.48) equal (10.49), or

$$D' = \frac{4W(1 - \bar{\alpha})e^{-4mV/S}}{cS[(\bar{\alpha} + 1) - e^{-4mV/S}]} \tag{10.50}$$

Approximately,

$$D' \doteq \frac{4W\left[1 - \left(\bar{\alpha} + \dfrac{4mV}{S}\right)\right]}{cS[\bar{\alpha} + (4mV/S)]} \tag{10.51}$$

Let

$$R'_T \equiv \frac{S[\bar{\alpha} + (4mV/S)]}{1 - [\bar{\alpha} + (4mV/S)]} = \frac{S\bar{\alpha}_T}{1 - \bar{\alpha}_T} \tag{10.52}$$
$$\equiv \text{total room constant, m}^2$$

where

$$\bar{\alpha}_T \equiv \bar{\alpha} + \frac{4mV}{S} \tag{10.53}$$

Alternatively,

$$R_T \equiv \frac{S\bar{\alpha}_T}{1 - \bar{\alpha}_T} \tag{10.54}$$
$$\equiv \text{total room constant, ft}^2$$

Then, in *metric units*,

$$\frac{|p_{\text{avg}}|^2_R}{\rho_0 c^2} = D' = \frac{4W}{cR'_T} \tag{10.55}$$

or

$$|p_{\text{avg}}|^2_R = \frac{4\rho_0 cW}{R'_T} \tag{10.56}$$

Equation (10.56) is identical in form to Eq. (10.44) except that the average absorption coefficient has been increased by $4mV/S$.

10.14. Total Steady-state Sound Energy Density. *Nondirectional Sound Source.* Near a point source of sound, the sound intensity is greater than at a distance. If the source is small enough and the room not too reverberant, the acoustical field very near such a source is independent of the properties of the room. In simpler terms, if one person's ear is only a few inches from another person's mouth, the room surrounding them has negligible effect on what the listener hears when the second person talks. At greater distances from the source, the direct sound field decreases in intensity, and, eventually, the reverberant sound field predominates.

If we are over one-third wavelength from the center of a point source, the energy density at a point r is

$$D_r \doteq \frac{|p_r|^2}{\rho_0 c^2} \tag{10.57}$$

where D_r = energy density at a point a distance r from the center of a spherical source in watt-seconds per cubic meter

p_r = magnitude of the rms sound pressure at the same point in newtons per square meter

The sound pressure due to a spherical source is related to the power radiated as

$$\frac{|p_r|^2}{\rho_0 c^2} = \frac{W}{4\pi r^2 c} \qquad (10.58)$$

Combining the equations for the direct and the reverberant sound energy densities (10.44) and (10.58), we get the mean-square pressure $|p|^2$ at any point a distance r away from the source in the room,

$$|p|^2 = W\rho_0 c \left(\frac{1}{4\pi r^2} + \frac{4}{R'} \right) \qquad (10.59)$$

Expressed as sound pressure level in decibels,

$$\text{SPL} = 10 \log_{10} W + 10 \log_{10} \rho_0 c + 94$$
$$+ 10 \log_{10} \left(\frac{1}{4\pi r^2} + \frac{4}{R'} \right) \quad \text{db} \quad (10.60)$$

where SPL = sound pressure level in decibels *re* 0.0002 microbar (0.00002 newton/m²) at a distance r from the source.

W = power emitted by the source in watts.

$\rho_0 c$ = characteristic impedance of air in mks rayls. At 22°C (71.6°F) and 751 mm Hg, $10 \log_{10} \rho_0 c = 10 \log_{10} 407 = 26.1$ db (mks units).

r = distance from the center of the source in meters.

R' = $S\bar{\alpha}/(1 - \bar{\alpha})$ = room constant in square meters. If there is air absorption, $\bar{\alpha}$ is replaced by $\bar{\alpha}_T$ of Eq. (10.53).

$\bar{\alpha}$ = average sound absorption coefficient for the room (dimensionless) [see Eq. (10.17)].

S = area of bounding surfaces of the room in square meters [see Eq. (10.18)].

Finally, remembering that the power level PWL is

$$\text{PWL} = 10 \log_{10} \frac{W}{10^{-13}}$$
$$= 10 \log_{10} W + 130 \text{ db} \qquad (10.61)$$

we get, in *metric units*,

$$\text{SPL} = \text{PWL} + 10 \log_{10} \rho_0 c - 36$$
$$+ 10 \log_{10} \left(\frac{1}{4\pi r^2} + \frac{4}{R'} \right) \quad \text{db} \quad (10.62)$$

In *English units*, SPL, W, $\rho_0 c$, and $\bar{\alpha}$ are the same as given after Eq. (10.60), but r^2, R, and S are in square feet. We have

$$\text{SPL} = \text{PWL} + 10 \log_{10} \rho_0 c - 25.6$$
$$+ 10 \log_{10} \left(\frac{1}{4\pi r^2} + \frac{4}{R} \right) \quad \text{db} \quad (10.63)$$

At normal temperature and pressure (71.6°F and 30 in. Hg),

$$SPL = PWL + 10 \log_{10}\left(\frac{1}{4\pi r^2} + \frac{4}{R}\right) + 0.5 \text{ db} \qquad (10.64)$$

where r = distance from the center of the source in feet
R = room constant in square feet [see Eqs. (10.46) and (10.54)]
Corrections for temperature and barometric pressure are given in Fig. 10.19. Those numbers are to be added to the sound pressure level given by Eq. (10.64).

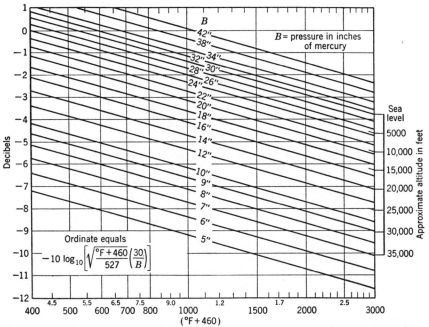

FIG. 10.19. Corrections for temperature and barometric pressure to be added to Eqs. (10.64), (10.66), (10.73), and (10.75). The temperature is in degrees Fahrenheit, and the atmospheric pressure is in inches of mercury, B. Zero correction is for 67°F and 30 in. Hg.

Values of the room constant R as a function of room volume for normally shaped rooms are shown in Fig. 10.20. The parameter in the curves is the average absorption coefficient $\bar{\alpha}$, or $\bar{\alpha}_T$ if there is air absorption.

The physical meaning of the equations above can be grasped more easily by a study of Fig. 10.21. This figure shows, for standard pressure and temperature, the sound pressure level in decibels relative to the power level in decibels, with the room constant R in square feet as the parameter.

Let us look in detail at the curve for $R = 1000$ (square feet). When the observing microphone is within 3 ft of the center of a continuously sounding spherical source, the sound pressure level is nearly the same as if the room were not there.   That is to say, the pressure is different by less than 1 db from that in open air.   On the other hand, when one is more than 20 ft from the center of the source, the sound pressure level on the average is constant at all positions in the room.   As we stated before, it is necessary in measuring the sound pressure to rotate or otherwise move the microphone over a wavelength of distance so as to obtain the rms average of the standing waves in the room.

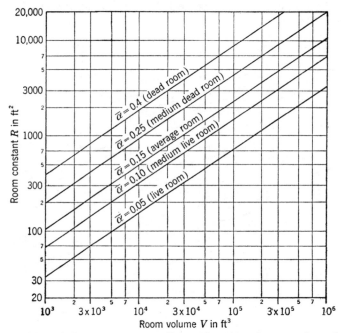

Fig. 10.20. Value of the room constant R as a function of room volume for rooms with proportions of about $1:1.5:2$.   These proportions give $S = 6.25V^{2/3}$.   The parameter is the average sound-absorption coefficient for the room.   The subjective ratings "dead," "live," and so forth, are the author's and are not necessarily in standard use.

As a practical example it is clear that a workman with his head very near a noisy machine will receive very little benefit from increasing the magnitude of the average absorption in the room, i.e., increasing R. However, workmen at a distance from the machine will be exposed to 3 db less sound pressure level for each doubling of the room constant R.

Small Directional Sound Source.   We have dealt with nondirectional sources in the preceding paragraphs, so that the direction in which r was measured did not matter.   The directivity of a source along any radial line (called an axis) may be specified by means of the directivity factor Q. The directivity factor is simply the ratio of the mean-square sound pressure p^2 produced by the sound source at a point distant from the source to the mean-square sound pressure that would be produced at that same point by a nondirectional source of the same power. This quantity, of course, is not only a function of the axis chosen but also a function of the position of the source in the room and, for most sources, a function of frequency.

The reverberant sound energy density is independent of position in the room, except for fluctuations due to standing waves, and is proportional to the total power radiated by the source. Hence, the directivity factor

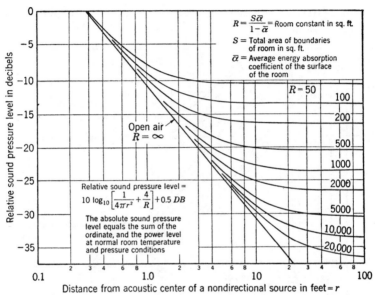

FIG. 10.21. Chart for determining the sound pressure level in a large, irregular room, produced by a nondirectional sound source. The ordinate equals the sound pressure level in decibels relative to the sound power level. The parameter is the room constant R in square feet as defined on the graph. For example, an R of 1000 at a distance of 9 ft from the center of the source gives an ordinate reading of -22.5 db. If the power level is 125 db, the sound pressure level at 9 ft will be 102.5 db.

has little effect on the reverberant sound energy density or the sound pressure level associated with it. An exception to this situation occurs when a highly directive source beams sound at a very highly absorbing surface in the room. Very little reflection then occurs to produce a reverberant sound field.

The direct sound energy density on any axis is proportional to the directivity factor Q. Hence, we simply have to make a correction to one part of Eq. (10.62). This equation in *metric units* becomes

$$SPL = PWL + 10 \log_{10} \rho_0 c - 36 + 10 \log_{10}\left(\frac{Q}{4\pi r^2} + \frac{4}{R'}\right) \quad db \quad (10.65)$$

where all symbols are defined after Eq. (10.60) and PWL is given by Eq. (10.61).

For normal temperature and pressure, with r^2 and R in *square feet*,

$$\text{SPL} = \text{PWL} + 10 \log_{10}\left(\frac{Q}{4\pi r^2} + \frac{4}{R}\right) + 0.5 \text{ db} \qquad (10.66)$$

Corrections for temperature and barometric pressure are given in Fig. 10.19. Those numbers are to be added to the sound pressure level given by Eq. (10.66).

Equation (10.66), for values of Q greater than unity, is plotted in Fig. 10.22 and for values of Q less than unity is plotted in Fig. 10.23. We see

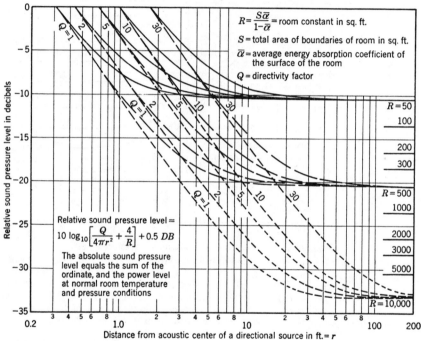

FIG. 10.22. Same as Fig. 10.19, except for a directional sound source with a directivity factor Q greater than unity.

clearly that even though the source is directive, the parts of the curves determined by the reverberant sound are nearly independent of Q. However, if Q is greater than unity, the direct sound field extends to a greater distance than if Q is unity or less.

For example, if $Q = 16$, the direct sound field for $R = 1000$ (square feet) predominates out to 10 ft, and the reverberant sound field does not predominate until a distance of 40 ft has been achieved. However, for $Q = \frac{1}{4}$, the direct sound field for $R = 1000$ predominates only out to 1.0 ft, and the reverberant sound predominates as early as 5.0 ft. In these examples, we have assumed that the sources are small enough so that we are in their "far-field."

Even when the source is nominally nondirectional, the directivity factor Q depends on the position of the source in the room. For example, we see from Fig. 4.20 that if a nondirectional source is located in a plane wall, $Q = 2$ because the power is radiated into hemispherical space only.

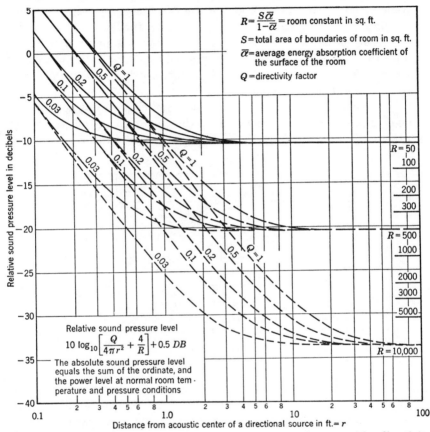

$R = \dfrac{S\bar{\alpha}}{1-\bar{\alpha}}$ = room constant in sq. ft.

S = total area of boundaries of room in sq. ft.

$\bar{\alpha}$ = average energy absorption coefficient of the surface of the room

Q = directivity factor

Relative sound pressure level

$10\,\log_{10}\left[\dfrac{Q}{4\pi r^2} + \dfrac{4}{R}\right] + 0.5\ DB$

The absolute sound pressure level equals the sum of the ordinate, and the power level at normal room temperature and pressure conditions

Distance from acoustic center of a directional source in ft.= r

FIG. 10.23. Same as Fig. 10.19, except for a directional sound source with a directivity factor Q less than unity.

In Table 10.2, we show values of Q for four typical positions of a small nondirectional source in a large room.

TABLE 10.2.   Values of Q for Small Nondirectional Sources Located at Typical Positions in a Large Rectangular Room

Position in room	Q
Near center	1
In center of one wall	2
At edge, halfway between floor and ceiling	4
At corner	8

We mentioned earlier that when the enclosure is small, it will react on the source to modify its radiated power. Whether an enclosure is to be regarded as small or not depends on two things. First, and the more important, is the ratio of the mean free path to the wavelength. For a room to be "large," the mean free path should be several wavelengths long. Second, the room either should be irregular in shape or else should contain irregular objects such as furniture, cabinets, recessed windows, large picture frames, free-standing screens, or the like. Many living rooms meet the second criterion fairly well, but the mean free path is smaller than several times a wavelength below about 200 cps. Such a room is a "small" room at lower frequencies below 200 cps and is a "large" room at higher frequencies.

If you wish a source of sound to radiate as much power as possible in a "small" near-rectangular room, the best location for a sound source is in a corner at either the floor or the ceiling level. Under this condition, the

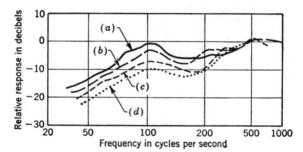

FIG. 10.24. Frequency response of a loudspeaker for four locations (a) at corner of a rectangular room; (b) at center of one wall at floor level; (c) at center of one wall halfway between floor and ceiling; and (d) suspended at exact center of room. The curves are smoothed versions of the original data.

acoustic impedance presented to the source is such as to draw more power from it at low frequencies than if it were elsewhere in the room. As an example, the average sound pressure levels in a near-rectangular room for a direct-radiator loudspeaker placed at each of four positions are shown in Fig. 10.24. It is seen that about eight times as much power (9 db more) is radiated from the loudspeaker at low frequencies if it is in the corner as compared with being in the exact center of the room. At frequencies above 500 cps, the reverberant sound pressure level is independent of the location of the loudspeaker, which indicates that the power output of this loudspeaker is not affected by the reaction of the room.

Finally, we make the observation that the average direct energy density D_D is related to the power radiated by the source regardless of its directivity factor by

$$D_D = \frac{Wt'}{V} \qquad (10.67)$$

where t' is the time for a mean free path. From Eq. (10.19), we have

$$D_D = \frac{4W}{cS} \tag{10.68}$$

Large Sound Source Covering One Wall of Room. When two rooms are contiguous and a source of sound exists in one of them, we can consider the sound that passes through the common wall between them as coming from a large source covering the entire wall of the second room. Let us consider the receiving room as being approximately rectangular and as having a volume $V = S_W L$, where S_W is the area of the radiating wall and L is the length of the room perpendicular to it. We wish to determine the sound energy density in the room as a function of the area of the wall and the room constant. We shall assume as before that the room has enough irregularities and is large enough to produce a diffuse reverberant sound field.

The sound energy density will be composed of two parts, the direct part and the reverberant part. The reverberant part is nearly independent of the shape, size, or location of the source in the room. The direct part will equal the sound energy density for the wave that leaves the radiating wall and travels a distance L across the room, whereupon it experiences its first reflection.

The reverberant part of the sound energy density is, from Eq. (10.43),

$$D' = \frac{4W}{cR'} \tag{10.69}$$

The direct part of the sound energy density equals the power per unit volume times the time t for the sound to traverse the length L of the room, that is, $t = L/c$, so that

$$D_D = \frac{WL}{Vc} = \frac{WL}{S_W L c} = \frac{W}{S_W c} \tag{10.70}$$

where S_W = area of the radiating wall in square meters

L = length of the room perpendicular to the radiating wall in meters

The total sound energy density D_T is the sum of Eqs. (10.69) and (10.70),

$$D_T = \frac{4W}{cS_W}\left(\frac{1}{4} + \frac{S_W}{R'}\right) \tag{10.71}$$

In terms of sound pressure, using Eq. (10.15), we have

$$|p|^2 = \frac{4W\rho_0 c}{S_W}\left(\frac{1}{4} + \frac{S_W}{R'}\right) \tag{10.72}$$

The validity of this equation in the frequency range above 150 cps for a room with dimension $L = 16$ ft and $S_W = 9 \times 12 = 108$ ft^2 has been established by London.[13]

In the derivation above, we have assumed that the direct sound is "beamed" across the room. This will not be true usually, except fairly near the wall. This assumption and, therefore, Eq. (10.72) hold in the "near-field" which exists within a distance from the wall about equal to one-half a panel width.

Expressed in *English units* and decibels, we get the sound pressure level in the *near-field*,

$$\text{SPL} = \text{PWL} + 10 \log_{10}\left(\frac{1}{S_W} + \frac{4}{R}\right) + 0.5 \text{ db} \qquad (10.73)$$

where PWL is as given in Eq. (10.61) and is for the power radiated by the wall, and

S_W = area of the radiating wall in square feet
$R = \bar{\alpha}S/(1 - \bar{\alpha})$
S = total area of all surfaces of the room in square feet
$\bar{\alpha}$ = average absorption coefficient of the room

In the "far-field," that is to say, at a considerable distance from the wall, the sound pressure level will be somewhere between that given by Eq. (10.73) and that given by Eq. (10.66), with $Q = 2$.

Equation (10.73) is written for normal room temperature and pressure conditions. For other temperatures and pressures add the corrections in Fig. 10.19 to the sound pressure levels calculated by this equation. Also, if there is air absorption replace R by R_T as given by Eq. (10.54).

10.15. Examples of Room Acoustics Calculations

Example 10.4. A large irregular room has the approximate dimensions of width 150 ft, length 200 ft, and height 40 ft. The wall areas are of four different types. The ceiling has an absorption coefficient $\alpha_1 = 0.3$; the side walls have $\alpha_2 = 0.4$; the rear wall has $\alpha_3 = 0.6$; and the front wall and floor have $\alpha_4 = 0.15$. A sound source with a power of 10 acoustic watts is put in the room. There are 100 people in it with an $S\alpha$ absorption of 3.5 ft^2 each. Determine (a) the mean free path; (b) the average $\bar{\alpha}$; (c) the total number of absorption units in sabins; (d) the reverberation time of the room; (e) the reverberant sound energy density; (f) the direct sound energy density; (g) the sound pressure level at 20 ft for $Q = 1$ and for $Q = 20$.

Solution. a. The volume of the room is

$$V = 1.2 \times 10^6 \text{ ft}^3 \ (3.4 \times 10^4 \text{ m}^3)$$

The total surface area is

$$S = 8.8 \times 10^4 \text{ ft}^2 (8.15 \times 10^3 \text{ m}^2)$$

$$\text{Mean free path} = \frac{4 \times 1.2 \times 10^6}{8.8 \times 10^4} = 54.5 \text{ ft } (16.6 \text{ m})$$

[13] A. London, Methods for Determining Sound Transmission Loss in the Field, *J. Research Natl. Bur. Standards*, **26**: 419–453 (1941).

b. $\bar{\alpha} = \dfrac{S_1\alpha_1 + S_2\alpha_2 + S_3\alpha_3 + S_4\alpha_4 + (S\alpha \text{ for people})}{S}$

$\qquad = \dfrac{0.9 \times 10^4 + 0.64 \times 10^4 + 0.36 \times 10^4 + 0.54 \times 10^4 + 0.035 \times 10^4}{8.8 \times 10^4}$

$\qquad = \dfrac{2.48}{8.8} \doteq 0.282$

c. The total number of absorption units in sabins is (see Fig. 10.15),

$$a = 8.8 \times 10^4 \left[-2.30 \log_{10} (1 - 0.282) \right]$$
$$= 8.8 \times 10^4 \times 0.33 = 2.9 \times 10^4 \text{ ft}^2$$

d. The reverberation time is

$$T = \dfrac{0.049 \times 1.2 \times 10^6}{2.9 \times 10^4} \doteq 2.0 \text{ sec}$$

e. The reverberant sound energy density is (see Eq. 10.43), in metric units,

$$D' = \dfrac{4W}{Sc} \dfrac{1 - \bar{\alpha}}{\bar{\alpha}} \text{ watts/m}^3$$
$$= \dfrac{4 \times 10 \times (0.72/0.28)}{8.15 \times 10^3 \times 344.8} = 3.66 \times 10^{-5} \text{ watt/m}^3$$

f. The direct sound energy density is (see Eq. 10.68), in metric units,

$$D_D = \dfrac{4W}{Sc} = \dfrac{4 \times 10}{8.15 \times 10^3 \times 344.8} = 1.42 \times 10^{-5} \text{ watt/m}^3$$

g. The power level is
$$\text{PWL} = 10 \log_{10} 10 + 130$$
$$= 140 \text{ db}$$

The room constant is

$$R = \dfrac{S\bar{\alpha}}{1 - \bar{\alpha}} = \dfrac{8.8 \times 10^4 \times 0.28}{0.72}$$
$$= 3.4 \times 10^4 \text{ ft}^2$$

The sound pressure level for $Q = 1$ is given by Eq. (10.64).

$$\text{SPL} = 140 + 10 \log_{10} \left(\dfrac{1}{4\pi \times 400} + \dfrac{4 \times 10^{-4}}{3.4} \right) + 0.5$$
$$= 140 - 35 + 0.5$$
$$\doteq 105.5 \text{ db } re \text{ 0.0002 microbar}$$

The sound pressure level for $Q = 20$ is given by Eq. (10.66).

$$\text{SPL} = 140 + 10 \log_{10} \left(\dfrac{20}{4\pi \times 400} + \dfrac{4 \times 10^{-4}}{3.4} \right) + 0.5$$
$$= 140 - 24 + 0.5$$
$$= 116.5 \text{ db } re \text{ 0.0002 microbar}$$

Example 10.5. The power level of a noise source in the adjoining room is attenuated by 40 db in passing through the common wall. If the noise source radiates 0.1 acoustic watt, what is the sound pressure level in the room of Example 10.4, if the common wall is 40 by 150 ft, (a) near the wall and (b) at a distance of 100 ft from the wall?

Solution.   a. From Eq. (10.73) we have near the wall

$$\text{SPL} = \text{PWL} + 10 \log_{10}\left[\frac{1}{S_W} + \frac{4}{R}\right] + 0.5 \text{ db}$$

$$\text{PWL} = 10 \log_{10}\frac{0.1}{10^{-13}} - 40 = 80 \text{ db}$$

$$S_W = 6000 \text{ ft}^2$$
$$R = 3.4 \times 10^4 \text{ ft}^2$$
$$\text{SPL} = 80 + 10 \log_{10}(2.93 \times 10^{-4}) + 0.5$$
$$= 80 - 35.5 + 0.5 = 45 \text{ db}$$

b. In Eq. (10.66), we find that the first term in the argument of the logarithm equals $2/(4\pi \times 10^4) = 0.159 \times 10^{-4}$. In (*a*), the first term equaled $\frac{1}{6000} = 1.67 \times 10^{-4}$. The true answer is somewhere in between, so let us assume arbitrarily the first term to equal 0.5×10^{-4}. This yields

$$\text{SPL} = 80 + 10 \log_{10}(0.5 \times 10^{-4} + 1.26 \times 10^{-4}) + 0.5$$
$$= 80 - 37.5 + 0.5 \doteq 43 \text{ db}$$

PART **XXV** *Sound Transmission through Walls between Enclosures*

10.16. Transmission Loss (TL) and Noise Reduction (NR). *Transmission Loss.*   The characteristics of a wall placed between two rooms is usually expressed in terms of the transmission loss (TL) in decibels. Transmission loss is defined as the ratio (expressed in decibels) of the acoustic energy transmitted through the wall to the acoustic energy incident upon it.   Because the area is the same for both the numerator and denominator, the transmission loss is said to apply to unit area.   In the United States, this is usually taken to be 1 ft². Mathematically,

$$\text{TL} \equiv 10 \log_{10}\frac{W_1}{W_2} \equiv 10 \log_{10}\frac{1}{\tau} \tag{10.74}$$

where W_1 = acoustic power in watts incident on the wall of area S_W (measured with perfectly absorbing wall)

W_2 = acoustic power in watts radiated by the wall into a perfectly absorbing space

$\tau = W_2/W_1$ = transmission coefficient

The transmission loss may be determined by an experimental arrangement like that shown in Fig. 10.25 and which is currently used at the National Bureau of Standards.   The wall under test is 7 ft 6 in. wide and 6 ft high, and the dimensions of the sound-transmitting area of the panel are 6 ft 6 in. wide by 5 ft.   The source of sound is placed in room 1, and the transmitted sound is measured in room 2.

The sound pressure level is measured in both rooms by moving microphones so as to measure the rms average in space. In room 1, the source is usually small, and the sound pressure level is measured in the reverberant region of the room, *i.e.*, where the room constant R is small compared to $16\pi r^2$ and Q is essentially unity.[14] In room 2, the sound is either measured by averaging the sound pressure level over the entire volume of the room or else averaging it in a plane parallel to and very near the wall. London[13] recommends the latter technique. He has shown, also, that if one makes measurements very near the wall in a highly reverberant room, the sound pressure levels are about 2.5 db higher than those obtained by averaging the sound field in the room. This

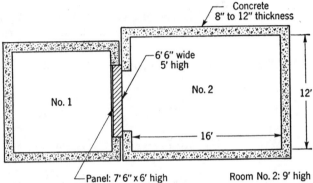

FIG. 10.25. Two-room arrangement used at the National Bureau of Standards for measuring the transmission of sound through panels. The panel under test is mounted between rooms 1 and 2. The source of sound is located in room 1.

number is as expected from the previous part and confirms the result obtained in Example 10.5.

The average reverberant sound pressure level in room 1 is related to the power level of the source in that room by [see Eq. (10.64)]

$$\text{SPL}_1 \doteq \text{PWL} + 10 \log_{10} \frac{4}{R_1} \qquad \text{db} \qquad (10.75)$$

where PWL $= 10 \log_{10} (W/10^{-13})$ db *re* 10^{-13} watt.

$\qquad$ $W =$ acoustic power in watts radiated by the source in room 1.

$\qquad$ $R_1 = [\bar{\alpha}S/(1 - \bar{\alpha})]_1 =$ room constant in square feet for room 1. If there is air absorption, replace $\bar{\alpha}$ by $\bar{\alpha}_T$ of Eq. (10.53).

Normal temperature and pressure are assumed, and the 0.5 db term of Eq. (10.64) has been neglected as it will cancel out in the manipulations later.

In room 1, the reverberant energy density near the common wall is proportional to the power $W(1 - \bar{\alpha})_1$, because an energy density propor-

[14] Beranek, *op. cit.*, Chap. 19, pp. 870–884.

tional to $\bar{a}W$ is associated with the direct sound field [see Eq. (10.41)]. The acoustic power W_1 that would be transmitted through the common wall if the wall were perfectly absorbent ($\alpha_W = 1$) is related to the reverberant energy in room 1 multiplied by the ratio $S_W\alpha_W/S\bar{\alpha}_1$, that is,

$$W_1 = W(1 - \bar{a})_1 \frac{S_W \cdot 1}{S\bar{\alpha}_1} = \frac{WS_W}{R_1} \qquad (10.76)$$

The subscripts 1 indicate that the quantities are for room 1.

The power W_2 transmitted into room 2 equals τW_1 [see Eq. (10.74)]; so

$$W_2 = \frac{WS_W\tau}{R_1} \qquad (10.77)$$

Expressing W_2 as power level in decibels gives us the source power level in room 2 in terms of the power level of the actual source in room 1.

$$\text{PWL}_2 = \text{PWL} + 10 \log_{10} \frac{S_W}{R_1} - \text{TL} \qquad \text{db} \qquad (10.78)$$

From Eq. (10.75), we get

$$\text{PWL}_2 = \text{SPL}_1 + 10 \log_{10} \frac{S_W}{4} - \text{TL} \qquad \text{db} \qquad (10.79)$$

Substituting PWL_2 into Eq. (10.73) yields the sound pressure level in room 2.

$$\text{SPL}_2 = \text{SPL}_1 - \text{TL} + 10 \log_{10} \frac{S_W}{4} + 10 \log_{10} \left(\frac{1}{S_W} + \frac{4}{R_2} \right) \qquad \text{db} \quad (10.80a)$$

In terms of the power level in room 1 (see Eq. 10.75),

$$\text{SPL}_2 = \text{PWL} - \text{TL} + 10 \log_{10} \left(\frac{S_W}{R_1} \right) + 10 \log_{10} \left(\frac{1}{S_W} + \frac{4}{R_2} \right) \qquad (10.80b)$$

Solving for the transmission loss from Eq. (10.80a) gives

$$\text{TL} = \text{SPL}_1 - \text{SPL}_2 + 10 \log_{10} \left(\frac{1}{4} + \frac{S_W}{R_2} \right) \qquad \text{db} \qquad (10.81)$$

Noise Reduction.   The *noise reduction* in decibels is defined as

$$\text{NR} \equiv \text{SPL}_1 - \text{SPL}_2 \qquad \text{db} \qquad (10.82)$$

Then, the noise reduction provided by a wall between two rooms is

$$\text{NR} = \text{TL} - 10 \log_{10} \left(\frac{1}{4} + \frac{S_W}{R_2} \right) \qquad \text{db} \qquad (10.83)$$

where NR = difference in sound pressure levels in decibels on the two sides of the wall determined by measuring the sound pressure level on the primary side with a microphone that is moved around in the reverberant sound field and then subtracting from it the sound pressure level with a microphone that is moved around in a region fairly near the surface on the secondary side.

TL = 10 times the logarithm to the base 10 of the ratio of the sound energy incident on the wall to the sound energy transmitted through the wall.

S_W = area of the transmitting wall either in square meters or in square feet.

R_2 = room constant for room 2 = $[S\bar{\alpha}/(1 - \bar{\alpha})]_2$, where S is the total area of the surfaces of the room on the secondary side and $\bar{\alpha}$ is the average absorption coefficient for room 2. S must have the same dimensions as S_W. If there is air absorption, replace $\bar{\alpha}$ by $\bar{\alpha}_T$ of Eq. (10.53).

Equation (10.83) has been verified experimentally by London (reference 13 of Part XXIV) for frequencies between 150 and 2000 cps, except for a small fixed additive quantity.

It should be noted especially that if the secondary side of the panel opens into a room with large absorption (R_2 large) or if it opens outdoors, the noise reduction equals

$$NR = TL + 6 \text{ db} \qquad (10.84)$$

As a final point, if one moves into the far-field of the radiating wall, which is approximately beyond a distance away equal to one wall width, the sound field will start to drop off with distance and, eventually, at a great enough distance from the wall, Eq. (10.80) becomes

$$SPL_2 = SPL_1 - TL + 10 \log_{10} \frac{S_W}{R_2} \qquad (10.85)$$

This situation would exist, for example, for a window in one wall of a room. Near the window, Eq. (10.81) would hold, and away from the window a considerable distance, Eq. (10.85) would hold.

10.17. Measured and Calculated Transmission Loss. *Average Transmission Loss.* The National Bureau of Standards and the Riverbank Laboratories at Geneva, Ill., are the accepted "official" laboratories in the United States for measuring the transmission loss of walls. These laboratories generally measure at frequencies of 125, 250, 500, 1000, 2000, and 4000 cps. The *average transmission loss* is determined by taking a simple arithmetic average of the transmission losses in decibels at 125, 250, 500, 1000, and 2000 cps.

In England, the measurements are officially taken by the National Physical Laboratory, except that the average transmission loss is determined from the data at 200, 300, 500, 700, 1000, 1600, and 2000 cps. The average transmission losses generally do not differ much between the two countries.

Single Walls. The transmission loss (TL) for single walls that are homogeneous in construction and are damped so that they do not ring when struck with a hammer is dependent primarily on the product of the surface density (pounds per square feet) and the frequency. The thickness is generally not important for walls that are under 12 in. except in so

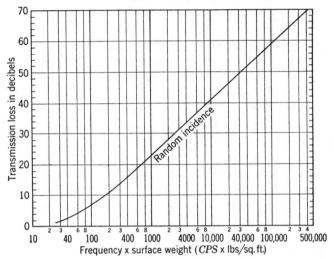

Fɪɢ. 10.26. Transmission loss (TL) for solid damped partitions. The average transmission loss may be determined from this graph by assuming a frequency of 500 cps. For undamped walls subtract about 5 db from the average transmission loss. For plane sound waves incident normal to the face of the panel, the transmission loss increases 6 db for a doubling of the abscissa rather than 5 db as is shown here for randomly incident sound.

far as it increases the surface density. The transmission-loss curve, for sound that is *randomly incident* on the primary side (measured with the arrangement of Fig. 10.25), is shown in Fig. 10.26. This curve indicates that, for damped single walls, the transmission loss increases about 5 db for each doubling of frequency or doubling of weight per square foot. In Table 10.3, we show the surface densities for a 1-in.-thick wall made from

TABLE 10.3.   Weights of Common Building Materials

Material	Surface density, lb/ft²/(inch thickness)
Brick	10–12
Cinder concrete	8
Dense concrete	12
Wood	2–4
Common glass	12.5–14.5
Lead	65
Aluminum	14
Steel	40
Gypsum	5

the materials tabulated. For a 2-in.-thick wall, the surface density doubles, and so forth.

Multiple Walls. When the average transmission loss must exceed 40 db, it is generally economical to use double walls. The average transmission loss through them is usually 5 to 10 db greater for a given weight of material than it is for a single wall using the same material. In Fig. 10.27, the average transmission loss for both single and double walls of various constructions is shown.

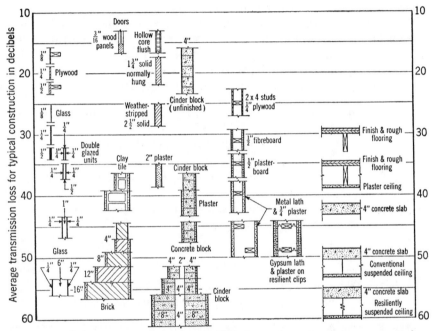

FIG. 10.27. Average transmission losses (TL) for various structures constructed as shown. In compiling this chart, it was assumed that no flanking paths exist. That is to say, all the sound arrives in the receiving room through the panel and not via alternate paths.

For example, the average transmission loss for a double cinderblock wall, each leaf 4 in. thick, separated by 4 in., and both leaves plastered on the outside, is about 55 db. A single cinder-block wall, 4 in. thick and plastered on both sides, yields a transmission loss of about 42 db. If the thickness of the single wall were increased to 8 in., we would expect that the transmission loss would increase about 5 db, as we found in the previous sections. Hence, the double wall is about 8 db better than the single wall for the same total weight.

The problem of incorporating double walls into building structures is beyond the scope of this book. The reader is referred to Cullum[15]

[15] D. J. W. Cullum, "The Practical Application of Acoustic Principles," pp. 53–61, Spon, Ltd., London, 1949.

for some architectural details. Extensive research has been done in England[16] and in Holland[17] on the construction of apartment and row houses with better noise isolation. These studies are available in technical reports that are issued from time to time.

10.18. Enclosures. When single or double walls are combined to produce an enclosure, the amount of noise reduction achieved is dependent upon the combined transmission losses. In fact, a small opening, cracks around a door, or thin windows may nullify a large part of the effectiveness of an otherwise good enclosure.

Let us imagine a case where a complete enclosure is placed in a uniform noise field. To determine the value of the transmission loss for insertion into Eq. (10.83) or Eq. (10.85) so as to find the total noise reduction for the enclosure, the areas and transmission coefficients for all the walls, doors, and windows must be considered in the equation

$$\text{TL} = 10 \log_{10} \frac{S}{S_1\tau_1 + S_2\tau_2 + S_3\tau_3 + \cdots} \quad \text{db} \qquad (10.86)$$

where S = total area of all surfaces of the enclosure either in square meters or in square feet

$S_1, S_2, S_3, \ldots$ = areas of particular surfaces of the room in the same units as S

$\tau_1, \tau_2, \tau_3, \ldots$ = transmission coefficients for those surfaces, respectively, found by use of

$$\tau_n = \frac{1}{\text{antilog}_{10}\,(\text{TL}_n/10)} \qquad (10.87)$$

where τ_n = transmission coefficient for a particular surface

TL_n = published transmission loss for that particular surface

Example 10.6. As a typical example, determine the total average transmission loss for a 4-in.-thick concrete block wall with a single ⅛-in.-thick glass window in it, if the areas are 200 ft² and 20 ft², respectively.

Solution. From Fig. 10.27 we see that, for glass, $\text{TL}_1 = 26$ db and, for 4-in. concrete block, $\text{TL}_2 = 46$ db. So

$$\tau_1 = (\text{antilog}_{10}\,{}^{26}\!/_{10})^{-1} = (4 \times 10^2)^{-1} = 2.5 \times 10^{-3}$$
$$\tau_2 = (\text{antilog}_{10}\,{}^{46}\!/_{10})^{-1} = (4 \times 10^4)^{-1} = 2.5 \times 10^{-5}$$

Hence,

$$\tau_{\text{avg}} = \frac{S_1\tau_1 + S_2\tau_2}{S} = \frac{0.05 + 0.005}{200} = 2.75 \times 10^{-4}$$

It is seen that the transmission through the small window ($S_1\tau_1$) is much greater than that through the remainder of the large wall ($S_2\tau_2$). The transmission loss for the combination is

$$\text{TL} = 10 \log_{10} \frac{1}{\tau_{\text{avg}}} = 10 \log_{10} 3.64 \times 10^3 = 35.6 \text{ db}$$

[16] Building Research Station, Watford, Herts., England.
[17] Research Institute for Public Health, Engineering TNO, The Hague, Holland.

Part XXV] SOUND TRANSMISSION THROUGH WALLS 331

Reference to Fig. 10.27 shows that if double-glazed windows were used with, say, two $\frac{1}{4}$-in.-thick panes separated 1 in., the transmission loss for the whole structure would equal that for the concrete block, that is, 46 db, instead of being 10 db less, as was the case for the example given here.

Example 10.7. An airplane engine is being tested in a cell alongside of which is a control room. The wall separating the control room from the test cell is a 12-in.-thick poured concrete wall with dimensions 40×20 ft. In this wall is a double window, made of two $\frac{1}{2}$-in.-thick panes separated by 6 in. of air and having dimensions of 10×3 ft. Determine the sound pressure level at 500 cps in the control room near the window, assuming that the room constant R is very large and that the sound pressure level in the test cell is 140 db.

Solution. The transmission loss for the wall is found from Table 10.1 and Fig. 10.26. At 500 cps,

$$\text{TL (concrete)} = 56 \text{ db}$$

The transmission loss for the window is found from Fig. 10.27.

$$\text{TL (window)} = 59 \text{ db}$$

In front of the window, the noise reduction is found from Eq. (10.83).

$$\text{NR} = 59 + 10 \log_{10} 4 = 65 \text{ db}$$

The sound pressure level equals

$$\text{SPL}_2 = 140 - 65 = 75 \text{ db}$$

To one side of the window, the sound pressure level equals

$$\text{SPL}_2 = 140 - 56 - 6 = 78 \text{ db}$$

CHAPTER 11

NOISE CONTROL

PART XXVI *Procedures and Sources*

11.1. Introduction. Noise is rapidly becoming a major national problem. People are living in more concentrated groups, and housing costs have risen so that building structures are becoming lighter and lighter in weight. Because of the absence of specifications for tolerable noise levels, building codes in many nations have permitted speculative builders to construct apartments and row houses with acoustically inadequate walls and resonant floors.

Noise from highways and airports has blighted many housing areas which would be assets otherwise to our cities. The airport noise problem has been a major deterrent to widespread private flying because the complaints of neighbors concerning the noise have made it necessary to locate flying fields in relatively inaccessible regions.

In the factory and office we have serious noise problems also. Drop forges, riveting machines, cutters, grinders, can-making, and weaving machines are a few examples of sources of noise that may affect hearing permanently and interfere with speech communication.

Passengers inside aircraft must be protected from the intense noise of the propellers and exhausts. This is accomplished partially by lining the fuselage with acoustical materials.

In recent years, the jet airplane has become one of the worst of our noise sources. The engines themselves must be tested in test cells that are rendered sufficiently quiet so as not to disturb neighbors. Jet aircraft not in flight must be backed against large mufflers located on the ramps at airports when warming up. As yet, there is no solution to the problem of exterior noise radiated in flight.

All these problems require the attention of acoustical engineers. A purpose of this chapter is to present the procedures by which one attempts to solve typical noise problems and to give performance data on some of the more common acoustic structures used in noise control.

The problem of noise reduction is given its proper meaning and context by the presence of people as listeners.   Criteria for acoustical design have been developed for factory spaces, offices, aircraft, and residential districts.   Some of these criteria will be presented in Chap. 13, which deals with psychoacoustical phenomena.   In this chapter, we shall confine ourselves to the physical side of the problem, leaving the amount of necessary noise reduction for a particular problem as something to be decided after a psychoacoustic criterion is selected.

11.2. Noise-control Procedures.   *Factors in Noise Control.*   The first rule of noise control is to design or to modify the construction of the machine or device so that it is less noisy.   Sometimes this can be done by such steps as replacing gears by a V-belt drive; using vibration mounts under the footings so that vibrations will not be transmitted into other structures to be radiated as sound; replacing one style of machine by another; using an abrasive cutting action in place of a toothed cutting action; and so forth.

Another approach to reducing the noise is to decrease the directivity index in the direction in which the listener is located.   This may sometimes be accomplished by rotating the source so that its principal axis of radiation points elsewhere.   In one example, the ventilating parts on a home motion-picture projector were modified so that the noise was radiated toward the operator rather than to the sides where the audience was seated.

Still another approach is to increase the distance between the source and the listeners.   For example, hospitals should not be built near airports, highways, and railroads.   If the lot is large, locate the building on it as far from a busy street as possible.

Even though every known procedure for modifying the basic design of an equipment is utilized, the device may still be too noisy for a particular location.   In that case, we must resort to one or more of the three standard techniques for noise control, *viz.*, relocating, muffling, or enclosing the device.   In order that the engineer may apply any one of these techniques quantitatively, we must consider the following factors if a satisfactory solution to the problem is to be achieved:

1. The *acoustic* power radiated by the source in each of a prescribed number of frequency bands should be determined.   Usually, noise measurements are made in one of the following ways:[1]

a. With an octave-band analyzer having eight frequency bands: 37.5 to 75, 75 to 150, 150 to 300, 300 to 600, 600 to 1200, 1200 to 2400, 2400 to 4800, and 4800 to 10,000 cps.

b. With a one-third octave-band analyzer having 25 frequency bands: 20 to 45, 45 to 57, 57 to 71, 71 to 90, 90 to 114, 114 to 142, 142 to 180, 180

[1] L. L. Beranek, "Acoustic Measurements," Chap. 12, pp. 516–592, John Wiley & Sons, Inc., New York, 1949.

to 228, 228 to 284, 284 to 360, 360 to 456, 456 to 568, 568 to 720, 720 to 912, 912 to 1136, 1136 to 1440, 1440 to 1824, 1824 to 2272, 2272 to 2880, 2880 to 3648, 3648 to 4544, 4544 to 5760, 5760 to 7296, 7296 to 9088, and 9088 to 11,520 cps.

c. With a narrow-band analyzer with a fixed bandwidth between 2 and 50 cps, or

d. With a narrow-band analyzer with a bandwidth equal to 2 to 20 per cent of the mean bandwidth.

The results obtained with one of these kinds of analyzer are reduced, sometimes, to what they would have been if they had been obtained by an analyzer with a 1-cps bandwidth.   If the plot of the data is in terms of power level in decibels (see page 14) as a function of frequency for a 1-cps band, we say that the plot gives the *power spectrum level* (see page 15).   Other bandwidths such as *critical bandwidths* sometimes are used also in presenting data.[1]

2. The *directivity characteristics* of the source giving the relative amounts of power radiated in different directions around the source at various frequencies should be measured.

3. The *transmission path* or *environment* coupling the source of noise to the listeners' ears should be studied.   This environment may be a room or a factory space, in which cases the formulas for the sound fields in large irregular rooms presented in the previous chapter are applicable. Those formulas give the sound pressure as a function of acoustic power, distance from the source and room characteristics.   The transmission path may be outdoors, where the sound pressure decreases linearly with distance, except for the additional effects of air and terrain absorption and temperature gradients.   Or the transmission path may involve intervening structures such as mufflers, ducts, walls, and hills.   These must be considered, whether near the source or near the listener, or both.

4. The *criteria* that are applicable to the problem at hand must be selected.   For example, if it were decided that workers near a factory should be able to converse and that distant neighbors should not be annoyed by the noise, the sound levels should be reduced below those shown by the appropriate curves in Part XXXII that express the maximum sound pressure levels for permitting achievement of these goals. Those curves are the criteria for design.

5. The *amount of noise reduction required*.   This is achieved by considering all four of the factors above.   The power level (in decibels) plus the directivity index minus the effects of environment and intervening structures give the sound pressure levels due to the source at the listener's position.   The difference between the sound pressure levels at the listener's position and the applicable criterion curve is the amount of noise reduction (in decibels) that must be achieved.

Vibration Isolation. Although this text is devoted entirely to the acoustics side of noise reduction, the importance of and the need for vibration isolation should not be overlooked.   Many times a device can be quieted adequately by merely mounting it on rubber isolators so that its vibrations do not excite surrounding structures that will in turn radiate noise.   Recent texts on vibration are referred to in the footnotes.[2,3]

Illustration of Noise Control. To aid in obtaining a feeling for the relative value of several types of noise-control possibilities, let us refer to Fig. 11.1.   Here is shown a hypothetical noise-producing machine. If it is mounted directly on a nonrigid floor, it produces the noise spectrum shown beside the first diagram (*a*).   The noise spectrum is measured in eight octave frequency bands at the position marked *M* on the first diagram.

The bar diagram at the right gives the calculated transfer function produced by the machine.   The transfer function is a quantity used to assist in the determination of the subjective magnitude *loudness level* of a complex noise, once its noise spectrum is known.   Until recently, the transfer function was believed to be proportional to the *loudness* of a sound, that is to say, if something were judged to be twice as loud, the transfer function would be double.   This concept is now open to question.   The transfer function is discussed in Part XXX.   The speech-interference level is shown by the height of the bar in the 1200 to 2400-cps band.   This concept is discussed in Part XXXII.   The effectiveness of sound control in relation to human beings is indicated not by changes in the spectrum per se, but specifically by such changes as reduce the transfer function and the speech-interference level.

In (*b*) we show the effect of putting soft vibration mounts beneath the machine.   The noise spectrum has decreased, particularly at the lower frequencies.   In this example, vibration isolation is of value but is not in itself sufficient.

In (*c*) we show that a small advantage results from erecting a short wall next to the machine.

In (*d*) we show that only slight advantage is to be gained from enclosing the machine in a porous acoustical blanket.   Such blankets are effective in controlling reverberation times and room constants but do not usually make good acoustic barriers.

In (*e*) we show the effect of enclosing the vibration-isolated machine in an airtight box.   The noise levels decrease a large amount, particularly at the higher frequencies.

[2] C. E. Crede, "Vibration and Shock Isolation," John Wiley & Sons, Inc., New York, 1951.

[3] H. M. Hansen and P. F. Chenea, "Mechanics of Vibration," John Wiley & Sons, Inc., New York, 1952.

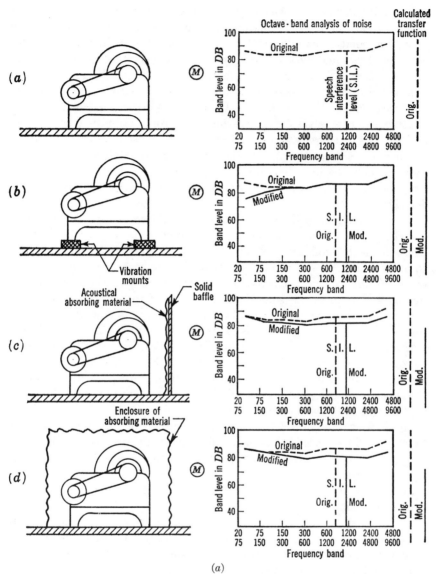

Fig. 11.1. Hypothetical examples of noise reduction showing relative values of vibration mounts, barriers, enclosures, and acoustic materials used in various combinations. The speech-interference level (see Part XXXII) is shown by the vertical bar marked

In (*f*) we show that the combination of vibration isolation and a tightly closed box yields a very substantial reduction of the noise spectrum, the transfer function, and the speech-interference level.

In (*g*) we see the effect of enclosing the machine in an airtight box that is lined with an acoustical absorbing blanket. Here, the noise control is seen to be very good.

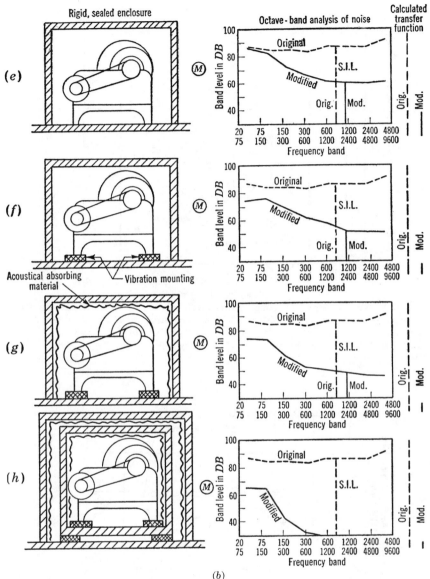

(b)

SIL. The calculated transfer function (loudness) before and after modification is shown by the bars to the right of the graphs.

In (h) we see the effect of a double enclosure with double vibration mounts and lined with an acoustical absorbing blanket. The noise levels are very low for this case.

An obvious difficulty with the solution of Fig. 11.1 is that no ventilation, or escape for the gases of combustion, is possible. To satisfy these needs, acoustic ducts or mufflers must be used. Attenuations of nearly

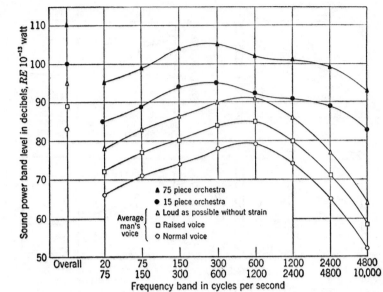

FIG. 11.2. Average long-time sound power levels relative to 10^{-13} watt are shown for average male voices and for 15- and 75-piece orchestras. For the human voice, the peaks as read on the fast C scale of a sound-level meter would be about 10 db higher than the over-all levels shown here. For orchestras the peaks would be about 15 db higher.

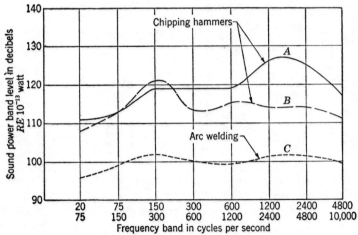

FIG. 11.3. Sound power levels as read on the slow meter scale of an octave-band analyzer for chipping hammers of various makes operating with 90 lb/in.[2] air. (A) Hammer operating on 18- by 18- by $\frac{3}{4}$-in. (thick) steel plate; (B) hammer operating in mid-air; (C) same for arc welding with 300 amp current.

any reasonable magnitude are possible by the correct choice of design of muffler.

11.3. Power Levels of Sources.   The power-level curve as a function of frequency for a source is the starting point in noise-control design. Most sources radiate a constant amount of acoustic power whether the

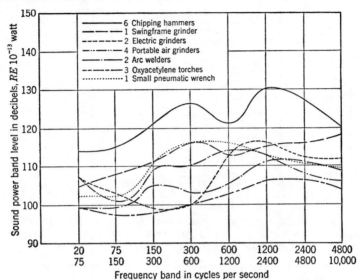

FIG. 11.4. Sound power levels as read on the slow meter scale of an octave-band analyzer for a number of industrial tools.

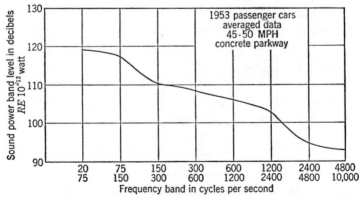

FIG. 11.5. Average sound power levels for passenger cars of 1953 traveling between 40 and 50 mph on a typical concrete highway; measurements made externally.

source is in a room or outdoors.   For example, a ventilating fan used in air-conditioning systems appears to radiate nearly the same acoustic power whether connected to a duct or not, and regardless of the range of back pressures usually encountered.   This happens because, even when the source is in a room, its radiation impedance is nearly like that for free space. We saw, in studying Fig. 10.21, that the sound field near the machine is nearly the same indoors as outdoors, unless the room constant is very small.

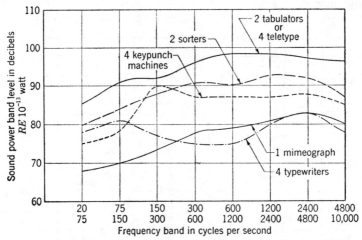

FIG. 11.6. Office machinery noise. Data were taken in large offices, and diffuse room conditions were assumed. Data were read from an average meter as 3 db below frequent peaks. All office machines were of a vintage prior to 1950.

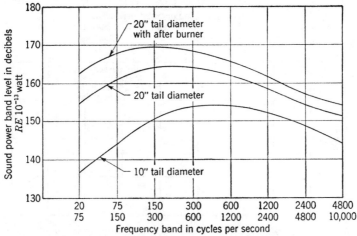

FIG. 11.7. Sound power levels produced by turbo-jet engines (without propellers) for three powers of jet.

There are cases, of course, where a small enclosure reacts on the source so strongly as to modify its radiated power. Such situations must usually be handled by exact mathematical analysis.

We shall divide sound sources into several general groups for analysis. These groups fall into three broad categories as shown in Table 11.1. Power levels, either directly measured or estimated from available data,

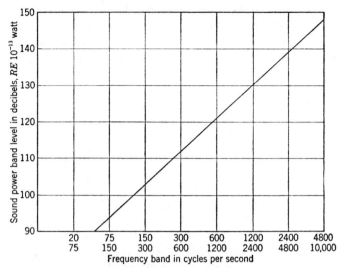

Fig. 11.8. Sound power levels produced by a ¾-in.-diameter steam hose used for cleaning meat-grinding equipment.

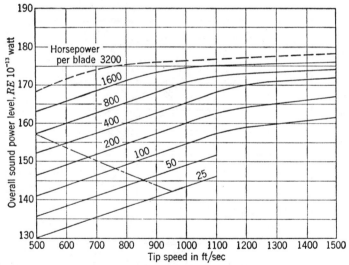

Fig. 11.9. Over-all sound power levels produced in flight by propellers of 1952 as a function of tip speed and horsepower per blade.

for the few sources listed in Table 11.1, are shown in Figs. 11.2 to 11.11. These graphs express the power level in each of eight frequency bands, where power level is defined as

$$\text{PWL} = 10 \log_{10} \frac{W}{10^{-13}} \quad \text{db} \qquad (11.1)$$

We usually describe PWL in the words "power level with reference to (re) 10^{-13} watt."

11.4. Directivity Patterns of Sources. The directivity pattern of a source of sound is a description, usually presented graphically, of the intensity levels at a given distance from the center of the source as a function of angle θ about the source on a designated plane passing through

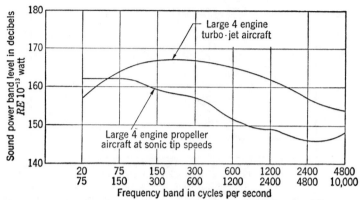

Fig. 11.10. Sound power band levels for two 4-engine aircraft at cruising speeds (1952 vintage).

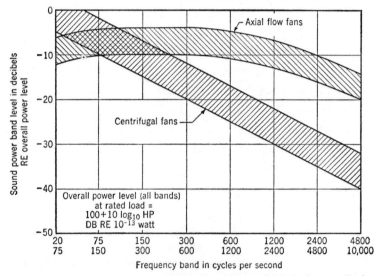

Fig. 11.11. Chart for determining the sound power band levels for ventilating fans of two types. The over-all power level is first determined from the formula given on the graph, where HP is the rated horsepower of the fan when it is operated as rated.

the center of the source. The directivity index was defined in Part XI as the ratio, expressed in decibels, of the intensity at a point on a designated axis of a source to the intensity that would exist at that same point if a non-directional source were radiating the same total acoustic power. The points of measurement for directivity patterns are far enough

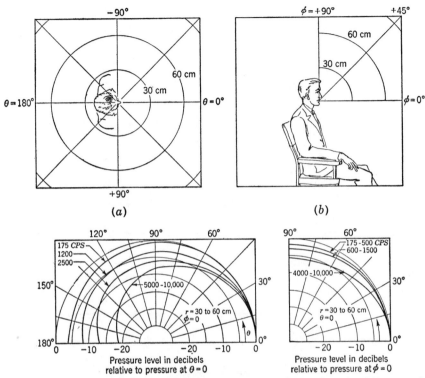

Fig. 11.12. Directivity patterns for the human voice. [*After Dunn and Farnsworth, J. Acoust. Soc. Amer.*, **10**: 184–199 (1939).] The directivity indexes are 175 cps = 1 db; 1200 cps = 3 db; 2500 cps = 4.5 db; and 5000 to 10,000 cps = 6 db.

TABLE 11.1.   Categories of Sound Sources†

A. Speech and music sources:
   Human voice
   Orchestras

B. Impact and vibration sources:
   Factory machinery
   Transportation machinery
   Office machinery

C. Sources with air streams:
   Jet-engine exhausts
   Air jets
   Propellers
   Aircraft in flight
   Ventilating fans

† Data are given in Figs. 11.2 to 11.11.

from the source so that they are in the far-field, *i.e.*, the sound pressure decreases linearly with distance along the designated axis.

In Figs. 11.12 to 11.14 we show the directivity patterns at selected frequencies for certain of the sources of Table 11.1. The directivity index for any particular axis may be determined from a directivity pattern by the methods of Part XI.

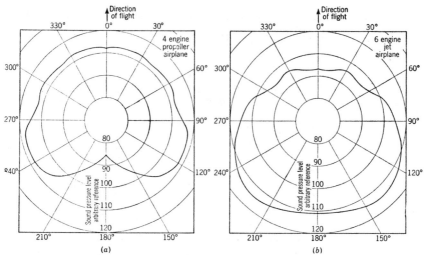

FIG. 11.13. Directivity patterns for cruising aircraft. (a) Propeller aircraft. (b) Jet aircraft.

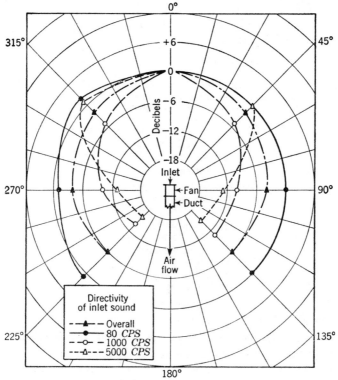

FIG. 11.14. Directivity patterns for inlet of a ventilating fan of the vane axial type with a case diameter of 16 in.

PART XXVII *Acoustic Transmission Paths*

The sound source and the listener are coupled by an acoustic transmission path. This path may be through the outdoor air or the air inside a room, or a combination of both. For example, the sound might originate outdoors and eventually arrive at a listener's ear inside a room. In handling a combination of both, the problem is usually broken down so that each part is handled separately, the basic fact being kept in mind that all the acoustic power radiated from the source must be accounted for. We shall treat the outdoor problem first.

11.5. Outdoors. *Spherical Radiation.* The sound pressure level in decibels at a point in free space (spherical radiation) is related to the power radiated and the directivity index by†

$$SPL = PWL + DI - 10 \log_{10} S_s + 0.5 \quad db \quad (11.2)$$

where PWL = power level in decibels *re* 10^{-13} watt [see Eq. (11.1)]

$DI = 10 \log_{10} Q$ = directivity index in decibels for the designated axis on which the sound pressure level is being determined

Q = directivity factor for the designated axis (see Part XI)

$S_s = 4\pi r^2$ = surface area in square feet of a sphere whose radius equals the distance r at which the sound pressure level is being determined

In writing Eq. (11.2) we have assumed that the ambient temperature is 67°F and the ambient pressure is 0.76 m (30 in.) Hg. Corrections to be added to the sound pressure level given by this equation are found in Fig. 10.19.

Hemispherical Radiation. Often, the sound source is located near the ground so that the power is radiated into "half space." In this case, at all points, the sound intensities will be doubled, which means that they are 3 db higher than those given by Eq. (11.2). The formula for this case is

$$SPL = PWL + DI - 10 \log_{10} S_h + 0.5 \quad db \quad (11.3)$$

where $S_h = 2\pi r^2$ = surface area in square feet of a hemisphere whose radius equals the distance r at which the sound pressure level is being determined. In this case also, the sound pressure level decreases 6 db for each doubling of distance.

Air Losses—Theoretical. If the frequency of the radiated sound is greater than 1000 cps, loss of acoustic energy in the air may become significant and must be taken into account. The loss expressed in decibels

† This equation comes directly from Eq. (10.66) with $R = \infty$.

per meter or per foot is [see Eq. (10.40)],

$$\Delta_m = 4.34(m_1 + m_2) \quad \text{db/m} \tag{11.4}$$

$$\Delta_m = 1.324(m_1 + m_2) \quad \text{db/ft} \tag{11.5}$$

where m_1 = loss in meter^{-1} due to viscosity and heat conduction as given by Eq. (10.38)

m_2 = loss in meter^{-1} due to molecular absorption and dispersion and given by Fig. 10.17

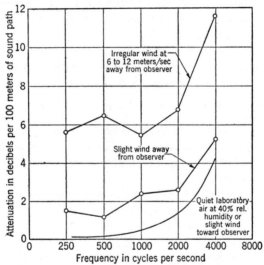

FIG. 11.15. The relative importance of wind in increasing the attenuation of sound over that measured under laboratory conditions. The bottom curve gives the attenuation as calculated for air under laboratory conditions at 40 per cent relative humidity. The two remaining curves show values of attenuation actually measured in outdoor experiments with the wind blowing from the observer toward the source. The effects of spherical divergence and of humidity (the lowest curve) were subtracted. The difference between the upper two curves represents the attenuation due to the gustiness (turbulence) of the wind and is seen to be of the order of magnitude of 4 to 6 db per 100 m. These results were obtained by measuring the sound pressure level at different distances from the sound source located in a 12-m- (39-ft) high tower. If the wind blows from the source toward the observer, the attenuations are nearly those of the bottom curve. (*From Ingard.*[4])

Air Losses—Measured.[4] Outdoors, the attenuation of sound will be much greater than that indoors, even on days when the air is "calm." The greater attenuation is caused by ever-present air movements and localized temperature gradients and rotational motion. In Fig. 11.15 we show a set of curves for outdoor-air attenuation.

[4] K. U. Ingard, Review of the Influence of Meteorological Conditions on Sound Propagation, *J. Acoust. Soc. Amer.*, **25**: 405–411 (1953). Much of the material in this chapter can be found in "Handbook of Acoustic Noise Control," Vol. I, Physical Acoustics, by Bolt Beranek and Newman, Inc. (1952). Order from Office of Technical Services, Department of Commerce, Washington, D.C., PB No. 111,200.

TERRAIN LOSSES.   In addition to the absorption in the air itself, sound is also attenuated in traveling over the terrain.   Wooded areas are more effective in attenuating sound than are grassy ones.   Hills also aid in scattering the sound energy.   Unfortunately, practical situations are so varied that it is not possible to discuss this problem here.

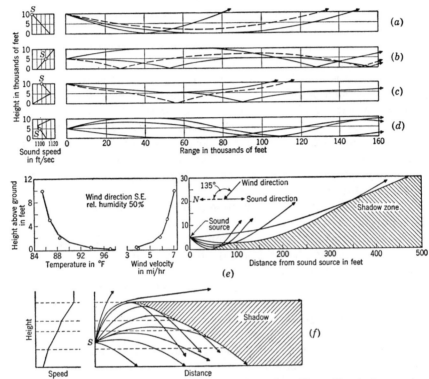

FIG. 11.16. Graphs showing the effects of sound-speed gradients (due to temperature and barometric pressure changes) as a function of distance from the source. (a) Uniform negative sound-speed gradient, airplane at 10,000 ft; (b) uniform positive sound-speed gradient, airplane at 5000 ft; (c) temperature inversion, airplane at 10,000 ft; (d) temperature inversion of opposite type, airplane at 5000 ft; (e) negative temperature gradient, source at 5 ft; (f) positive temperature gradient, source at 5 ft.

TEMPERATURE GRADIENTS.   Temperature gradients produce pronounced effects on sound propagation, although wind generally influences the propagation of sound to a greater extent.   A positive temperature gradient with height above the earth's surface produces a downward refraction of sound waves and, hence, an enhancement of their levels. The reverse effect will be observed on a hot summer's day when the surface of the earth is heated to a temperature much higher than that of the air a dozen or more feet above it.   In this case, the sound may seem to disappear entirely in a distance of only 100 ft.

Six examples of diffraction caused by temperature gradients are shown in Fig. 11.16. In (a) and (e) the temperature gradient is negative, and the sound is audible to a limited distance and then bends upward. In (b) and (f) the gradient is positive, and a large part of the energy radiated upward is returned to the earth. Clearly, in the latter case the sound pressure will not decrease as much as the normal 6 db for each doubling of distance. In (c) and (d) temperature inversions are assumed to occur at about 5000 ft. The sound at the ground for these cases is refracted in a manner similar to the refraction of radio waves in the Kennelly-Heaviside layer of ionized gas in the upper atmosphere. Note that where the speed of sound is lowest, the density of the air is greatest so that a temperature inversion produces the effect of a layer of dense air.

For the situations of Fig. 11.16, the sound does not decrease according to the simple relations of Eqs. (11.2) and (11.3). In noise-reduction problems, the effects of temperature gradients must be borne in mind, because a noise-reduction treatment that is satisfactory most days and nights may, on certain occasions, be inadequate because of the failure of the inverse-distance law to hold.

Barriers. Many times the question is asked, If a wall is built between the source of noise and my home, will the noise be significantly reduced? The answer to this question is not simple. It depends on the height of the wall, on the wavelength of the sound, on the distances the source and the listener are from the wall, and on the heights of the source and listener above the ground. Unless both the source and listener are at ground level and unless the noise source is reasonably near the wall so that its height need not be great, a barrier of this type is generally not feasible from an economical standpoint.

A simplified solution to this problem, made under the assumption that the source and the microphone are near the ground, is shown in Fig. 11.17.[5] The abscissa to this figure is given by

$$N = \frac{2}{\lambda}\left[R\left(\sqrt{1 + \frac{H^2}{R^2}} - 1\right) + D\left(\sqrt{1 + \frac{H^2}{D^2}} - 1\right)\right]$$
$$\doteq \frac{H^2}{\lambda R} \quad \text{if } D \gg R \geq H \tag{11.6}$$

where H = height of barrier above the line of sight between source and observer

R = horizontal distance from source to barrier

D = horizontal distance from barrier to observer

Measured data are plotted on this graph for comparison. It is seen that the attenuations are limited in practice to about 15 to 20 db.

[5] R. O. Fehr, The Reduction of Industrial Machine Noise, *Proc. 2d Natl. Noise Abatement Symposium* (1951).

11.6. Enclosing Walls. A very common procedure for noise control is to enclose completely the source of noise. If this is done, the considerations shown in Fig. 11.1 apply basically. First, the enclosure must be nearly hermetically sealed, except for planned ventilation. Second, the mass and structure of the walls, windows, and doors must be so chosen that the desired attenuation is achieved. Third, the source of noise should be adequately vibration-isolated from the floor and from the walls and ceiling of the enclosure so that sound does not arrive on the outside by a structure-borne path.

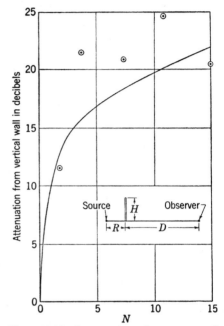

Average Transmission Loss. The attenuations of walls and windows were tabulated in Figs. 10.26 and 10.27. We showed in Eq. (10.86) and Example 10.6 how to determine the average transmission coefficient for the entire enclosure.

Doors. Doors are a particularly troublesome problem because of the difficulty of achieving adequate seals around the edges and at the threshold. Many nationally advertised doors, claimed to give 35 to 45 db average attenuation, have been found to give only 20 to 25 db of attenuation when installed, simply owing to the lack of proper seals.

The primary problem with seals is that nearly all doors are irregular

Fig. 11.17. Attenuation due to a wall erected to a height H above the line of sight between sound source and observer. [*After Fehr The Reduction of Industrial Machine Noise, Proc. 2d Natl. Noise Abatement Symposium* (1951).] The circles are measured data for $H = 30$ ft, $R = 20$ ft, $D = 100$ ft, $N \doteq 40/\lambda$. Beyond $N = 15$, A was never greater than 20 db and dropped to 12 db at 3500 cps.

enough so that some parts of the seal must be compressed as much as ⅛ in. in order for other parts of the seal to come in contact with the door. Because of the large area involved at the points of pressure and because of the stiffness of the rubber or felt used, the force necessary to produce adequate compression of the seal is very large. It is difficult, therefore, for a person to push the door shut. Also, a heavy strain is put on the fittings.

One simple, and generally satisfactory, type of seal is shown in Fig. 11.18. The door closes against an extruded rubber gasket, 0.5 to 1.0 in. wide. The gasket can be run continuously around the door. The principal disadvantage is that a raised sill is necessary, with the necessity for a sloping threshold on one side of it.

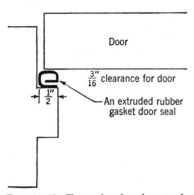

Door

$\frac{3}{16}''$ clearance for door

$\frac{1}{2}''$

An extruded rubber gasket door seal

FIG. 11.18. Example of a door seal made from an extruded rubber gasket. This seal runs around all four edges of the door.

Two possible seals at the threshold are shown in Fig. 11.19a and b.[6] The draft excluder is a device that is actuated by a plunger on the inside edge of the door. When the door is closed, the seal is forced down against the floor. This type of device can be used without a raised sill, although it is much less effective than the arrangement of Fig. 11.18.

The door must have a single panel of sufficient mass to prevent sound transmission, or else it must have heavy panels on each side floated in rubber with the air space in between filled with a mineral wool or glass blanket.

11.7. Mufflers and Ducts. Whenever a continuous flow of air accompanies a noise, it is not possible to erect a completely closed room around

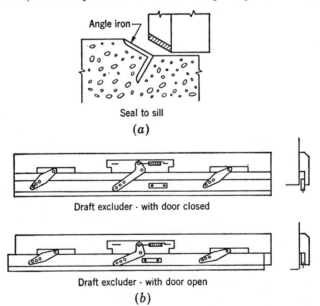

Angle iron

Seal to sill

(a)

Draft excluder · with door closed

Draft excluder · with door open

(b)

FIG. 11.19. Two possible means for providing seals at the threshold of a door. (*After Cullum, "The Practical Application of Acoustic Principles," Spon, Ltd., London,* 1949.)

the source of noise in order to reduce its disturbing effects. Instead, a muffler or a sound-attenuating duct must be used.

[6] D. J. W. Cullum, "The Practical Application of Acoustic Principles," Spon, Ltd., London, 1949.

In a muffler or duct, sound attenuation may occur by the reflection of energy back into the source or by absorption of the energy. Reflection is accomplished through the use of acoustic filters or ducts with bends in them. Absorption occurs in structures that dissipate the energy in porous blankets or tiles or in damped vibrating members. A few examples of each type of noise-reducing element are discussed here.

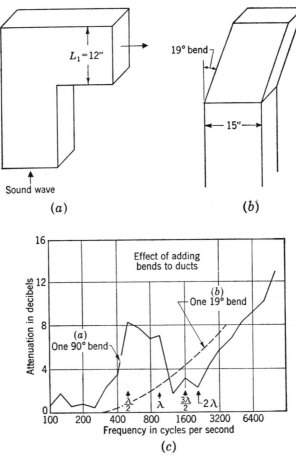

FIG. 11.20. Attenuation of sound by unlined bends in ventilating ducts. (a) Dimensions of unlined 90° bend. (b) Dimensions of unlined 19° bend. (c) Measured data for (a) and (b).

Ventilating Ducts. BENDS. Noise produced by ventilating fans may be reduced by the introduction of bends into the duct system. The performances of unlined 19° and 90° bends are shown in Fig. 11.20. It is seen that the attenuation is high at the basic resonance of the duct side walls (900 cps), at $L_1 = \lambda/2$ and $3\lambda/2$ and at high frequencies.

If the bends are lined with an acoustic material with an absorption coefficient in excess of 0.8, the losses are given approximately by the curves of Fig. 11.21.

LININGS. Linings in ducts are commonly used to reduce noise emanating from ventilating fans. The measured performance of four ducts each about 8 by 11 in. inside dimensions and each lined with a different material is shown in Fig. 11.22. The data are given as decibels per foot

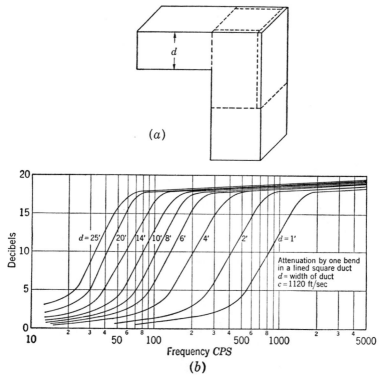

FIG. 11.21. (a) Physical dimensions of 90° lined bend in a duct. (b) Approximate curves for estimation of attenuation at the bend.

as a function of frequency. In Fig. 11.23, there are shown similar data for six sizes of ducts lined with Celotex rigid rock-wool sheet, 1.0 in. in thickness. The ordinate of this figure is decibels per foot, but the abscissa is the ratio of the perimeter P to the open cross-sectional area A in inches^{-1}. Frequency is given as the parameter. From these data, it is concluded that an approximate formula for the sound attenuation is[7]

$$\text{db/ft} = \frac{12.6\bar{\alpha}^{1.4}P}{A} \tag{11.7}$$

[7] H. J. Sabine, The Absorption of Noise in Ventilating Ducts, *J. Acoust. Soc. Amer.*, **12**: 53 (1940).

where $\bar{\alpha}$ = average absorption coefficient for the acoustic material lining
the duct

P = perimeter of the inside of the duct in inches

A = cross-sectional area of the inside of the duct in square inches

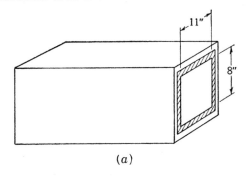

(a)

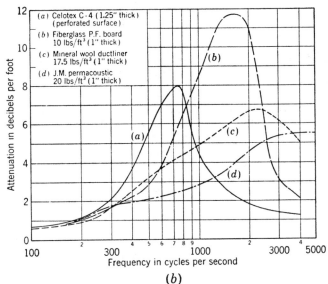

(b)

FIG. 11.22. (a) Sketch of a lined duct; and (b) measured attenuations in decibels per
foot for the duct of (a) with four different materials.

Acoustic Filters. Acoustic filters are used in the exhaust pipe of
every automobile and diesel-engine locomotive and on many industrial
machines. If a filter has no openings to the outside air except for the
necessary outlet opening, it is easy to make it reject high-frequency
sound.[8] It is much more difficult to make a filter reject low-frequency
sound while passing high-frequency sound. In general, there must be a

---

[8] A discussion of classic acoustic wave filters is given by W. P. Mason, "Electromechanical Transducers and Wave Filters," pp. 126–143, D. Van Nostrand Company,
Inc., New York, 1942; and by H. F. Olson, "Dynamical Analogies," pp. 52–104,
180–182, D. Van Nostrand Company, Inc., New York, 1943.

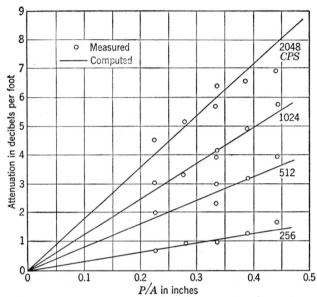

FIG. 11.23. Attenuation in decibels per foot for ducts lined on four sides with acoustic material. The solid lines are computed from Eq. (11.7), and the circles were measured by Sabine.[7]

TABLE 11.2

Type	Filter	Analogous circuit	Attenuation characteristic
Low pass No. 1	S_2 (closed) l_2 l_1 l_1	M_{AH} M_{A1} M_{A1} M_{AH} C_{A2}	Attenuation $f \rightarrow$
Low pass No. 2	S_2 (closed) l_2 S_1 l_1 l_1	M_{A1} M_{A1} M_{AH} C_{A2}	Attenuation $f \rightarrow$
Low pass No. 3	S_2 (closed) l_2 S_1 l_1 l_1 l_1	M_{A1} M_{A1} M_{A1} C_{A2} C_{A2}	Attenuation $f \rightarrow$
High pass No. 1	Diaphragm S_2 (open) l_2 l_1 l_1	S_2 (open) M_{A1} M_{A1} C_A C_A M_{A2}	Attenuation $f \rightarrow$
High pass No. 2	S_2 (open) l_2 l_1 l_1	M_{A1} M_{A1} M_{A2}	Attenuation $f \rightarrow$

separate space into which one can bleed the low-frequency sound if a high-pass filter is to be achieved, or else series diaphragms must be used. Obviously, if continuous air flow through the filter is desired, series diaphragms are not practical.

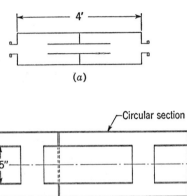

(a)

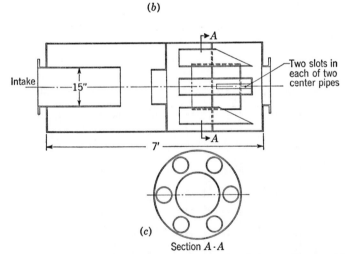

(b)

(c)

Section A-A

Fɪɢ. 11.24. Cross-sectional sketches of three nondissipative-type filters. (a) Simple two-chamber filter with connecting tube. (b) and (c) Filters for large diesel engines.

The basic information needed to select components for acoustic filters is given in Chap. 5 on Acoustic Components. In Table 11.2, we list a series of filter constructions along with approximately equivalent circuits and theoretical attenuation characteristics. The theoretical attenuation characteristics shown are based on the assumption that the filters are terminated at both ends in their image impedances. In practice, this will not be true, and the exact performance will depend on the nature of the termination. Also, all dissipation has been neglected in the circuits.

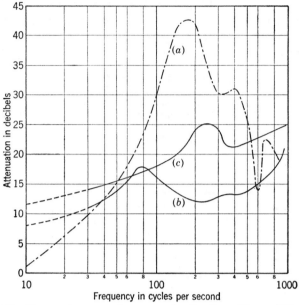

FIG. 11.25. Measured attenuation curves for the three filters of Fig. 11.24.

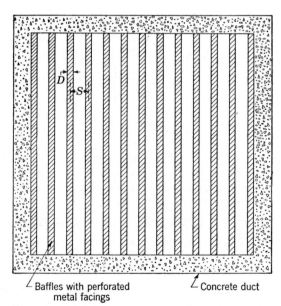

Baffles with perforated   Concrete duct
metal facings

FIG. 11.26. Elevation view of the end of a large duct containing parallel baffles. The percentage open area is $100(S - D)/S$. Each baffle is usually constructed from two parallel sheets of perforated metal filled with glass or mineral wool.

For low-pass filter 1, the holes on either end are acoustic masses, and the side protuberance is an acoustic capacitance. In the impedance analogy one obtains the equivalent circuit shown. For low-pass filter 2, the tubes on either end serve as acoustic masses. The side protuberance is a series acoustic mass provided by the holes and an acoustic compliance C_{A2}. In low-pass filter 3, two side protuberances and three sections of

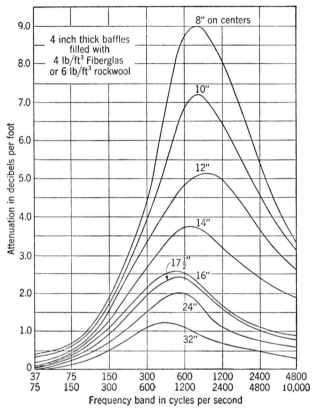

FIG. 11.27. Attenuation in decibels per foot for baffles that are 4 in. thick and mounted with the on-center spacings indicated on the curves. The baffles usually are filled with 4 lb/ft³ Fiberglas or 6 lb/ft³ rock wool. A continuous-spectrum noise is assumed. Smoothed curves were drawn through the experimental points.

tube are shown. These produce two shunt acoustic compliances and three series acoustic masses.

In the case of the high-pass filters, the shunt branch is simply a tube to the open air. If l_1 is made short, the filters will not cut off on the high-frequency end until the frequency is great.

Examples of three practical filters are shown in Fig. 11.24. The measured attenuations are shown in Fig. 11.25. Notice that the average attenuations for a filter of this construction becomes as high as 25 db at higher frequencies. More sections will increase the high-frequency attenuation, and a larger size will lower the lowest frequency of high attenuation.

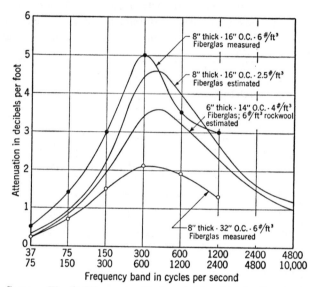

FIG. 11.28. Same as Fig. 11.27 except with 6- and 8-in.-thick baffles filled with absorbing material as indicated.

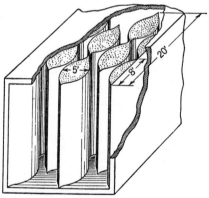

FIG. 11.29. Isometric drawing for standard Soundstream absorber. Each column has a maximum width of 5 ft and a length of 8 ft, measured from thin portion to thin portion. The columns are separated so that the lateral open distance between them is 4 ft 10 in.

Parallel Absorbent Sheets. Many applications, such as air-conditioning cooling towers and aircraft-propeller test cells require the movement of large quantities of air at high speeds with low back pressures. For these applications, a series of parallel baffles are frequently used as in Fig. 11.26, where parallel baffles of thickness D are shown with an on-center spacing S.

The performance of such baffles with thicknesses of 4 and 8 in. for a number of on-center spacings is shown in Figs. 11.27 and 11.28.   The data are given in terms of decibels per linear foot and are measured by moving a microphone through the structure (*i.e.*, into the page for Fig. 11.26) and plotting the sound pressure level as a function of distance. The slope of this curve gives the attenuation in decibels per foot.

The amount of attenuation for a structure of a given length is found by multiplying the attenuation per foot by the length in feet.   However, very high attenuations (50 db or more) are very difficult to achieve because the sound will by-pass the absorbing material by traveling along the metal supporting structure or the outside walls unless frequent vibration breaks and heavy walls are employed.

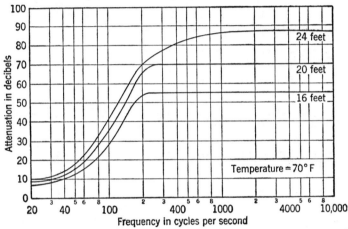

FIG. 11.30. Performance curves for three lengths of standard Soundstream absorbing units constructed and mounted as shown in Fig. 11.29.

Sinusoidal Absorbent Sheets.   A recently developed material for attenuating sound in air-conducting passages is shown in isometric section in Fig. 11.29.†   The attenuating properties of a structure of this type for units that have a "half wavelength" of 8 ft and a thickness (*i.e.*, double amplitude) of 5 ft, as shown in the figure, are given in Fig. 11.30 for 16-, 20-, and 24-ft lengths of the treatment.

Absorbent Mufflers.   Mufflers are also made with absorbent linings. The two most common constructions are shown in Fig. 11.31.   They are usually made either circular or square in cross section.   The approximate performance data for these two types of mufflers are given in Fig. 11.32 for the special case of $d/\lambda = 1$ at 370 cps.   These curves are valid approximately for other sizes of mufflers if the abscissa is replaced by the scale given at the top of the graph.

Flanking Transmission.   In conclusion, it should be emphasized that a sound-attenuating structure is no better than the most transmissive of its component parts.   If a concrete test cell for a propeller is to be used

† This material is sold under the trade name of Soundstream Absorber.

in a location where the criterion requires a noise reduction of 60 db at a particular frequency, it does no good to provide this amount of attenuation with the intake and exhaust treatments of the stacks and then to choose concrete walls that attenuate the sound only 45 db. Also, it must be remembered that sound will be transmitted as vibrations along

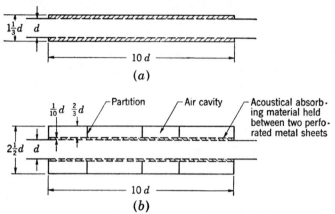

(a)

(b)

Fig. 11.31. Construction of two types of dissipative mufflers: (a) solid packed; (b) airspaces and compartmentation behind absorptive layer.

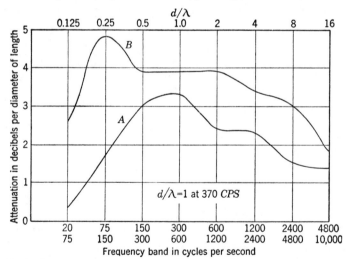

Fig. 11.32. Performance of the mufflers of Fig. 11.31, for the special case of $d/\lambda = 1$ at 370 cps.

mechanical members and be reradiated as sound. Hence, vibration breaks must be provided at sufficient intervals to preserve the full value of the treatment. Usually, vibration breaks are necessary in concrete structures and metal mufflers whenever the attenuation exceeds 25 db at frequencies below 100 cps or exceeds 50 db at frequencies above 1000 cps.

CHAPTER 12

ACOUSTIC MEASUREMENTS

PART XXVIII *Measurement of Acoustic Levels*

The student does not work long in the field of acoustics without realizing that in nearly every study measurements of sound pressure and particle velocity are necessary. Here is a brief introduction to the broad subject of acoustic measurements. More complete treatments have been given elsewhere.[1-3]

This part discusses the technique of measuring sound pressure levels, sound levels, and power levels in frequency bands of various widths. It presents the techniques that one must use when measuring fluctuating sounds and describes the effect the angle of incidence has on the readings of the sound-level meter. It tells how to choose microphones, analyzers, and recorders for particular purposes. The effect of the observer's presence on measured data is discussed, as well as the effect of the meter case itself. Finally, this part discusses briefly the use of the American Standard sound-level meter.

12.1. Sound Pressure and Power Levels. In most cases, the parameter of a sound field that is measured most commonly and expeditiously is the sound pressure, rather than the particle velocity or other quantities. This is so, in part, because stable pressure-actuated microphones that will operate over wide ranges of pressures and frequencies are much easier to build and to calibrate than are velocity microphones or intensity-measuring devices. Further, the ear, which is so often the ultimate receptor of the desirable or undesirable sounds being measured, is a pressure-operated instrument.

[1] Leo L. Beranek, "Acoustic Measurements," John Wiley & Sons, Inc., New York, 1949.

[2] A. P. G. Peterson and L. L. Beranek, "Handbook of Noise Measurement," General Radio Company, Cambridge, Mass., 1953.

[3] E. G. Richardson, "Technical Aspects of Sound," Elsevier Publishing Company, Inc., New York, 1953.

Because of the wide range of sound pressures encountered in practice, measurements are usually expressed in terms of the sound pressure level, in decibels, defined as

$$\text{SPL} = 20 \log_{10} p + 74 \text{ db } re \text{ } 0.0002 \text{ microbar} \qquad (12.1)$$

where p is the sound pressure in microbars (0.1 newton/m^2 or 1 dyne/cm^2).

The basic equipment for measurement of sound pressure level comprises a calibrated microphone, a calibrated amplifier, and a calibrated indicator such as a meter or oscillograph or graphic recorder, as shown in Fig. 12.1. If the sound to be measured is anything other than a pure tone (single-frequency) it is necessary that the basic equipment have a uniform response vs. frequency characteristic over the entire frequency range of interest in the complex sound. Such an instrument is referred to as a "sound-pressure-level meter."

Measurement of Sound Pressure Level as a Function of Frequency. When the sound to be measured is complex, consisting of a number of tones, or is a noise with a continuous spectrum, the single value of over-all sound pressure level given by the basic equipment often is not sufficient for analytical purposes. One may wish to know also the sound pressure level as a function of frequency. In that case band-pass filters whose mid-band frequency is either continuously or stepwise variable must be added to the basic equipment, so that the sound pressure may be measured in known bands of frequencies.

In general, three types of analyzers are used, (1) constant bandwidth, (2) constant-percentage narrow bandwidth, and (3) octave, half-octave, or third-octave bandwidth. The first of these, the constant-bandwidth type, has a pass band that is a fixed number of cycles wide. Common bandwidths range between 5 and 200 cps. This instrument is used mostly for determining the harmonic components of a sound, when the frequencies are sufficiently stable so that the components will not shift in and out of the narrow pass band during measurement.

The second type, the constant-percentage narrow-bandwidth analyzer, has a bandwidth that is a fixed percentage of the mid-band frequency. Thus, at high frequencies, the pass band covers a wider range in cycles per second than at low frequencies, so that one needs to take fewer readings at high frequencies with this type than with the constant-bandwidth type. The constant-percentage bandwidth analyzer is particularly useful in measuring the harmonic components of a wave whose fundamental frequency fluctuates somewhat. If the band is wide enough to contain the varying fundamental, it will also be wide enough at the higher frequencies to contain any of the harmonics.

A third commonly used instrument, the octave-, half-octave-, or third-octave-bandwidth analyzer, has many applications in situations where great detail is not required in the analysis of continuous-spectrum noises.

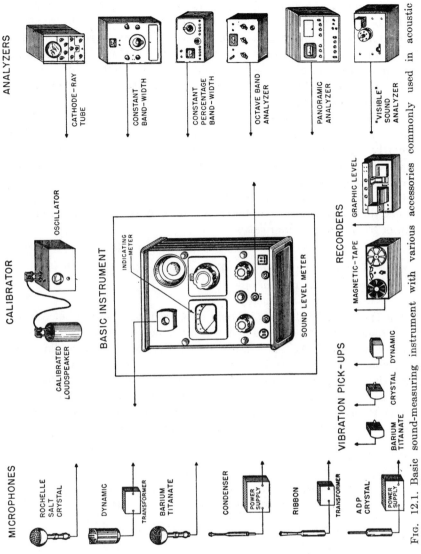

MICROPHONES

ROCHELLE
SALT
CRYSTAL

DYNAMIC

TRANSFORMER

BARIUM
TITANATE

CONDENSER

POWER
SUPPLY

RIBBON

TRANSFORMER

ADP
CRYSTAL

POWER
SUPPLY

CALIBRATOR

OSCILLATOR

CALIBRATED
LOUDSPEAKER

BASIC INSTRUMENT

INDICATING
METER

SOUND LEVEL METER

VIBRATION PICK-UPS

BARIUM
TITANATE

CRYSTAL

DYNAMIC

RECORDERS

MAGNETIC-TAPE

GRAPHIC LEVEL

ANALYZERS

CATHODE-RAY
TUBE

CONSTANT
BAND-WIDTH

CONSTANT
PERCENTAGE
BAND-WIDTH

OCTAVE BAND
ANALYZER

PANORAMIC
ANALYZER

"VISIBLE"
SOUND
ANALYZER

Fig. 12.1. Basic sound-measuring instrument with various accessories commonly used in acoustic measurements.

363

As indicated by the name, the upper cutoff frequency of each pass band of an octave-bandwidth analyzer is twice the lower cutoff frequency. For a half-octave analyzer, the upper cutoff frequency is $\sqrt{2}$ times the lower, and for a third-octave analyzer the ratio is the cube root of 2. A commonly used set of octave pass bands in commercial equipment is 37.5 to 75, 75 to 150, 150 to 300, 300 to 600, 600 to 1200, 1200 to 2400, 2400 to 4800, and 4800 to 9600 cps.

With any of these types of analyzers, care must be taken that attenuation of the filter outside the pass band is adequate. This becomes particularly important in cases where the spectrum of the noise slopes markedly or where one is measuring low-level sounds in the presence of very high level sounds.

Once the sound pressure level is known as a function of frequency bands, the *spectrum level* $S(f)$ for each band may be determined as a function of frequency by the relation

$$S(f) = \text{SPL} - 10 \log_{10} \Delta f \qquad \text{decibels} \qquad (12.2)$$

where SPL = sound pressure level measured with a bandwidth Δf
$\quad \Delta f$ = bandwidth in cps
$\quad f$ = center frequency

The spectrum level is what would have been measured if the analyzer had a bandwidth of 1 cps and if the spectrum level were uniform throughout the bandwidth Δf. It is a useful quantity for comparing data taken with analyzers of different bandwidths. Obviously, reduction of wideband data to spectrum levels has meaning only if the spectrum of the noise is continuous and does not contain prominent pure-tone components.

To convert octave-band readings in decibels (for a continuous-spectrum noise) to spectrum levels in decibels, use the values in Table 12.1.

TABLE 12.1.   Conversion of Octave-band Levels to Spectrum Levels†

Octave band, cps	Number of decibels to be subtracted from octave-band levels	Plot at the frequency, cps
(20–75)	(17)	(39)
37–75	16	52
75–150	19	105
150–300	22	210
300–600	25	420
600–1200	28	850
1200–2400	31	1700
2400–4800	34	3400
4800–9600	37	6800

† To nearest 0.5 db.

The measurements of sound pressure level made with an analyzer having a bandwidth Δf may be converted to the sound pressure level in a critical bandwidth Δf_c (as plotted in Fig. 12.2 and discussed in Part **XXX**) by means of the following formula,

$$\text{SPL}_c = \text{SPL} - 10 \log_{10}\left(\frac{\Delta f}{\Delta f_c}\right) \qquad \text{db} \qquad (12.3)$$

where SPL_c is the sound pressure level in the critical bandwidth. This formula should be applied only to noises with a continuous spectrum. If a noise being measured is a combination of a continuous-spectrum noise and pure tones and if the purpose of the measurements is to indicate

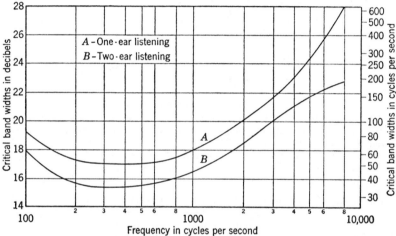

Fig. 12.2. Critical bandwidths for listening. (A) One-ear listening. (B) Two-ear listening. The right-hand ordinate is Δf in cycles per second. The left-hand ordinate is $10 \log_{10} \Delta f$ in decibels.

the manner in which a listener will hear the noise, the data are plotted as shown in Fig. 12.3. Here the sound pressure levels of the continuous-spectrum portion of the noise are for critical bandwidths. The pure tones are plotted as actually measured. This graph then directly indicates the number of decibels the pure tone must be reduced if it is to become inaudible in the background noise. To convert from octave-band readings in decibels (for a continuous-spectrum noise) to critical band levels in decibels, one may make use of the values given below in Table 12.2.

Measurement of Total Power and Directivity of a Source. In many situations, it is necessary to take measurements of sound pressure level around a machine or other noise-producing device and to derive from these measurements the properties of the source defined by the following quantities or functions:

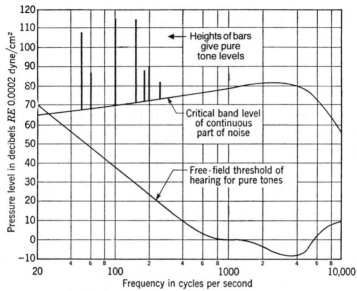

FIG. 12.3. Typical method for plotting a combined continuous-spectrum and line-spectrum noise. The continuous part is expressed in decibels for critical bandwidths referred to a reference intensity. The discrete components are plotted directly as measured. The resulting graph yields the number of decibels that any line component must be reduced either to become inaudible in the presence of the continuous-spectrum noise, or if there is no noise, to fall below the threshold curve.

TABLE 12.2.   Conversion of Octave-band Levels to Critical Band Levels†

Octave band, cps	Number of decibels to be subtracted from octave-band levels	Plot at the frequency, cps
(20–75)	0 (or −1)	(39)
37–75	0 (or −1)	52
75–150	1.0	105
150–300	6.0	210
300–600	9.0	420
600–1200	11.5	850
1200–2400	13.0	1700
2400–4800	13.5	3400
4800–9600	14.0	6800

† To nearest 0.5 db.

1. The total sound power W in watts radiated by the source, as expressed by the power level

$$PWL = 10 \log_{10} W + 130 \text{ db} \qquad (12.4)$$

2. The directional characteristics of the source, as expressed by the directivity patterns or, at particular angles, the directivity factors Q (or directivity indexes DI).

3. The frequency characteristics of the source, as expressed by power levels as a function of frequency defined in discrete frequency bands.

Nondirectional Sound Sources.   If a sound source is known to be nondirectional and is located in free space, measurements as a function of frequency of the sound pressure level at one location at a distance of several source diameters are sufficient to determine the sound power level of the source as a function of frequency.   The power level is related to the measured sound pressure level by the formula

$$\text{PWL} = \text{SPL} + 20 \log_{10} r + 10.5 \text{ db} \tag{12.5}$$

where r is the distance of the microphone from the source in feet.   This relation and Eq. (12.6) are correct at 68°F and 30 in. Hg.   At other temperatures and pressures it is necessary to *subtract* the correction of Fig. 10.19 from the power level.

When a nondirectional source is located indoors, a single measurement is again sufficient.   That is to say, the measurements are made in the reverberant field at a distance r from the source and if $16\pi r^2$ is large compared with the room constant R, then the equation below is used to determine the sound power level for a given frequency band,

$$\text{PWL} = \text{SPL} + 10 \log_{10} \frac{R}{4} - 0.5 \text{ db} \tag{12.6}$$

where R is the room constant in square feet.   To evaluate R, the average absorption characteristics of the room as a function of frequency must be known.   If the approximation given above is not satisfied, then Eq. (10.64) should be used to determine the sound power in a given frequency band.   Otherwise, the source should be moved into anechoic space.

Directional Sound Sources.   More extensive measurements must be made to evaluate the characteristics of a directional sound source.   When the source is located outdoors, sound pressure levels at many points on a sphere surrounding the source must be measured or estimated.   The radius of the sphere should be several source diameters so that the data are taken in the far-field.   In practice, measurements are made only at a limited number of locations around the machine (see Fig. 4.19), and the sound levels at other locations are estimated by interpolation between the measured levels or by making use of any geometrical symmetry in the source.

Measurements should be made at a sufficient number of positions such that the maximum variation between measurements is no greater than about 6 db.   In reading the sound level meter at a measuring point, the mean reading of the meter should be recorded if the total fluctuation is 6 db or less.   Otherwise, the region surrounding the source should be broken up into more areas.

To determine the total acoustic power at a given frequency, the sphere (or hemisphere) over which measurements have been taken should be divided into the same number of areas as measurement positions used to obtain data. The power in watts, W_S (at normal temperature and barometric pressure) that passes through each area, at a given frequency, is found approximately from

$$W_S = 10^{-13}S\left(\text{antilog}_{10}\frac{\text{SPL}}{10}\right) \qquad \text{watts} \qquad (12.7a)$$

where S = area of the incremental surface in *square feet*
    SPL = sound pressure level in decibels
or from

$$W_S = 10^{-12}S'\left(\text{antilog}_{10}\frac{\text{SPL}}{10}\right) \qquad \text{watts} \qquad (12.7b)$$

where S' = area of the incremental surface in *square meters*.† The sum of the powers so determined for all the areas represents the total acoustic power W, in watts, radiated by the source at a given frequency or in a given octave band.

The calculations can be greatly simplified by computing an average sound pressure level, $\overline{\text{SPL}}$, in decibels, as the arithmetic average of the SPL's in equal-area measuring sections. Of course, some error will be present if this is done, except for the case in which all SPL's are equal. The value of $\overline{\text{SPL}}$ obtained by arithmetic averaging of decibels is always too low. The maximum error for the case where there is a 10-db spread is -2.6 db for either four or eight measuring stations. Thus, the following rule of thumb may be adopted: If the spread of measured values of SPL at four or eight measuring stations is around 10 db and if the SPL's in decibels are averaged, add 1 db to the average to get the $\overline{\text{SPL}}$. The resulting $\overline{\text{SPL}}$ will usually be off not more than ± 1 db. If the spread of measured values is 5 db or less, add no correction. Here also, the error will always be less than 1 db.

Methods for dividing a spherical surface into 10 or 20 sections of equal area were given in Part XI (page 110). In practice, it is often not possible to obtain readings over a surface completely surrounding the source, at a distance of several source diameters. However, the radiation pattern produced by the sound source frequently will exhibit some form of symmetry such that the sound pressure level can be assumed constant along a given line. For example, a loudspeaker that is placed symmetrically in a large baffle often will produce a directivity pattern that is symmetrical with respect to circles around the main radiation axis. In a similar manner, one often assumes that the radiation pattern produced by a noise source, situated on a plane surface, is symmetrical about some axis. Measurements in a horizontal plane along a circle surrounding the source will then suffice. If a directional sound source is located

† These equations come from Eqs. (1.18) to (1.21).

indoors, one should make certain that the measurements of the direct air-borne sound are without any appreciable contribution from the reflected sound. To ensure that the reflected sound does not contribute appreciably to the readings, the average sound pressure level measured at the selected measuring points should be at least 8 db higher than the average sound pressure level measured at more distant points, where reflected sound predominates. However, the measuring point should still be several source diameters distant from the source in order to be certain that one is in the far-field.

Calculation of Directivity Factor. The directivity factor also can be calculated after the average sound pressure level $\overline{\text{SPL}}$ has been determined. The $\overline{\text{SPL}}$ can be determined by the approximate method given above or by use of

$$\overline{\text{SPL}_r} = \text{PWL} - 20 \log_{10} r - 10.5 \text{ db} \tag{12.8}$$

where PWL = power level for the source in the given frequency band
  r = radius in feet for which the average sound pressure level is determined

The directivity index $\text{DI}_{(\theta,\phi)}$ in a particular direction given by the angles θ and ϕ is related to the sound pressure level $\text{SPL}_{(r,\theta,\phi)}$ measured at distance r by the formula

$$\text{DI}_{(\theta,\phi)} = \text{SPL}_{(r,\theta,\phi)} - \overline{\text{SPL}_r} \tag{12.9}$$

The directivity factor $Q_{(\theta,\phi)}$ is obtained from the directivity index by converting the value of DI in decibels into a power ratio, *i.e.*,

$$Q_{(\theta,\phi)} = \text{antilog}_{10} \frac{\text{DI}_{(\theta,\phi)}}{10} \tag{12.10}$$

Choice of Microphone. The microphone used for measuring sound pressure levels should be selected for its suitability for the particular sounds to be measured. If extremely low sound levels are to be measured, only certain types of microphones are suitable. If measurements are desired at very low frequencies or at very high frequencies, other types of microphones are best suited to the task. If durability is an important consideration at the expense of optimum frequency response, a still different selection of microphones should be made.

In the paragraphs that follow, we shall discuss separately the problems of measurement of low sound levels, high sound levels, low-frequency noise, high-frequency noise, and the effects of varying temperature, severe handling, and long cables.

Low Sound Levels. A microphone that is used to measure low sound levels must have low "self-noise," and it must produce an output voltage sufficient to override the noise at the grid of the first amplifier tube in the sound-level meter. Rochelle salt and moving-coil microphones are ideally suited to measurements of low sound levels. For both these types the output voltage is sufficiently high to permit measurement of sound levels down to approximately 20 db *re* 0.0002 microbar. The ambient noise level of the crystal microphone is extremely low if the frequency range is restricted to the region above 20 cps. The ambient noise level for the dynamic microphone is equivalent to a sound pressure level of 15 db. Therefore, by using the dynamic microphone it is possible to measure sound levels as low as 20 db with fairly good accuracy.

High Sound Levels. Ordinary Rochelle salt, moving-coil, and capacitor microphones may be used for the measurement of sound pressure levels that do not exceed approximately 140 db. For higher levels, specially designed microphones of the crystal or capacitor type with stiff diaphragms can be obtained commercially.

Low-frequency Noise. Crystal and capacitor microphones are most suitable for measuring noise at low frequencies. With either of these two types, measurements may be made down to fractions of a cycle if special amplifiers are used. The limitation at low frequencies, if any, is entirely in the amplifiers, since both types respond to static as well as alternating pressures.

High-frequency Noise. For measurements up to about 12,000 cps a capacitor microphone gives satisfactory results. For measurements above 12,000 cps the most suitable types of microphones are very small crystals. The primary requirements for accurate measurement of high-frequency sounds are small size and freedom from resonance peaks in the microphone itself. These restrictions usually mean a sizable reduction in sensitivity.

Varying Temperature. If a microphone must operate in an environment where the temperature varies radically from one time to another, its response and other relevant properties should not be a strong function of temperature. A typical Rochelle salt crystal microphone has the temperature characteristics shown in Fig. 12.4. It requires temperature corrections of significant magnitude if used with a cable. The outputs of moving-coil, ribbon, and capacitor microphones are less variable with temperature and serve a variety of applications where the Rochelle salt microphone would not be suitable because of variation in ambient temperature. Many of these types will operate in temperatures up to about 200°F, although accurate data on their performance in this range are not available. Rochelle salt, in particular, is completely destroyed at 130°F and should not be used at temperatures above 113°F.

Humidity. In general, most modern microphones are relatively unaffected by large changes in relative humidity. However, extended periods of operation in relative humidities greater than 85 per cent should be avoided, and if necessary, some means of desiccation should be provided. Desiccation is particularly important for capacitor microphones.

Long Cables. If long cables must be used, the output of the microphone should be at low impedance. This may be accomplished either by using an impedance-matching transformer or a suitable vacuum-tube circuit at the microphone or by selecting a microphone with a low-impedance transducing element. A moving-coil microphone may be used with cables of nearly any length provided the loop resistance is

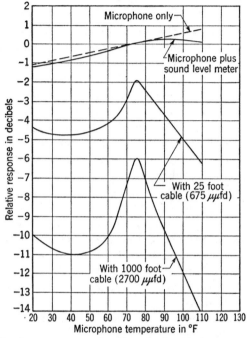

Fig. 12.4. Variation in response as a function of microphone temperature for a typical Rochelle salt crystal microphone with a capacitance of 1600 mmfd at 70°F, for the Rochelle salt crystal microphone alone and with various lengths of cable between the microphone and the grid of the input stage of the sound-level-meter amplifier. (*Courtesy of General Radio Company, Cambridge, Mass.*)

known and the output of the microphone is corrected to account for the losses in the cable.

Capacitor or crystal microphones usually may be used with cables up to 200 ft in length provided the cable corrections are applied to the measured data.

Measurement of Fluctuating Sounds. When a sound being measured fluctuates in intensity, the values of sound pressure that are measured will depend upon the dynamic characteristics of the indicator used. The type of indicator chosen will depend on the attribute of the sound pressure level which is to be measured. Ordinarily, if one wishes to measure instantaneous peak pressures, an oscilloscope will be necessary. "Fast" meters, or high-speed graphic level recorders, may be used for investigating relatively rapid changes in sound pressure level. For obtaining long-time average values, "slow" meters, or recorders, are used whose response is intentionally made very sluggish by appropriate design of their electrical or mechanical circuits. The long-time average value and the range of instantaneous peaks and dips of the sound pressure are usually sufficient statistics to characterize the sound for practical purposes.

Sounds at Grazing and Random Incidence. For the proper measurement of sound pressure levels it should be remembered that the response of most microphones varies somewhat with direction of the sound impinging upon it. For most diaphragm types of microphones, the "flattest" response characteristic is obtained for randomly incident sound. This response is nearly the same as that for grazing-incident sound. However, both the random and the grazing-incidence responses differ greatly from the response for normally incident sound for frequencies where the microphone dimensions are no longer small compared with the wavelength of sound. Thus, the microphone should be located so that when the sound level meter is held in the usual manner, the sound being measured impinges on the diaphragm of the microphone at grazing incidence. In a reverberant room, the sound is approximately randomly incident and the orientation of the microphone is uncritical. If by any chance considerable reflected sound at the high frequencies is incident normally on the diaphragm of the microphone, serious errors may result.

12.2. Sound Level and the Sound-level Meter. *Characteristics.* In measurements of noise, it is often desirable to use equipment whose indications are weighted in accordance with the way in which the human ear would respond to the noise. No such meter has been invented, but a meter has been standardized that was originally thought to measure something akin to loudness level. Readings from this meter are called "sound levels." The American Standards Association has published a set of specifications in which the requirements for a standard sound-level meter are given.[4] Three weighting networks are used to simulate the response of the hearing mechanism to low-, medium- or high-level *pure tones*. The *A*, *B*, and *C* networks give response characteristics that are approximately the inverse of the 40-, 70-, and 100-phon equal-loudness contours *for pure tones*, respectively, as shown in Fig. 12.5. *The sound level of a noise is, by definition, the reading of a sound-level meter built to conform to the American Standard requirements and operated about as follows:* Use the *A* network if the readings lie between 24 to 55 db; *B* network for readings between 55 to 85 db; *C* network for readings above 85 db. Readings are usually taken with each of the three weighting networks. From these readings, information regarding the frequency distribution of the noise can be obtained. If the sound level is the same on each of the networks, the sound energy is probably predominant at frequencies above

[4] American Standard Z24.3—1944, "Sound Level Meters for Measurement of Noise and Other Sounds," American Standards Association, Inc., New York.

500 cps. If the sound level is the same for networks B and C, but less with network A, the sound probably predominates in frequencies between 150 and 1000 cps. Finally, if the sound level is greatest with network C, the sound predominates in frequencies below 150 cps. It is emphasized that although these weighting networks are useful in giving the loudness level of *pure tones*, they are not able to give the loudness level of complex noises. This subject will be discussed more in Chap. 13.

Since the C network has a uniform response with frequency, it may also be used to read sound pressure levels.

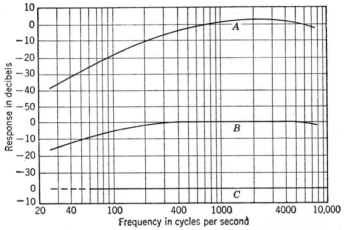

Fig. 12.5. Frequency-response characteristics taken from the American Standard for sound-level meters, Z24.3–1944.

Effect of Observer and Meter Case on Measured Data. If the sound-level meter is held in the hands of an observer, the readings at the higher frequencies will be disturbed by his presence. The magnitude of the error for one type of sound-level meter can be evaluated from comparison of the dashed and solid lines shown in Fig. 12.6. These data show the difference between the readings of the meter with and without the observer present, as compared with the response of the microphone alone when used on the end of a long cable. Two locations are shown: (1) the sound-level meter is between the observer and the noise source, and (2) the noise source is located to one side of the observer, and the sound-level meter is held in front of the observer. The graphs show that the error is smallest when the sound-level meter is held in front of the observer with the sound coming from his side (0°).

The meter case itself may also disturb the sound field at the microphone as shown by the solid lines of Fig. 12.6. The effect of the meter case is greater at high frequencies than at low frequencies, and it varies considerably with angle of incidence.

When very accurate data are required, the microphone should be connected to the meter by an extension cable.   In this case, the microphone should be held so that the sound radiated directly from the machine approaches the microphone from the side at grazing incidence, as we discussed before.

12.3. Effect of Background Noise.   The sound pressure level produced by a given machine must often be measured in a location where it is impossible to eliminate completely the noise from other sources.   Ideally, the measurement should determine only the direct air-borne sound from the machine, without any appreciable contribution from ambient noise from other sources.   If the ambient noise at the test location is to be

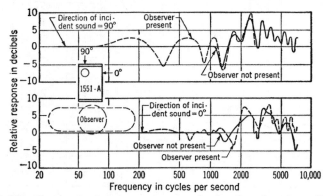

FIG. 12.6. The effect on the frequency response of using the microphone directly on a General Radio type 1551-A sound-level meter with and without an observer present. The decibel readings were obtained using a single-frequency plane acoustic wave in an anechoic chamber, and they are the difference between the response under the condition shown and the response of the microphone alone.   (*Courtesy of General Radio Company, Cambridge, Mass.*)

neglected, its level should be at least 8 db lower than the sound pressure level produced by the machine being tested.   When this condition is attained, the measured sound pressure level is in error by only a fraction of a decibel owing to the presence of the background noise.

If the ambient noise level is steady, a correction may be applied to the measured data according to the information given in Fig. 12.7.   For example, if the combination of ambient and apparatus noise is 4 db greater than the ambient noise alone, then from Fig. 12.7 we find that 2.2 db should be subtracted from the reading for the total noise to obtain the sound level due to the apparatus alone.

12.4. Information to Be Recorded.   The observer should indicate as part of his results all conditions under which the data were taken.   He should record:

1. Description of the space in which the measurements were made: Nature and dimensions of floor, walls, and ceiling.   Description and location of nearby objects and personnel.

2. Description of device under test (primary noise source): Dimensions, name-plate data, and other pertinent facts including speed, power rating, pulley or gear sizes, etc. Kinds of operations and operating conditions. Location of device and type of mounting.

3. Description of secondary noise sources: Location and types. Kinds of operations.

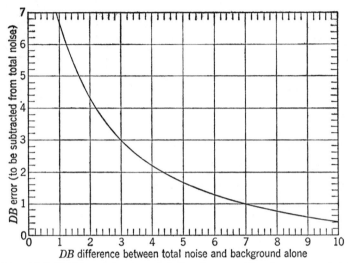

FIG. 12.7. Background-noise correction for sound-level measurements.

4. Serial numbers and type numbers on all microphones, sound level meters, and analyzers used. Length and type of microphone cables used.

5. Positions of observer.

6. Positions and orientations of the microphone during the tests.

7. Weighting scales and meter speed used on the sound level meter.

8. Temperature of the microphone.

9. Ambient noise spectrum with the device being tested not in operation.

10. Amount of fluctuation, plus or minus, of the meter.

11. Results of maintenance and calibration checks.

12. Name of the observer.

13. Date and time of test.

When the measurement is being made to determine the extent of noise exposure of personnel, the following items are also of interest:

1. Personnel exposed—directly and indirectly.

2. Time pattern of the exposure.

3. Attempts at noise control and personnel protection.

4. Audiometric examinations.

One suggested two-page data sheet is shown on page 376.

PAGE 1

SOUND SURVEY

DATE _____

ORGANIZATION _____

ADDRESS _____

INSTRUMENTS USED
SOUND-LEVEL METER – TYPE _____ MODEL # _____

MICROPHONE _____ TEMPERATURE _____ CABLE (Length) _____

ANALYZER – TYPE _____ MODEL # _____

OTHERS _____

NOTE: If noise is directional, record – Distance of the source, microphone position, incidence on microphone (Normal, Grazing, Random).

_____

_____

INDUSTRY _____ TYPE OF MACHINE _____

MACHINE MODEL # _____ NUMBER OF MACHINES _____

LOCATION OF MACHINE IN ROOM _____

ENVIRONMENT (Type of building, walls, ceiling, etc.; other operations, any attempts at sound control)

_____

_____

_____

PERSONNEL EXPOSED – DIRECTLY _____ INDIRECTLY _____

EXPOSURE TIME PATTERN _____

ARE EAR PLUGS WORN _____ TYPE _____

ARE THERE AUDIOMETRIC EXAMINATIONS _____

PREPLACEMENT _____ PERIODIC _____

Note information as to who makes these examinations, conditions under which they are made, time of day they are made, where records are kept.

_____

_____

_____

_____

_____

Engineer _____

PAGE 2

DATE _____

		FREQUENCY RANGES (Cycles per Second)										
	Flat 0	4800–10,000	2400–4800	1200–2400	600–1200	300–600	150–300	75–150	37.5–75	Flat 0	Time
LOCATION											

SOUND LEVEL VALUES (DECIBELS)

NOTE: Record A, B and C Networks Readings from the Sound-Level Meter

PART XXIX *Reciprocity Calibration of Transducers*

Laboratory standard microphones are calibrated by a primary technique, *i.e.*, one that depends on measurements only of length, mass, time, and ratios of other quantities.   There are several primary techniques for microphone calibration, but, except for the reciprocity technique, these generally are based on assumptions difficult to verify or are limited to a particular frequency range.

12.5. The Reciprocity Principle.   The principle of reciprocity is expounded in texts on linear-circuit theory.   Stated in one of its simpler forms, it says that if a constant-current generator i is connected to two accessible terminals of a linear passive electrical network and an open-circuit voltage e at two other accessible terminals is measured, then, as a second experiment, if the same constant-current generator i is connected to the second set of terminals, the same open-circuit voltage e will be measured at the first two terminals.

In a circuit that consists of two linear passive reversible electro-mechano-acoustic transducers, coupled by a gaseous medium (see Fig. 12.8), we say similarly: Assume that a constant-current generator i_1 is connected to the terminals of transducer 1 and, as a result of this action, an open-circuit voltage e_2 is produced at the terminals of transducer 2. Then, as a second experiment, assume that a constant-current generator i_2 is connected to the terminals of transducer 2 and an open-circuit voltage e_1 is measured at the terminals of transducer 1.   By the reciprocity theorem,

$$\frac{e_2}{i_1} = \frac{e_1}{i_2} \tag{12.11}$$

Referring to Fig. 12.8c, d, and e, we can, by extension of the reciprocity theorem, write that

$$\frac{e_2'}{i_1} = \frac{e_1}{i_2''} \tag{12.12}$$

where i_2'' = short-circuit current for a generator of internal impedance R and open-circuit voltage e_2''.   This generator is used in Fig. 12.8d to produce e_1.

As proof of Eq. (12.12), let us assume that the internal impedance at the terminals of transducer 2 is Z_2.   Then e_2' (see Fig. 12.8c) is related to the open-circuit voltage e_2 (see Fig. 12.8a) by

$$e_2 = \frac{R + Z_2}{R} e_2' \tag{12.13}$$

Also, the current i_2 (see Fig. 12.8b and d) is related to the voltage e_2'' by

$$i_2 = \frac{e_2''}{R + Z_2} \qquad (12.14)$$

From Fig. 12.8e,

$$i_2 = \frac{Ri_2''}{R + Z_2} \qquad (12.15)$$

Substitution of Eqs. (12.13) and (12.15) into Eq. (12.11) yields Eq. (12.12).

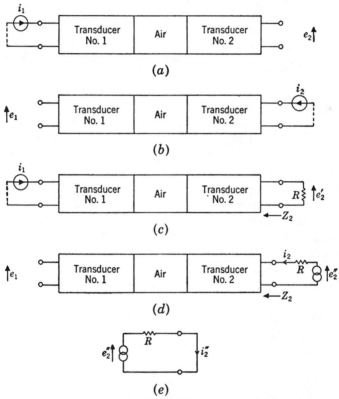

FIG. 12.8. Block diagrams for the two steps of an experiment demonstrating the reciprocity theorem using two reversible linear passive transducers connected together by an elastic medium (air). Reciprocity says that $e_2/i_1 = e_1/i_2$ or $e_2'/i_1 = e_1/i_2''$.

The equality given in Eq. (12.12) makes it possible to calibrate a transducer by the so-called reciprocity technique, as we shall show shortly.

12.6. Directivity of Reversible Transducers. As a further extension of the reciprocity theorem, it can be said that the directivity patterns of a given transducer at the same distance r (Fig. 12.9) are the same whether it is used as a microphone or as a loudspeaker. For large values of r the directivity patterns are independent of r because the sound pressure decreases linearly with r if r is large enough.

In Fig. 12.9, a pair of transducers is shown, one of which can be rotated in a plane at angles θ. If the open-circuit voltage of transducer 2 varies as some function of θ, say $e_2 = E_2 f_2(\theta)$, when a constant current i_1 is applied to the terminals of transducer 1, and if the open-circuit voltage e_1

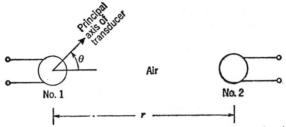

FIG. 12.9. Two electro-mechano-acoustic transducers in air.

for a constant current i_2 applied to the other transducer equals $E_1 f_1(\theta)$, then, from Eq. (12.11),

$$\frac{E_2 f_2(\theta)}{i_1} = \frac{E_1 f_1(\theta)}{i_2} \tag{12.16}$$

If $f_1(\theta) = f_2(\theta) = 1$ for $\theta = 0$, we have from Eq. (12.11) that

$$\frac{E_2}{i_1} = \frac{E_1}{i_2} \qquad \text{for } \theta = 0 \tag{12.17}$$

So, at all angles,

$$f_2(\theta) = f_1(\theta) \tag{12.18}$$

12.7. Plane-free-wave Response of a Microphone. Usually, we are interested in determining the open-circuit voltage response of a microphone e_{oc} when the microphone is placed in a plane free sound wave that has a sound pressure p_{ff} prior to insertion of the microphone. Expressed in decibels,

$$\Re = 20 \log_{10} \frac{e_{oc}}{p_{ff}} + 20 \log_{10} \frac{p_{ref}}{e_{ref}} \qquad \text{db} \tag{12.19}$$

where p_{ref} and e_{ref} are reference pressure and voltage, respectively. These quantities are often taken as 1 dyne/cm^2 and 1 volt, respectively, so that the second term on the right-hand side of Eq. (12.19) is zero. It is the quantity $\Re$ that we wish to determine by the reciprocity technique.

12.8. Reciprocity Theorem for Electro-mechano-acoustical Systems. Referring to Fig. 12.9, let us consider transducer 1 to be the one for which the response $\Re$ as a function of frequency is desired, and let us consider that transducer 2 is an auxiliary transducer. To apply the reciprocity principle without error, the sound field radiated by transducer 2 must meet certain conditions.   We shall see that these conditions can be met using any linear passive reversible transducer provided its presence in the experiment does not affect the value of $\mathfrak{R}$ for the other transducer and provided that the transducers are located in the far-fields of each other. The far-field of a transducer was defined in Part X as that part of the medium beyond a distance r_c from the transducer such that the sound pressure along a radial line passing through its acoustic center varies in inverse proportion to distance r.

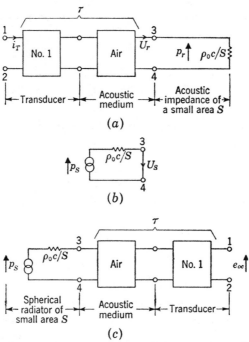

Fig. 12.10. Block diagrams for a transducer and the acoustic medium in the imped-ance-type analogy.

When transducer 1 is driven by an electric current i_T, it will radiate sound.   Let us select a very small area S, in the far-field of the trans-ducer, perpendicular to the radiated wave front at the position of trans-ducer 2, but without No. 2 present.   Let this small area S be represented by an acoustic impedance.   Let us represent transducer 1 by a four-pole network as shown in Fig. 12.10.   The air lying between it and the very small area S is represented by a second four-pole network connected in tandem and terminated by the acoustic impedance of the area S.

The small area S is assumed to be far enough away from the transducer so that the acoustic impedance for it is $\rho_0 c/S$.   Then, by the extension of the reciprocity theorem proved in Par. 12.5,

$$p_r = \tau i_T \qquad \text{from Fig. 12.10}a \qquad (12.20)$$

$$e_{oc} = \tau U_s \qquad \text{from Fig. 12.10}b \text{ and } c \qquad (12.21)$$

where i_T = electric current in amperes supplied to transducer 1 when it is operated as a loudspeaker.

e_{oc} = open-circuit voltage in volts produced by transducer 1 when it is operated as a microphone in the sound field of transducer 2. The free-field sound pressure at distance r, p_{ff}, is measured prior to insertion of the transducer.

U_r = volume velocity in cubic meters per second of the air particles normal to the very small area S located a distance r from transducer 1 with No. 2 absent. This velocity is produced by a current i_T in the coil of transducer 1.

p_r = sound pressure in newtons per square meter at this very small area S located a distance r from transducer 1 with No. 2 absent. Note that $p_r = \rho_0 c U_r / S$.

p_s = sound pressure in newtons per square meter that would be produced at the surface of an acoustic generator of very small area S and internal acoustic impedance $\rho_0 c/S$ if it were terminated in infinite acoustic impedance.

U_s = volume velocity in cubic meters per second that would be produced at the surface of an acoustic generator of area S and internal acoustic impedance $\rho_0 c/S$ if it were terminated in a zero acoustic impedance. Note that $U_s = p_s S / \rho_0 c$.

τ = reciprocity constant for transducer 1 plus the acoustic medium with an acoustic-impedance termination $\rho_0 c/S$ and an open-circuit electrical termination.

12.9. General Calibration Equation. In order that the reciprocity theorem hold, transducer 2 must produce the same sound field as though it had the same internal acoustic impedance as the very small area S presents to a plane wave, namely, $\rho_0 c/S$. We now determine this sound field. At points in the far-field ($r > r_c$) of the small source of area S, the source may be treated as though it were a simple spherical radiator of area $S = 4\pi a^2$ with an internal impedance $\rho_0 c/S$. For such an idealized source, the free-field (undisturbed) sound pressure p_{ff} at distance r is

$$p_{ff} = \frac{p_0 a}{r} e^{-jkr} \qquad (12.22)$$

and, from Eq. (2.64),

$$Z_{AS} = \frac{j\omega \rho_0 a}{S} \qquad (12.23)$$

where Z_{AS} = acoustic radiation impedance in mks acoustic ohms for a spherical source of very small radius of a meters

p_0 = sound pressure at the surface of the sphere in newtons per square meter

The acoustical circuit for our small spherical transducer is given by

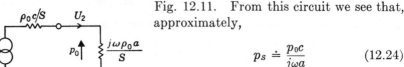

Fig. 12.11. From this circuit we see that, approximately,

$$p_s \doteq \frac{p_0 c}{j\omega a} \qquad (12.24)$$

FIG. 12.11. Acoustical circuit for a small spherical source with an internal impedance $\rho_0 c/S$ and a radius a.

where p_s was defined after Eq. (12.21). This equation is valid if $|\omega a| \ll c$.

Substitution of (12.24) into (12.22), remembering that $S = 4\pi a^2$ and $p_s = U_s \rho_0 c/S$, gives

$$p_{ff} = \frac{j\omega p_s S}{4\pi r c} e^{-jkr} = \frac{U_s \rho_0 c}{2\lambda r} e^{j(\pi/2 - kr)} \qquad (12.25)$$

When Eq. (12.25) is substituted in Eq. (12.21), we get

$$e_{oc} = \tau \frac{p_{ff} 2\lambda r}{\rho_0 c} e^{j(kr - \pi/2)} \qquad (12.26)$$

From (12.20) and (12.26),

$$\frac{e_{oc}}{p_{ff}} = \frac{p_r}{i_T} \frac{2\lambda r}{\rho_0 c} e^{j(kr - \pi/2)} \qquad (12.27)$$

The ratio on the left side of this equation is the desired calibration of the microphone. We must now devise a means for determining the mag-

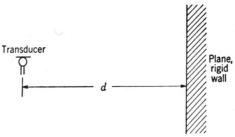

FIG. 12.12. Experimental arrangement for calibrating a transducer by the self-reciprocity method.

nitude of the ratio at the start of the right side without resorting to measurement of p_r. There are two methods for doing this. One is called the *self-reciprocity method* and the other, *the auxiliary-transducer method*.

12.10. Self-reciprocity Method for Free-field Calibration (see Fig. 12.12). The essential features of this method are the transducer to be calibrated, a large rigid plane wall, and an electrical apparatus for disconnecting the transducer from a source of electrical power and connecting it to an audio amplifier in a small fraction of a second.

In performing the experiment, the microphone is connected to a constant-current generator that generates a train of waves of the frequency at which a calibration is desired and of sufficient length to establish a quasi-steady-state condition in the middle part of the wave train. The quasi-steady-state current i_T into the microphone is measured. The train of waves travels outward and is reflected back from the large wall toward the microphone.

While the wave is in transit, the electrical circuit disconnects the terminals of the transducer from the generator and connects them to an amplifier so that the transducer operates as a microphone. When the reflected wave train reaches the transducer, an open-circuit voltage e_{oc} is measured. The data are expressed as a ratio, e_{oc}/i_T.

The quantity r in Eq. (12.27) is equal to $2d$, the total distance the wave has traveled. The sound pressure p_{ff} arriving at the microphone in the reflected wave is exactly the same as p_r, that sent out by the transducer when it was acting as a loudspeaker. Hence, we can write

$$\frac{e_{oc}}{i_T} = \frac{p_r}{i_T} \frac{e_{oc}}{p_{ff}} \tag{12.28}$$

Substitution of Eq. (12.28) in (12.27) yields

$$\frac{e_{oc}}{p_{ff}} = \sqrt{\frac{e_{oc}}{i_T} \frac{4\lambda d}{\rho_0 c}} \, e^{-i(\pi/4 - kd)} \tag{12.29}$$

where e_{oc} = open-circuit voltage of the transducer, in volts

p_{ff} = free-field sound pressure acting to produce the open-circuit voltage e_{oc}, in newtons per square meter

$k = 2\pi/\lambda$, in meters^{-1}

$\rho_0 c$ = characteristic impedance of air, in mks rayls (newton-seconds per cubic meter)

In the measurement, the absolute value of e_{oc} and i_T is not required; their ratio only, as determined by a calibrated potentiometer, is necessary. In addition, the relative phase of the two is needed if a phase calibration is desired. An accurate measurement of d and of frequency is also required, and because $\rho_0 c$ is a function of both temperature and barometric pressure, T and P_0 must be determined.

12.11. Auxiliary-transducer Method for Free-field Calibration. The self-reciprocity method has two important disadvantages in practice. A complicated electronic apparatus is necessary to perform the switching operation, and a large, rigid reflecting wall in a space that is otherwise nonreflecting is required. Actually, a smooth concrete floor in the center of a very large open space outdoors fits the requirements of a reflecting wall in an otherwise anechoic space. However, a "radar" type of electronic apparatus is always necessary, and the vagaries of weather must be contended with.

With the use of an auxiliary transducer and an anechoic chamber, the switching apparatus and the necessity for performing measurements outdoors is obviated.   By this method, the output of the transducer being calibrated (No. 1) is measured by an auxiliary transducer (No. 2) at a distance d (see Fig. 12.13a).   Then the sensitivity of the No. 1 transducer is compared with that of No. 2 by placing them successively in the same sound field and finding the ratio of their complex voltage outputs (see Fig. 12.13b and c).   In the (b) and (c) tests, the distance d' must be great enough so that no interaction between the loudspeaker and the two transducers occurs, and it need not equal d.

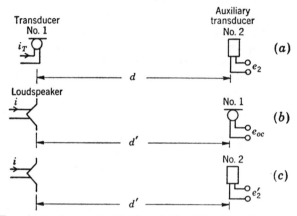

FIG. 12.13. Experimental steps in the free-field calibration of a transducer by the auxiliary-transducer method.

From these measurements the first ratio beneath the radical of Eq. (12.29) becomes

$$\frac{e_{oc}}{i_T} = \frac{e_2}{i_T}\frac{e_{oc}}{e_2'} \tag{12.30}$$

where i_T = current in amperes supplied to transducer 1 when it operates
         as a source
    e_2 = open-circuit voltage in volts produced by transducer 2 in
         test (a)
    e_{oc} = open-circuit voltage in volts produced by transducer 1 in
         test (b)
    e_2' = open-circuit voltage in volts produced by transducer 2 in
         test (c)

Substitution of (12.30) in (12.29) gives

$$\frac{e_{oc}}{p_{ff}} = \sqrt{\frac{e_2}{i_T}\frac{e_{oc}}{e_2'}\frac{2\lambda d}{\rho_0 c}}\, e^{-j(\pi/4 - kd/2)} \tag{12.31}$$

All quantities are in practical electrical and in mks units.   Note here that the total distance the wave travels is d instead of $2d$ as was the case for the self-reciprocity method.

The desired response $\mathcal{R}$ in decibels is found by inserting Eq. (12.31) into Eq. (12.19).

12.12. Pressure Response of a Microphone. Sometimes, we are interested in determining the open-circuit voltage response of a microphone e_{oc} due to an average sound pressure p_d over the diaphragm. Expressed in decibels,

$$\mathcal{R}_d = 20 \log_{10} \frac{e_{oc}}{p_d} + 20 \log_{10} \frac{p_{\text{ref}}}{e_{\text{ref}}} \tag{12.32}$$

The quantities p_{ref} and e_{ref} were discussed after Eq. (12.19).

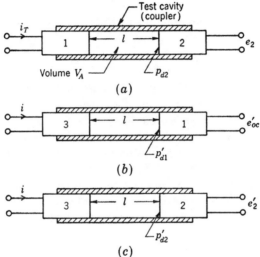

FIG. 12.14. Experimental steps in the pressure calibration of a transducer by the auxiliary-transducer method.

12.13. Auxiliary-transducer Method for Pressure Calibration. By a pressure calibration of a microphone we mean the determination of the ratio of the open-circuit voltage to the sound pressure at the diaphragm producing it. We shall designate the pressure at the diaphragm as p_d and the open-circuit voltage as e_{oc}.

This type of calibration is generally performed on microphones which are small in size, whose diaphragm is exposed to the air on only one side, and for which the mechanical impedance of the diaphragm is high. Microphones meeting these qualifications are commonly of the piezo-electric or electrostatic types, although microphones of the electromagnetic type might in special designs be calibrated.

With this type of calibration two auxiliary transducers are always used. The arrangements of the transducers during the three parts of the test are shown in Fig. 12.14.

We designate the transducer being calibrated as No. 1 and the two auxiliary transducers as Nos. 2 and 3. During each of the three parts of the tests two of the transducers are coupled together by a rigid-walled cavity having (after the microphones are inserted) an enclosed volume V_A. The diaphragms of the transducers 1 and 2 have acoustic impedances Z_{AD1} and Z_{AD2}, respectively. The diaphragm impedance of transducer 3 is of no particular importance.

In order to simplify the analysis, we shall make the following assumptions:

$$Z_{AD1} = Z_{AD2} \tag{12.33}$$

$$|Z_{AD1}| \gg \frac{1}{\omega C_A} \tag{12.34}$$

where C_A is the acoustic compliance of the coupling cavity,

$$C_A = \frac{V}{\gamma P_0} \quad \text{m}^5/\text{newton} \tag{12.35}$$

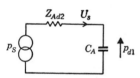

FIG. 12.15. Acoustical circuit for transducer with internal acoustic impedance Z_{Ad2} operating into a cavity with an acoustic compliance C_A.

where $V = Sl$, l is the length of the cavity, S is the cross-sectional area, P_0 = atmospheric pressure, and γ = ratio of specific heats. The length l must be less than $\frac{1}{20}$th to $\frac{1}{50}$th wavelength depending on the accuracy desired (see Part XIII, Par. 5.5).

Analogous to the case for the free-field calibration of a microphone [see Eqs. (12.20) and (12.21)], we have the reciprocity relationships

$$p_{d2} = \tau i_T \tag{12.36}$$
$$e_{oc} = \tau U_s \tag{12.37}$$

where p_{d2} = sound pressure in newtons per square meter at the face of the diaphragm of the No. 2 transducer in experiment (a) of Fig. 12.14

p_s = sound pressure in newtons per square meter that would be developed in the cavity by the No. 2 transducer acting as a source if the cavity impedance were infinite

U_s = volume velocity in cubic meters per second of the diaphragm of the No. 2 transducer acting as a source if the cavity impedance were zero

When the No. 2 transducer acts as a source, as is required by Eq. (12.37), we have the circuit of Fig. 12.15. From it we see that

$$p_{d1} = \frac{p_s}{j\omega C_A Z_{AD2} + 1} \doteq \frac{p_s}{j\omega C_A Z_{AD2}} \tag{12.38}$$

or

$$p_{d1} \doteq \frac{U_s}{j\omega C_A} \tag{12.39}$$

The calibration that we desire is e_{oc}/p_{d1}, so that

$$\frac{e_{oc}}{p_{d1}} = e_{oc}\frac{j\omega C_A}{U_s} \tag{12.40}$$

From Eqs. (12.36), (12.37), and (12.40) we get

$$\frac{e_{oc}}{p_{d1}} = j\omega C_A\frac{p_{d2}}{i_T} \tag{12.41}$$

But, we see that

$$\frac{p_{d2}}{i_T} = \frac{e_2}{i_T}\frac{p_{d2}}{e_2} \tag{12.42}$$

The ratio e_2/i_T we get from experiment (a) of Fig. 12.14.   From (b) and (c) of Fig. 12.14 we get e'_{oc}/e'_2.

Now p_{d2}/e_2 must equal p'_{d2}/e'_2 because transducer 2 is assumed linear, and p_{d1}/e_{oc} must equal p'_{d1}/e'_{oc} for the same reason.   Hence, we may rewrite Eq. (12.42) as

$$\frac{p_{d2}}{i_T} = \frac{e_2}{i_T}\frac{p'_{d2}}{e'_{oc}}\frac{e'_{oc}}{e'_2} \tag{12.43}$$

But, in (b) and (c) of Fig. 12.14, $p'_{d1} = p'_{d2}$ because $Z_{AD1} = Z_{AD2}$.   Hence we have

$$\frac{p_{d2}}{i_T} = \frac{e_2}{i_T}\frac{p_{d1}}{e_{oc}}\frac{e'_{oc}}{e'_2} \tag{12.44}$$

Substitution of (12.44) in (12.41) gives us the desired equation,

$$\frac{e_{oc}}{p_{d1}} = \sqrt{\frac{e_2}{i_T}\frac{e'_{oc}}{e'_2}}\, j\omega C_A \tag{12.45}$$

Finally, the pressure response $\mathfrak{R}_d$ is found from Eq. (12.32).

CHAPTER 13

HEARING, SPEECH INTELLIGIBILITY, AND PSYCHOACOUSTIC CRITERIA

PART XXX *Hearing*

The hearing mechanism is the final recipient of sounds produced by audio systems. Designers of audio systems must know the range of frequencies and the sound pressures to which this mechanism responds and the manner in which speech sounds and music must be presented to the listener if he is to gain a satisfactory amount of information and pleasure from the audio signal. Some of the basic characteristics of hearing and speech are covered in this chapter. The problem is much more complex than there is space to indicate here, so that we strongly recommend that you refer to more specialized texts for further study.[1-3]

The physicist or engineer is likely to approach the study of hearing as though he were considering a well-behaved system whose *modus operandi* was well understood. In the physical sciences quantities such as mass, wavelength, voltage, and intensity are measured. We are guided in our choice of which quantities to study by knowledge gained from theory and from related measurements. Furthermore, it is generally possible to hold constant all but a few of the independent variables and to measure the behavior of the system as a few of the variables are systematically maneuvered. In many systems, it is possible to superimpose the results of varying quantities individually to obtain the behavior under conditions where several quantities are changing simultaneously.

[1] S. S. Stevens *et al.*, "Handbook of Experimental Psychology," Chaps. 25–28, John Wiley & Sons, Inc., New York, 1951.

[2] W. A. Rosenblith, K. N. Stevens, and the staff of Bolt Beranek and Newman, "Handbook of Acoustic Noise Control," Vol. II, Noise and Man, *WADC Tech. Rept.* 52–104 (June, 1953). Order from Office of Technical Services, Department of Commerce, Washington, D.C., PB No. 111,274.

[3] I. J. Hirsh, "The Measurement of Hearing," McGraw-Hill Book Company, Inc., New York, 1952.

In psychological studies we can also measure, or at least rate, quantities such as just perceptible excitations, just noticeable differences, onset of pain, increased nervous activity, personal likes and dislikes, and so forth. However, the situation is different from that in physics, because there exists almost no theoretical body of knowledge to guide us as to what constitutes a valid measurement or, for that matter, as to which variables are important and which are ancillary. In psychological measurements, it is very difficult to hold part of the independent variables constant, and generally the principle of superposition does not hold. Experimenters soon learn that responses obtained as a result of a stimulus presented to a listener under one set of conditions are not the same as the responses obtained to the same stimulus after the listener has acquired a different mental attitude toward the experiment, the experimenter, or some other thing. In network theory we would speak of "different initial conditions."

In reading the material which follows you should bear in mind that the data presented were obtained by particular experimenters using particular stimuli, presented to listeners with particular mental biases, under particular ambient conditions. As a result, other experimenters ostensibly repeating the same experiment may obtain substantially different results unless great care is used to repeat all factors involved in the experiment. For a detailed discussion in nontechnical terms of psychological meaurements see Hirsh.[3]

In this chapter we shall study the sensation of hearing which results from a stimulation of the hearing mechanism. The hearing mechanism comprises the mechanical parts of the ear, the auditory nervous system including the brain, and the indicator of a response that the whole man represents.

No small physical apparatus possesses properties any more remarkable than those of the ear. It not only can withstand the most intense sounds produced in nature, which have sound pressures of 10^3 to 10^4 dynes/cm^2, but at the other extreme it responds to sound pressures, at some frequencies, which are as small as 10^{-4} dyne/cm^2. These very small sound pressures produce a displacement of the eardrum that is of the order of 10^{-9} cm for frequencies near 1000 cps. This distance is less than one-tenth the diameter of a hydrogen molecule!

The hearing mechanism is more than an extremely sensitive microphone. It functions also as an analyzer with considerable selectivity. Sounds of particular frequencies can be detected in the presence of interfering background noises; i.e., the hearing mechanism operates as though it were a set of contiguous "filter" bands. Even more remarkable is the ability of the auditory system to judge loudness, pitch, and musical quality, functions that are performed in some manner in association with the brain.

13.1. Mechanical Properties of the Ear. In simplified form, the ear is something like the representation of Fig. 13.1. The sound enters it through the *auditory canal (external ear).* This canal has a diameter of about 0.7 cm and a length of about 2.7 cm. The *eardrum* is a thin membrane terminating this canal and has an area of about 0.8 cm². Three small bones, or *ossicles,* joined together, called the hammer, anvil, and stirrup, are located in the *middle ear.* The first of these small bones connects to the main eardrum, and the third connects with a second mem-

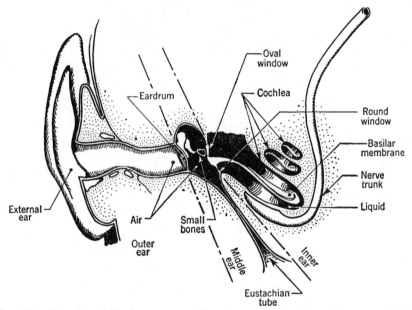

FIG. 13.1. Drawing of the human ear. Sound waves enter the auditory canal and move the eardrum, which sets the three ossicles in motion. When the oval window moves, it causes a motion of a colorless liquid inside the cochlea, with a resulting movement of the basilar membrane. [*After Wiener, Physics Today,* **4**: 13 (*December,* 1951).]

brane called the *oval window.* The oval window forms the entrance to the *inner ear* for the normal passage of sound.

The oval window is located at one end of the cochlea. The *cochlea* is a hollow, snail-shaped member formed from bone and filled with a colorless liquid. It is spiral-shaped with a length of about 35 mm and with a cross-sectional area of about 4 mm² at the stirrup end, decreasing to about one-fourth that size at the far end. It is divided down the middle by the *cochlear partition* (see Fig. 13.2*a* and *c*), which extends along the length of the cochlea. This partition is partly bony and partly a gelatinous membrane called the *basilar membrane* (see Fig. 13.2*b*). Surprising as it may seem, the basilar membrane is smaller at the larger end of the cochlea (see Fig. 13.2*b*). On the surface of the basilar membrane there terminate about 25,000 nerve endings of the main auditory nerve. The liquid-filled chamber is further divided into two parts by a very thin membrane called *Reissner's membrane*. In addition to the oval window, which communicates with the liquid-filled chamber above the cochlear partition (see Fig. 13.2a), there is a round window which communicates

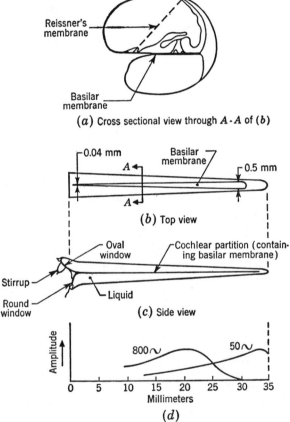

FIG. 13.2. Top, side, and cross-sectional sketches of the cochlea, stretched out to snow the liquid-filled spaces above and below the basilar membrane. At the bottom is a graph of the amplitudes of vibration of the basilar membrane at 50 and 800 cps.

with the lower liquid-filled chamber and which acts as a pressure release. The area of the oval window is about 3 mm², and that of the round window is about 2 mm².

The volume of air contained in the middle ear is about 2 cm³. The mechanical advantage of the system of three small bones (ossicles) in transferring vibrations from the eardrum to the oval window is about 1.3 to 1. However, the effective ratio of areas of the eardrum and of the oval window times this mechanical advantage provides an increase in sound pressure from the eardrum to the liquid of the cochlea of a factor of about 15. This transformation is advantageous to the transfer of the vibrations of the air to the cochlear fluid.

A motion of the oval window produced by a motion of the eardrum sets up one or more waves that travel along the membrane and through the liquid, with the result that for each frequency there is a point of maximum excitation on the membrane (see Fig. 13.2d). The end of the membrane nearest the oval window "resonates" at the higher frequencies, and that at the far end of the spiral "resonates" at the lower frequencies. The nature of the vibration pattern has been explored, and the results are shown in Figs. 13.3 and 13.4. In Fig. 13.3, measurements of the relative amplitude of motion of the different parts of the membrane are shown for seven different frequencies. These and other data yield a curve of position of maximum vibration against frequency as shown in Fig. 13.4. In other words, the basilar membrane is a wide-band mechanical filter which partially separates a complex sound into its components. As a result, a particular group of nerves is excited more vigorously by a particular frequency than by other frequencies.

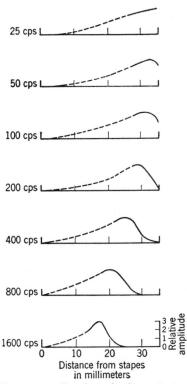

FIG. 13.3. Graphs of the relative amplitudes of vibration of the basilar membrane at various frequencies. [After v.Békésy, Akust. Z., 8: 66–76 (1943).]

It might be assumed from Fig. 13.3 that the hearing mechanism is a rather coarse sort of analyzer because the widths of the resonance curves are so broad. This conclusion is not borne out by actual tests. Apparently the nervous system sharpens up this "resonance."

We shall conveniently select a mathematical model to explain the ability of the ear to detect one tone in the presence of equally intense or more intense tones of other frequencies. It does not follow that this mathematical model will hold for other situations. This model is the same one as we would use in describing the properties of an electrical filter for separating one frequency component from an assemblage of such components. In describing electrical filters, one speaks of "bandwidths." We shall do the same here.

The bandwidths of the hearing process, as measured by a person's ability to detect a pure tone in the presence of a white, random noise,† are commonly called *critical bandwidths*.‡   In detecting a tone in the presence of noise, the hearing mechanism appears to reject the noise outside the critical band centered on the pure tone, thereby making it appear to behave as a filter set.   These critical bandwidths are shown plotted as a

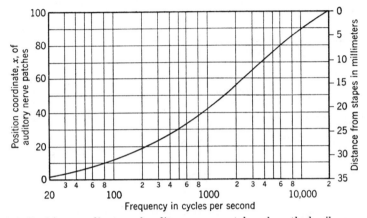

Fig. 13.4. Position coordinate x of auditory-nerve patches along the basilar membrane for various frequencies.   Zero on the left-hand ordinate is located at the widest part of the basilar membrane and 100 at the stapes.   The distance (measured from the stapes) of the position of maximum vibration is given on the right-hand ordinate. [*After v.Békésy, Akust. Z.*, **7**: 173–186 (1942).]

function of frequency in Fig. 13.5.   The critical bandwidths for one-ear listening are slightly different from those for two-ear listening, so that two curves are shown, one for each case.   In Par. 13.7, we shall see how Fig. 13.5 is useful in evaluating the effectiveness of noise in masking out pure-tone sounds.

From Fig. 13.4 we see that the frequency region below 250 cps seems to occupy a very small portion of the basilar membrane.   This would

† Random noise is an acoustical quantity (*e.g.*, sound pressure) or an electrical quantity (*e.g.*, voltage) whose instantaneous amplitudes occur, as a function of time, according to a normal (Gaussian) distribution curve.   A common random noise is that resulting from the random motion of molecules of the air.   Another is that produced by the random motion of electrons in an electrical resistance.   Random noise need not have a flat (uniform) frequency spectrum.

White noise is a sound wave or electrical wave having a continuous and uniform distribution of energy as a function of frequency in the audible frequency range. White noise need not be random.

‡ The concept of critical bandwidths was proposed by H. Fletcher, Auditory Patterns, *Revs. of Mod. Phys.*, **12**: 47–65 (1940).   Later work was done by J. P. Egan and H. W. Hake, On the Masking Pattern of a Simple Auditory Stimulus, *J. Acoust. Soc. Amer.*, **22**: 622 (1950) and by T. H. Schafer *et al.*, Frequency Selectivity of the Ear as Determined by Masking Experiments, *J. Acoust. Soc. Amer.*, **22**: 490 (1950).

make it seem that low-frequency sounds are less important or less easily separated than those in the middle-frequency range. Nevertheless, as listeners, we particularly enjoy low-frequency tones in our music, and we can easily tell one from another. There is evidence that low-frequency stimuli excite all parts of the basilar membrane. It is also possible that the low-frequency sounds are sent to the brain for analysis rather than being analyzed at the basilar membrane, as is the case for the higher-frequency sounds.

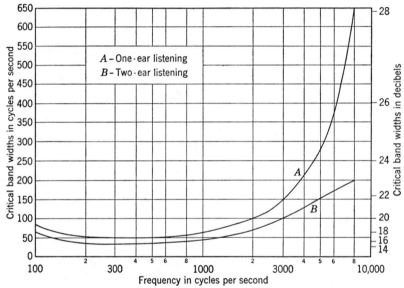

Fig. 13.5. Critical bandwidths for listening as determined by detecting pure tones in the presence of white, random noise. (A) One-ear listening; (B) two-ear listening. The left-hand ordinate is Δf_c in cycles per second, and the right-hand ordinate is 10 $\log_{10} \Delta f_c$ db. [After French and Steinberg, J. Acoust. Soc. Amer., **19**: 90–119 (1947).]

13.2. Thresholds of Hearing. *Thresholds of Audibility.* A threshold of audibility for a specified signal is the minimum effective sound pressure of that signal that is capable of evoking an auditory sensation (in the absence of any noise) in a specified fraction of the trials. It is usually expressed in decibels *re* 0.0002 microbar.

The American Standard threshold of audibility for pure tones, for a listener with acute hearing seated in an anechoic (echo-free) chamber facing a source of sound at a distance greater than 1 m, is shown as curve 2 of Fig 13.6. The sound pressure is measured before the listener enters the sound field at a point where the center of the head will be located. A similar curve, except that the sound is supplied by earphones and the sound pressure is measured at the entrance to the ear canal, is shown as curve 1 of Fig. 13.6.

Two significant differences are seen between curves 1 and 2 of **Fig. 13.6.**
First, they are displaced from each other by an average of about 10 db,
and, second, curve 2 has more irregularities in it.   The relative displacement is a combination of two factors.   The binaural (two-ear) threshold
would be displaced downward about 2 db from the monaural (one-ear)
threshold.   The remaining difference of about 7 to 8 db is believed at low
frequencies to be caused by noise produced under the earphones by the
irregular twitching of the muscles against which the earphones rest.
Because of this, for the same pressure *at the eardrum* in each case, a sound
produced by a loudspeaker is audible at a pressure 5 to 10 db *lower* than is

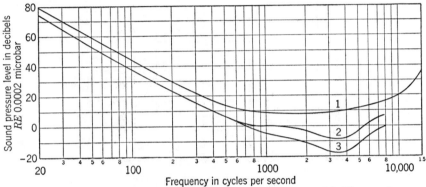

Fig. 13.6. Thresholds of audibility determined in three ways.   (1) Monaural curve;
signal presented by an earphone and sound pressure measured at the entrance to the
ear canal.   [*After Dadson and King, J. Laryngol. Otology*, **46**: 366–378 (1952); *v.Békésy,
Ann. Physik*, **26**: 554–566 (1936).]   (2) Binaural curve; pressure presented by a
single source at a distance in front of the listener; sound pressure was measured in the
field before an observer entered it.   (*American Standard for noise measurement,*
Z24.2–1942.)   (3) Binaural curve; signal presented by a number of small loudspeakers located randomly in a horizontal plane about the listener's head and sound
pressure measured in field before observer entered it.   [*After Wiener and Ross, J.
Acoust. Soc. Amer.*, **18**: 401–408 (1946).]

a sound produced by earphones.   The greater differences in curve 2 at
frequencies above 800 cps are the result of diffraction of the sound around
the head and acoustial resonances in the outer-ear canal.
    The question is sometimes asked as to how low or how high a frequency
a person can hear.   Tests have revealed that curve 1 of Fig. 13.6 may be
extrapolated downward to show the level required to produce audibility
at 2 cps and that listeners appear to hear at frequencies that low.   Note,
however, that at 2 cps a sound pressure level of about 135 db would be
necessary to produce audibility.
    The upper limit of hearing is quite variable from person to person.
It is generally found that young people hear up to 20,000 cps if the
tone is fairly intense.   Middle-aged people usually hear up to 12,000 to
16,000 cps.   Again, the level at which the tone is presented is important.

The threshold of audibility varies with a great many factors. It is different from person to person. Even for the same person, it varies from day to day and hour to hour. After exposure to even a moderate noise level, temporary, though slight, deafness occurs which shifts the threshold upward.

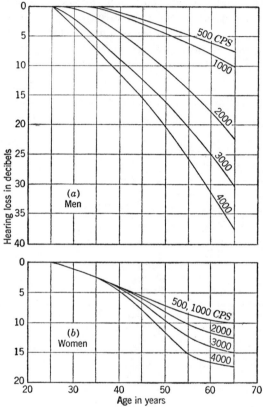

FIG. 13.7. Shift in average threshold of hearing with age as a function of frequency for men and women.

One of the principal factors affecting the threshold of audibility is age. In Fig. 13.7 we show the results of studies of progressive loss of aural sensitivity with increasing age.†

Thresholds of Tolerance. At the other extreme of the hearing range we are interested in the maximum sound levels which the ear can stand without discomfort, tickle, or pain. Listeners wearing earphones report that they begin to experience discomfort when a pure tone (sine wave) reaches

† These data are part of the results of a study by a group under the chairmanship of W. A. Rosenblith appointed by American Standards Association Sectional Committee Z-24. The results are reported in "The Relations of Hearing Loss to Noise Exposure," American Standards Association, Inc., New York.

levels greater than 110 db *rc* 0.0002 microbar (see Table 13.1).   A tickling sensation is aroused in the ear when the levels are greater than 130 db. Definite pain may occur when the levels exceed 140 db.   These values seem to be nearly independent of frequency in the range between 50 and 8000 cps.   Listeners who are exposed to high levels daily can stand about 10 db more in two of the three categories just listed (see Table 13.1). Similar results are reported for noises which have a continuous spectrum, such as has white noise or random noise.   In the case of a continuous wide-band spectrum, however, the threshold is reached when the energy in any one critical bandwidth at some point along the frequency scale reaches the levels given in the previous paragraph.

TABLE 13.1.   Threshold of Tolerance†

| | Pure tones | |
Threshold	Naïve ears	Exposed ears
Discomfort................	110	120
Tickle....................	132	140
Pain.....................	140	
Immediate damage..........	150 to 160‡	

† Sound supplied by earphones.   Frequencies between 50 and 8000 cps.
‡ Not accurately known.

13.3. Pitch.   Pitch is defined as that aspect of auditory sensation in terms of which sounds may be ordered on a scale extending from "low" to "high," such as a musical scale.   Pitch is a subjective quantity.   It is chiefly a function of the frequency of a sound, but it is also dependent upon the sound pressure level and the composition.   The unit is the *mel*.

Frequency is a physical quantity that is measured with physical apparatus; the unit is cycles per second.   Two tones of the same frequency, but with different sound pressure levels, will sound different in pitch.   That is to say, a 200-cps tone of one level will sound as though it had a different frequency from a 200-cps tone of another level.   We also find that a listener does not usually consider an octave in frequency as a doubling of pitch.

To measure pitch, we set up an experiment in which we give an observer two separate tone generators (oscillators).   These generators are so arranged that they may be connected alternately to a loudspeaker or to earphones.   The observer is asked to adjust the frequency of one of the oscillators until it seems to have twice the pitch of the other.   This procedure is repeated for a number of settings of the reference oscillator extending over the frequency range.   The observer is also asked to increase and to decrease the pitch by other factors.   From these data, a scale of pitch in mels is developed. A reference pitch of 1000 mels was chosen as the pitch of a 1000-cps tone with a sound pressure level of 60 db *re* 0.0002 microbar. A complete curve of pitch vs. frequency is shown in Fig. 13.8.

An interesting feature of the curve of Fig. 13.8 is that it has very nearly the same shape as does the curve of Fig. 13.4. This similarity leads one to believe that pitch determination is a judgment that is based on the location of the point of excitation along the basilar membrane.

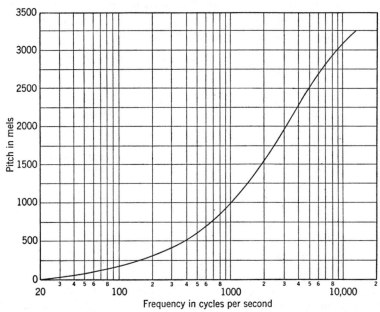

FIG. 13.8. Relation between subjective pitch expressed in mels and frequency. Note that subjective pitch increases more and more rapidly as frequency is increased logarithmically. The musical scale, by comparison, is a logarithmic scale; that is to say, an octave is a doubling of frequency. [*After Stevens and Volkman, Am. J. Psychol.*, **53**: 329–353 (1940).]

In conclusion, we should note that we know very little about the pitch of complex sounds and some interesting contradictions exist, as we shall show in Par. 13.6.

13.4. Loudness Levels and Loudness. *Loudness Levels.* When we hear a sound, we often make a judgment of its "loudness." For example, we say that a crash of thunder is "extremely loud," while the singing of a person in the distance is "not very loud." These qualitative expressions of "very loud," "less loud," or "soft" have been made quantitative for some kinds of sound.

The simplest way to speak quantitatively of the loudness of a sound is to compare it with some standard sound. This standard sound has been chosen to be a 1000-cps tone. The *loudness level* of any other sound is defined as the sound pressure level of a 1000-cps tone that sounds as loud as the sound in question. The unit of loudness level is the *phon*. For example, if a 1000-cps tone with a sound pressure level of 70 db *re* 0.0002 microbar sounds as loud as a certain square wave (regardless of the sound pressure level of the square wave), the square wave is said to have a *loudness level* of 70 *phons*.

Extensive measurements have been made to determine the loudness levels of pure tones and narrow bands of noise as a function of frequency and sound pressure levels. These results are shown in Figs. 13.9 and

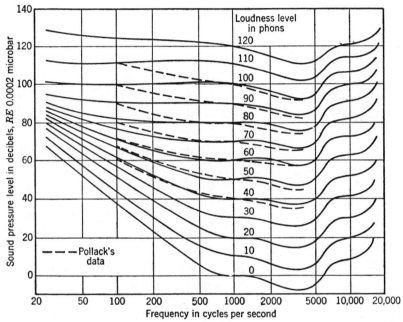

FIG. 13.9. Equal-loudness contours for pure tones by Fletcher and Munson. (*American Standard for noise measurement*, Z24.2–1942.) The dotted lines are equal-loudness contours for bands of noise, 250 mels wide as determined by Pollack. [*Pollack, J. Acoust. Soc. Amer.*, **24**: 533–538 (1952). *See also Beranek, Peterson, Marshall, and Cudworth, J. Acoust. Soc. Amer.*, **23**, 261–269 (1951).]

13.10.   On each of the graphs the ordinate gives the sound pressure level of the sound. The numbers appearing on the contours are the loudness levels of the sound and are the sound pressure levels of a 1000-cps tone that is equally loud. Three sets of graphs are given in the two figures; those shown by the dotted curves are repeated twice.

The first of these three graphs (solid lines of Fig. 13.9) is the American Standard set of loudness contours for pure tones and is for observers listening in an anechoic chamber with the source of sound located at a distance of more than 1 m from the listener. These contours are usually called the Fletcher-Munson contours. The sound pressure levels of the tones were measured with the listener out of the sound field at a point corresponding to the center of his head.

The second set of graphs (solid lines of Fig. 13.10) is also for pure tones, the data being taken in England under similar circumstances to those of Fig. 13.9. These contours are usually called the Churcher-King contours. It is not known whether the curves of Fig. 13.9 or those of Fig. 13.10 are more typical of the population at large. In this country, at present, we generally use the set recommended by the American Standards Association.

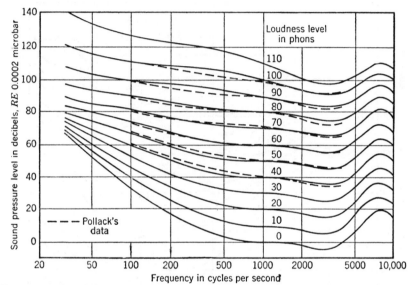

FIG. 13.10. Equal-loudness contours for pure tones by Churcher and King. [*J. Inst. Elec. Engrs. (London)*, **81**: 57–90 (1937).] The dotted curves are the same as for Fig. 13.9.

The third set of graphs (dotted lines on Figs. 13.9 and 13.10) is for bands of noise, 250 mels wide (see Par. 13.3 *Pitch*), supplied by an earphone. The sound pressure levels were measured just beneath the cushion of the earphone supplying the sound. These contours resemble closely the shape of the Churcher-King contours.

The equal-loudness curves are of particular significance in the design of radio receiving equipment. In concert halls, music is usually played at quite high levels. When this music is reproduced in our homes via the radio or phonograph, we often listen to it at greatly reduced levels. Because the ear discriminates against low-frequency tones when the levels are low, as we can see from the equal-loudness curves, the music appears to have lost its bass quality. In order to give the illusion that the music is being reproduced with the same tonal content as it had in the concert hall, radio designers install compensating networks which change the response of the audio system in the radio automatically when the volume control is adjusted, in the opposite manner to the way the ear changes its response as the signal level is varied. Such compensation is called *frequency weighting*, and the electrical device that modifies the frequency response is called the *weighting network* or the *tone control*.

Loudness. The concept of loudness level, which we have just learned, is very useful, but it does not give the whole story. From the loudness-level contours of Figs. 13.9 and 13.10 it is not possible to say *how much* louder one sound is than another. For example, is a sound with a loudness level of 100 db twice, three times, or four times as loud as one with a loudness level of 80 db? To answer this question several extensive sets of measurements have been made to determine a scale of loudness. But first let us define loudness.

Loudness is defined as the intensive attribute of an auditory sensation, in terms of which sounds may be ordered on a scale extending from "soft" to "loud." Loudness is chiefly a function of the sound pressure, but it is also dependent on frequency and wave form. The unit is the *sone*. By definition, a pure tone of 1000 cps, 40 db above a normal listener's threshold, produces a loudness of 1 sone. In the measurements that led to a loudness scale, subjects were asked to change the loudness by factors of 2, 10, 0.5, and 0.1. From these data a curve of loudness vs. loudness level was constructed, as shown in Fig. 13.11.

We see from Fig. 13.11 that, for higher sound pressure levels, 10 db in loudness level corresponds roughly to a doubling of the loudness in sones. At lower levels, 10 db correspond to a change in loudness amounting to a factor of about 3. At the very lowest levels, 10 db change in loudness level corresponds to a change in loudness by a factor of nearly 20.

This curve has been under review recently by various experimenters, and there is good evidence that it should have a smaller slope. It is well established, however, that this curve is useful in the determination of the loudness level of a complex sound given the loudness level of its components or the spectrum level. Therefore, we shall call the curve of Fig. 13.11 the *transfer function*.

Calculation of Loudness Level of Combinations of Pure Tones. The data of Figs. 13.9 to 13.11 provide us with information from which it is possible to estimate the loudness level of a combination of pure tones if we know the sound pressure level of each. The method is simple if the tones are spaced no closer together than about 450 mels (see Fig. 13.8 for the relation between frequency and mels). If their frequencies are closer than about 450 mels of each other, their weighted intensities must be added and the sum treated as a single pure tone, as we shall now describe.[4]

[4] L. L. Beranek, A. P. G. Peterson, J. L. Marshall, and A. L. Cudworth, Calculation and Measurement of the Loudness of Sounds, *J. Acoust. Soc. Amer.*, **23**: 261–269

For a combination of N tones, five steps are necessary in the calculation of the combined loudness level.

1. Divide the N tones according to frequency to form 7 to 10 groups, each group preferably having the frequency limits shown in Table 13.2,† and determine the weighted rms sound pressure level of each group. By the "weighted" sound pressure level we mean that the tones are all

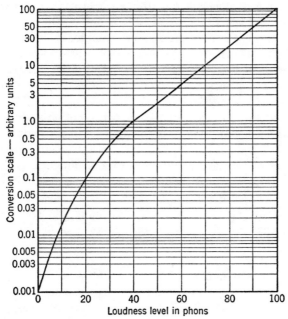

Fig. 13.11. Plot of the transfer function (loudness in sones) as a function of loudness in phons. The loudness level of a particular sound is equal to the sound pressure level of a 1000-cps tone that sounds equally loud. The sounds are assumed to be presented to an uncovered ear. [*After Fletcher and Munson, J. Acoust. Soc. Amer.,* **9,** 1–10 (1937).]

converted to the middle frequency of the band with the aid of equal-loudness contours. Suppose that we have two tones of 100 and 200 cps with sound pressure levels of 60 and 50 db, respectively. Then, from Fig. 13.9, we see that the 100-cps tone of 60 db is equivalent in loudness to a 150-cps tone of 55 db and the 200-cps tone of 50 db is equivalent in loudness to a 150-cps tone of 55 db. Combination of the two 55-db levels on an energy basis gives an equivalent-tone level of 58 db.

2. The loudness level in phons of each of the 10 groups is determined from Fig. 13.9.

(1951). As can be seen from this reference, calculation of loudness has only limited meaning.

† Eight octave-frequency bands may be used instead of the 10 of Table 13.2.

3. Each of the loudness levels in phons is converted to transfer numbers with the aid of Fig. 13.11.

4. The 10 different transfer numbers are added together to obtain the total transfer number.

5. The total transfer number is converted into the desired loudness level in phons with the aid of Fig. 13.11.

TABLE 13.2.   Frequency Bands of Equal Width of 300 Mels

Band No.	Pitch limits, mels	Mean pitch, mels	Frequency limits, cps	Mean-pitch frequency, cps
1	0– 300	150	20– 200	94
2	300– 600	450	200– 500	340
3	600– 900	750	500– 860	670
4	900–1200	1050	860–1330	1080
5	1200–1500	1350	1330–1900	1600
6	1500–1800	1650	1900–2570	2230
7	1800–2100	1950	2570–3450	2960
8	2100–2400	2250	3450–4660	4000
9	2400–2700	2500	4660–6300	5400
10	2700–3000	2850	6300–9000	7500

For example, assume we have four tones in a free field as follows: a 100-cps tone with a sound pressure level of 47 db; a 180-cps tone with a sound pressure level of 35 db; a 600-cps tone with a sound pressure level of 40 db; and a 3000-cps tone with a sound pressure level of 30 db. These fall into three bands, the lowest band having a weighted sound pressure level of 44 db. From Fig. 13.9 we see that the loudness levels are, respectively, 20 phons, 38 phons, and 33 phons. The transfer numbers are determined from Fig. 13.11 and are, respectively, 0.09, 0.95, and 0.55. Summing these gives us a total transfer number of 1.70. From Fig. 13.11 we see that this corresponds to a loudness level of 46 phons.

Calculation of Loudness Level of Continuous-spectrum Noises.   The loudness level of a continuous-spectrum noise may be calculated in a manner similar to that outlined in the preceding section.   In this case, the noise is first analyzed into sound pressure levels by a series of contiguous band-pass filters such as the 10 given in Table 13.2.   The level in each band is then treated as though it were the level of a pure tone whose frequency lay at the mean-pitch frequency of the band.   The same procedure as that given for calculating the loudness level of a group of pure tones is followed [see steps 2 to 5 in the previous paragraph].

13.5. Differential Sensitivity to Sound Pressure and Frequency.
Minimum Perceptible Changes in Sound Pressure Level.   A person is able to detect a change in sound pressure level of about 1 db for any tone between 50 and 10,000 cps if the level of the tone is greater than 50 db

above the threshold for that tone. Under ideal laboratory listening conditions, with signals supplied by an earphone, changes in level of as little as 0.3 db can be detected by the ear in the middle-frequency range. For sound pressure levels less than 40 db, changes in level of 1 to 3 db are necessary in order to be perceptible.

Minimum Perceptible Changes in Frequency. For frequencies above 1000 cps and pressure levels in excess of 40 db, the minimum perceptible change in frequency which the ear can detect is of the order of about 0.3 per cent.† At frequencies below 1000 cps and for the same range of pressure levels, the ear can detect a change in frequency of as little as about 3 cps.† At low pressure levels and particularly at low frequencies the minimum perceptible change in frequency may be many times these values.

13.6. The Case of the Missing Fundamental. Harmonic distortion at the lower frequencies in a small radio receiving set or loudspeaker may have some beneficial effects. Because it is impossible to radiate a sizable percentage of the available power from a small loudspeaker at low frequencies, one finds it difficult to understand why some small radio sets and loudspeakers sound reasonably well. One reason is that the lowest bass notes are actually supplied either physiologically or psychologically because several of their harmonics are present in the signal. The pitch of a sequence of frequencies such as 400, 600, and 800 cps is apparently that of a 200-cps tone. However, if the frequencies 500 and 700 cps are added, the pitch will drop to that of a 100-cps tone. The perceived pitch of a combination of tones spaced equally in frequency is usually not that of the mean frequency of the combination, but rather that of the constant difference frequency.

Even when all the frequencies of a musical composition below a certain value, say 300 cps, are removed, the quality of the music remains the same to a surprising degree. This happens even without harmonic distortion in the sound-reproducing system because most music is relatively rich in harmonic content, but it is especially true if just the right amount of harmonic distortion is produced at the lower frequencies.

13.7. Masking. The acoustical engineer is often asked the question, To what level do I need to reduce a tone so that it will become inaudible in the presence of background noise? In automobiles, for example, a tone may be emitted from the loudspeaker of the car radio at the frequency of the vibrator in the power supply. It may not be economical to eliminate this tone completely, but it could conceivably be reduced until it is lost in the general noise inside the automobile. To establish the tolerable level for this tone, we need to know to what extent one sound hides, or *masks*, another.

Quantitatively, we define a term, *masking*, as the number of decibels

† See reference 2, pp. 30–37.

by which a listener's threshold of audibility is raised (changed) by the presence of another (masking) sound.   As an example, let us suppose that we have a tone whose frequency is 1500 cps.   Let us assume that it can be heard until its level becomes less than +1 db *re* 0.0002 microbar. That is, the threshold of audibility for this tone is +1 db.   Next, let us turn on a second tone with a frequency of 1200 cps and a level of, say, 80 db.   Now we find that the first tone of 1500 cps is not audible until its level has been increased to a value of 54 db.   We say that the *masking* of the second tone by the first is 54 minus 1 db, which equals 53 db.

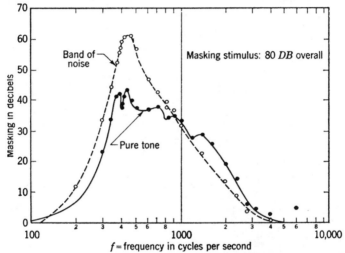

Fig. 13.12. Masking produced by two stimuli, one a pure tone at 400 cps and the other a narrow band of noise centered at 410 cps, as a function of frequency of the tone being masked.   [*After Egan and Hake, J. Acoust. Soc. Amer.*, **22** : 622–630 (1950).]

The masking effect is always greater above the frequency of the tone than below.   For example, the masking of a pure tone of frequency f produced by a narrow band of noise or by a pure tone of 400 cps with an rms sound pressure level of 80 db has the values shown in Fig. 13.12.   Beats in the perceived sound occur whenever the frequencies of the two tones approach each other.

When the noise has a continuous spectrum with no drastic peaks or dips, the masking effect on a pure tone may be determined with the aid of Fig. 13.5.   For this case, only the noise in a narrow frequency band on either side of a pure tone serves to mask the tone.   In other words, the properties of the ear which make it perform like a filter come into action. Let us explore this fact further.   If the level of the noise is expressed in terms of its *spectrum level*, *i.e.*, rms sound pressure levels in bands 1 cps in width, a pure tone to be audible must have a level which is greater than the spectrum level of the noise by the number of decibels shown in Fig.

13.5.   Another way of saying this is that the level of the pure tone must exceed slightly the total rms level of the noise in a *critical bandwidth* if the tone is to be heard.   Critical bandwidths Δf_c expressed in cycles per second and in decibels ($10 \log_{10} \Delta f_c$) are shown on the two ordinates of Fig. 13.5.

The meaning of a critical bandwidth can be clarified further by referring to Fig. 13.13.   Let us assume that we have a pure tone of some frequency and a continuous-spectrum noise with a flat spectrum whose extent in frequency is from f_a to f_b cps.   Also, let us assume that the bandwidth of this noise $f_b - f_a = \Delta f$ cps, can be varied over a wide range

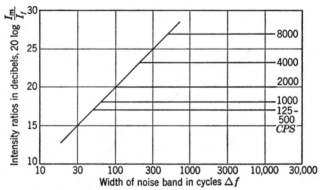

Fig. 13.13. Ratio of the intensity of the masked tone I_m to the intensity per cycle of the noise I_f plotted against the width of the noise band Δf in cycles.   [*After Fletcher, Revs. Modern Phys.*, **12**: 47–65 (1940).]

centered around the frequency of the pure tone.   When the bandwidth is zero, the masking will obviously be zero.   As the bandwidth is increased, the masking M, in decibels, will increase in direct proportion to the logarithm of the bandwidth up to its critical value Δf_c.   Above that width no further masking of the pure tone occurs.   Hence, if a person is listening to a pure-tone signal in the presence of background noise, there is no advantage to be gained from the use of a filter to remove the background noise, unless its width is *less* than that of a critical bandwidth.

PART XXXI *Speech Intelligibility*

The subject of speech communication is too large to be treated here, but a few of the simple attributes of speech signals and a method for estimation of syllabic intelligibility will be given.   References for further reading are given in the footnotes.[1,3,5–10]

[5] L. L. Beranek, W. H. Radford, J. A. Kessler, and J. B. Wiesner, Speech-reinforce-

13.8. Speech Spectrum. Speech is a succession of utterances that produces a wave whose frequencies and amplitudes change rapidly with time. We have already shown in Part XXVI (page 338) that the human voice has, on the average, a power spectrum that peaks (for men) at about 500 cps and a spectrum (in octave bands) that drops off above 1000 cps at a rate of about 8 db per octave. We also showed that at high frequencies the voice is directional.

Each syllable of speech lasts about ⅛ sec, and the average interval between syllables is about 0.1 sec. Some sounds, the vowels for example, are produced at the vocal cords. Other sounds are produced by the noises of air movement through the mouth and over the tongue and lips. The frequency spectrum of either type of sound is shaped by the resonant cavities formed by the throat, mouth, teeth, and lips.

The vowel sounds are not as critical to speech intelligibility as the consonant sounds. It is unfortunate that the consonant sounds are so weak and, therefore, are easily masked by noise. In some languages, such as Hebrew, no vowels are written, only consonants.

When the long-time average speech spectrum is plotted in terms of its spectrum level (rms sound pressure in 1-cps bands) as measured 1 m in front of the talker, it appears as shown in Fig. 13.14 by the curve marked "average level of speech." Tests reported by French and Steinberg[7] and Beranek[8] indicate that the useful dynamic range of speech in each frequency band appears to be about 30 db. Of this number, the rms peaks lie about 12 db above the average level, and the weakest syllables lie about 18 db below the average level.

13.9. Estimation of Speech Intelligibility. In attempting to estimate speech intelligibility we must also consider the properties of the hearing mechanism. Tests by French and Steinberg[7] have shown that the proper frequency scale to use as a base for calculations is one that is nearly proportional to the pitch scale discussed in the previous part. We must also bring into the calculations the threshold of hearing and the "overload" point of the average hearing mechanism. By overload point we mean the level at each frequency above which the hearing mechanism no longer seems to respond to the stimulus.

ment System Evaluation, *Proc. IRE*, **39**: 1401–1408 (1951).

[6] R. H. Bolt and A. D. MacDonald, Theory of Speech Masking by Reverberation, *J. Acoust. Soc. Amer.*, **21**: 577–580 (1949).

[7] N. R. French and J. C. Steinberg, Factors Governing the Intelligibility of Speech Sounds, *J. Acoust. Soc. Amer.*, **19**: 90–119 (1947).

[8] L. L. Beranek, Design of Speech Communication Systems, *Proc. IRE*, **35**: 880–890 (1947).

[9] G. A. Miller, "Language and Communication," McGraw-Hill Book Company, Inc., New York, 1951.

[10] R. K. Potter, G. A. Kopp, and H. C. Green, "Visible Speech," D. Van Nostrand Company, Inc., New York, 1947.

Articulation Index. All these factors are combined into the one graph of Fig. 13.14. The abscissa of this graph is frequency in cycles per second, plotted on the experimentally determined articulation-index scale of French and Steinberg.[7] On this graph are plotted as *spectrum levels* (see Part II) (1) the threshold of hearing for continuous-spectrum sounds;

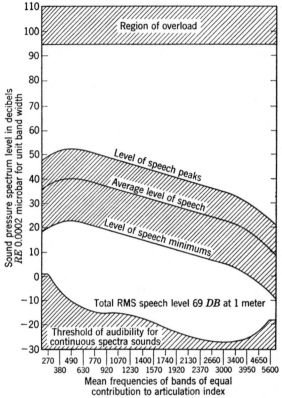

FIG. 13.14. Plot, on a spectrum level basis, of (1) the speech area for a man talking in a *raised* voice; (2) the region of "overload" of the ear of an average male listener; and (3) the threshold of audibility for young ears. All curves are plotted as a function of frequency on a distorted frequency scale. [*After Beranek, Proc. IRE,* **35**: 880–890 (1947).]

(2) the peak, average, and minimum levels of speech for a *raised* man's voice measured at a distance of 1 m directly in front of him; and (3) the "overload spectrum level" for the ear.

According to the work of French and Steinberg and of Beranek, if the spectrum levels of speech at a listener's ear are such that the center shaded region of Fig. 13.14 lies above the threshold of hearing of the listener and above the ambient noise, but below the overload line, all syllables of the speech will be audible to the listener and the speech intelligibility will be nearly perfect.   This corresponds to an *articulation* index of 100 per cent.

On the other hand, if noise covers part of the shaded speech region or if part of the region falls below the threshold of hearing or above the overload level for the hearing mechanisms, the articulation index is less than unity.

The percentage articulation index is defined as the ratio (times 100) of the speech area not covered over by the items named in the previous sentence to the total speech area as shown in the center of Fig. 13.14.

Procedure for Calculation of Articulation Index.   To calculate the speech intelligibility for a given situation, the following steps must be followed.

1. The *orthotelephonic gain* of the system should be determined as a function of frequency.   The orthotelephonic gain is defined as

$$\text{Orthotelephonic gain} = 20 \, \log_{10} \frac{p_2}{p_1} \tag{13.1}$$

where p_1 = long-time average of the sound pressure produced at the listener's ear by an average male talker at 1 m distance in an anechoic chamber, the talker using a raised voice (6 db above a normal voice).   This is the reference condition.

p_2 = long-time average of the sound pressure due to the direct speech produced at the listener's ear by the same talker using a raised voice but with the actual conditions of communication present.   That is to say, p_2 includes the effects of amplification, focusing, distance, and barriers between talker and listener, but not the effects of noise, reverberation, or distortion.   A narrow-band continuously variable filter should be used for measuring both p_1 and p_2.

The directivity characteristics of a speech-reinforcing system or of a reflecting surface or canopy near the talker are treated as part of the orthotelephonic gain.   For example, if a reflecting canopy boosts the *direct* speech level by 3 db, it adds 3 db to the directivity index for the voice itself.   Hence it adds 3 db to the orthotelephonic gain.

In free (anechoic) space the orthotelephonic gain is found from Fig. 13.15.   That is to say, the voice level decreases uniformly with distance at the rate of 6 db for each doubling of distance, so that the amount by which the center shaded region is shifted upward or downward for talkers and listeners in a nonreverberant space is given in Fig. 13.15.   The effects of voice strength are also included.   This chart also holds for the *direct portion* of the speech in a reverberant room, provided that reflections, from focusing surfaces or from reflecting canopies, that combine within 0.05 sec with the direct sound are treated as additions to the direct portion.

2. The shaded region of Fig. 13.14 is moved upward or downward and its shape is changed as a function of frequency by adding to it the ortho-

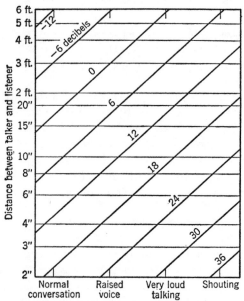

FIG. 13.15. Contours for proper location of the shaded articulation-index area of Fig. 13.14. This chart gives the orthotelephonic gain for two people facing each other in free space at the distances shown. It also takes into account the effect of voice level. Zero decibel corresponds to a raised voice at 3 ft. In free space, the orthotelephonic gain is a function of distance alone and not of frequency.

telephonic gain (in decibels). If the person is talking at other than the indicated voice level, the shaded region is further shifted upward or downward by the amounts given in Table 13.3.

TABLE 13.3. Changes in Sound Pressure Level of Voice with Voice Condition

	Normal voice	Raised voice	Loud as possible without straining vocal cords	Shouting
Shift in level, db...	−6	0	+6	+12

3. The long-time average spectrum level of the ambient noise arriving at the listener's ear is determined as a function of frequency and is plotted on the same graph.

4. Reverberant sound is also treated as ambient noise. Bolt and MacDonald[6] have presented data which, when modified to include the increase in the directivity index of the direct speech due to the addition of a sound system or a reflecting canopy, permit one to plot a reverberant "speech" spectrum on the graph of Fig. 13.14. The shape of the reverberant speech spectrum as a function of frequency is determined as follows:

a. Determine the reverberation time in seconds and the room constant *R* in square feet at each frequency.

b. Determine the differential directivity index at each frequency, defined as the difference between the directivity index for the actual situation and the directivity index for the voice alone.† Convert to directivity factor [see Eq. (4.19)].

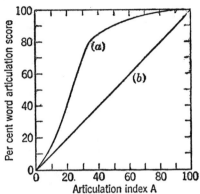

c. Using the reverberation time, the differential directivity factor *Q*, the room constant *R*, and the distance between the source and the listener *r*, determine $10 \log (QR/r^2)$. Then enter Table 13.4 to determine at each frequency the number of decibels *N* shown.

Fig. 13.16. Percentage word-articulation score for phonetically balanced word lists read by experienced announcers to (*a*) trained listeners familiar with the content of the word lists and (*b*) naïve listeners selected from the general population with no familiarity with the word lists. These relations are approximate and will differ widely for different listeners, word lists, and talkers. (*From unpublished data of the author.*)

d. To obtain the reverberant speech spectrum, subtract, at each frequency, the number of decibels *N* from the upper edge of the shaded speech region of Fig. 13.14. This gives a curve that always moves up or down with the shaded speech region as long as the directivity index and the distance *r* between the talker and listener remain constant.

TABLE 13.4.   Table for Calculating Approximately the Effect of Reverberation on Speech Intelligibility‡

Reverberation time, sec	*N* Direct speech peak level minus reverberant speech level, db						
	$10 \log (QR/r^2)$						
	0	5	10	15	20	25	30
0.5	15	20	25	30	35	40	45
1	10	15	20	25	30	35	40
2	5	10	15	20	25	30	35
3	3	8	13	18	23	28	33
4	1	6	11	16	21	26	31

‡ *R* is the room constant as defined in Part XXIV, *r* is the distance between the listener and the source, and *Q* is the differential directivity factor of the source.

† The directivity index for the voice alone is determined from the directivity patterns given on p. 343.

5. The *articulation index* is determined by finding the percentage of the finally plotted center-shaded region of Fig. 13.14 that does not lie below

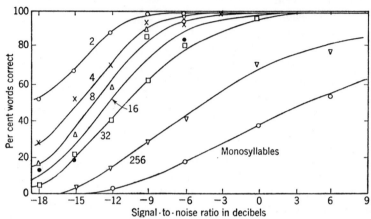

Fɪɢ. 13.17. Relation between articulation-test score and the difference in level between speech signal and noise for test list of different size. The listeners knew the vocabulary for each test. [*From Miller, Heise, and Lichten, J. Exptl. Psychol.*, **41**: 329–335 (1951).]

the noise level, or below the threshold of hearing, or below the reverberant speech level, or above the overload line at 95 db.

6. Estimate the word articulation from the graph of Fig. 13.16.

13.10. Psychological and Linguistic Factors Affecting Sentence and

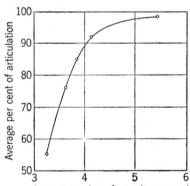

Fɪɢ. 13.18. The improvement of articulation as the number of sounds per word is increased. Five groups of 20 words, having different average numbers of sounds per word, were read 20 times to a group of listeners. [*After Egan, OSRD Rept.* 3802 (*Nov.* 1, 1944).]

the results of an articulation test.

Word Intelligibility. The prediction of sentence or word intelligibility is very difficult because it depends on many factors other than noise level, reverberation time, and level of the signals. In Fig. 13.17 we show the relation between the per cent words correct and the signal-to-noise ratio for word lists that vary in size from 2 words to 256 and for monosyllables. The listener knew the vocabulary and had only to choose among the words to make his response. In Fig. 13.18 we show the improvement in articulation as the number of sounds per word is increased. Different talkers and different listeners yield different scores, as can be seen from Figs. 13.19 and 13.20. Learning is also a big factor in the results of an articulation test. For example, in Fig. 13.21 we show a

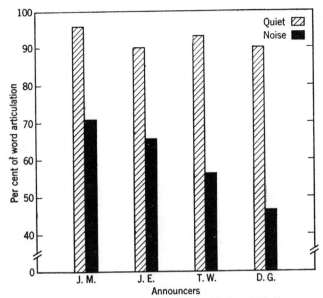

Fig. 13.19. Bar graph showing the differences in word-articulation scores obtained with four different announcers.   [*After Egan, OSRD Rept.* 3802 (*Nov.* 1, 1944).]

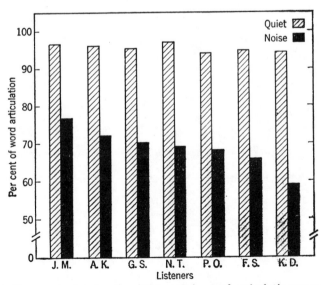

Fig. 13.20. Bar graph showing the differences in word-articulation scores obtained with a typical group of listeners.   [*After Egan, OSRD Rept.* 3802 (*Nov.* 1, 1944).]

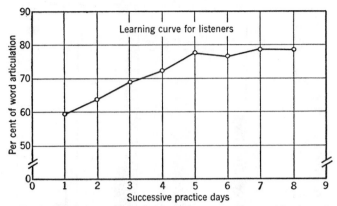

FIG. 13.21. Typical learning curve obtained for a test crew with the same speech-communication system. Each point on the curve represents the average score on 12 tests for a crew of 10 listeners. The tests were read over the interphone system by three well-practiced announcers. [*After Egan, OSRD Rept.* 3802 (*Nov.* 1, 1944).]

TABLE 13.5. Characteristics of Auditorium, Voice, and Reinforcing System

Frequency, cps	Reverberation time T, sec, full audience	Room constant R, ft²	Speech peaks, raised voice, at 1 m, db	Orthotelephonic gain at 60 ft, db	Differential directivity factor Q of system	Ambient noise in ⅓ octave bands, db re 0.0002 microbar
(1)	(2)	(3)	(4)	(5)	(6)	(7)
270	2.0	2.5×10^4	50	−10	10	42
380	1.8	2.8	51	− 2	16	36
490	1.6	3.1	52	0	20	36
630	1.5	3.3	52	2	20	33
770	1.4	3.6	51	3	20	33
920	1.3	3.8	50	3	20	32
1070	1.2	4.2	48	2	20	30
1230	1.2	4.2	47	3	20	29
1400	1.2	4.2	45	4	20	26
1570	1.1	4.6	44	5	20	25
1740	1.1	4.6	42	7	20	24
1920	1.1	4.6	41	6	20	24
2130	1.1	4.6	40	6	20	24
2370	1.1	4.6	38	5	20	24
2660	1.0	5.0	37	3	20	24
3000	1.0	5.0	36	2	20	24
3400	1.0	5.0	34	0	20	24
3950	1.0	5.0	31	− 2	20	24
4650	1.0	5.0	27	− 5	20	25
5600	1.0	5.0	22	−10	20	26

typical learning curve for a listening test crew, with all other factors except their learning held constant.

These various factors reveal that speech-intelligibility tests should be undertaken only after careful planning and that the statistical nature of the results should be fully appreciated.[11]   The factors discussed above

TABLE 13.6.   Calculation of Articulation Index

Frequency, cps	Speech peaks plus ortho-tele-phonic gain, db	Spectrum level of ambient noise, db	$\frac{QR}{r^2}$	$10 \log \frac{QR}{r^2}$, db	N, db	Reverberation speech spectrum, db	Energy summation of Cols. (3) and (7)	Articulation index, Col. (2) minus Col. (8) divided by 600
(1)	(2)	(3)	(4)	(5)	(6)	(7)	(8)	(9)
270	40	21	70	18	23	17	22	0.030
380	49	29	125	21	26	23	30	0.032
490	52	30	172	22	30	22	31	0.035
630	54	32	183	23	31	23	33	0.035
770	54	31	200	23	31	23	33	0.035
920	53	29	210	23	32	21	30	0.038
1070	50	26	232	24	33	17	27	0.038
1230	50	25	232	24	33	17	26	0.040
1400	49	24	232	24	33	16	25	0.040
1570	49	23	255	24	34	15	24	0.042
1740	48	22	255	24	34	14	23	0.042
1920	47	20	255	24	34	13	21	0.043
2130	46	19	255	24	34	12	20	0.043
2370	43	15	255	24	34	9	16	0.045
2660	40	12	280	24	34	6	13	0.045
3000	38	9	280	24	34	4	10	0.047
3400	34	5	280	24	34	0	6	0.047
3950	29	−1	280	24	34	−5	0	0.048
4650	22	−10	280	24	34	−12	−8	0.050
5600	12	−20	280	24	34	−22	−18	0.050
								0.825

indicate that absolute predictions of articulation scores are not possible. However, one can say that if the calculated articulation index exceeds 60 per cent, a speech-communication system is probably satisfactory. If the articulation index is less than 30 per cent, the system is probably unsatisfactory.   Between 30 and 60 per cent, the system should be viewed with suspicion and detailed articulation tests performed if possible.

[11] L. L. Beranek, "Acoustic Measurements," pp. 625–635, 761–792, John Wiley & Sons, Inc., New York, 1949.

Example. A speech-reinforcing system is to be used in an auditorium. The characteristics of the auditorium, the voice, and the reinforcing system are tabulated in Table 13.5. Calculate the articulation index at a seat 60 ft from the loudspeaker for a person talking in a raised voice. Comment on the adequacy of the system. The volume of the auditorium is 1 million ft³, and the auditorium seats 4000 people.

Solution. Following the procedure that was given in Par. 13.9, we first add the orthotelephonic gain to the speech peak levels. This is shown in column 2 of Table 13.6 and amounts to shifting upward and to distorting the shape of the shaded region of Fig. 13.14. Next, the ambient noise spectrum, as given in third-octave bands, is converted to spectrum level by subtracting 10 times the logarithm of the bandwidth from each number (see column 3). Determine $10 \log_{10} (QR/r^2)$ (see columns 4 and 5). Enter Table 13.4 to obtain N, and subtract this number of decibels from column 2. This yields column 7.

We must add, on an energy basis, the figures that appear in columns 3 and 7, as these two columns give us the energy of the total interfering noise. The result is in column 8. To obtain the articulation index for each band, subtract column 8 from 2 (calling all differences greater than 30 db equal to 30), and divide by 20×30 because there are 20 bands and each band has a maximum contribution of 30 db. The result is shown in column 9. The total articulation index equals the sum of the figures in column 9, namely, 0.82. The approximate percentage word intelligibility from Fig. 13.16 is 98 per cent for trained listeners, who know the complete vocabulary, and 80 per cent for naïve listeners. This applies to a person 60 ft from the loudspeaker. Nearer, the articulation index will be greater, and farther away it will be less.

This system should be perfectly satisfactory when there is a full audience. When there is no audience, however, the reverberation time for this particular auditorium is about 2 sec higher, so that N in column 7 of Table 13.6 is about 13 db less (because both R and T change), and column 7 will be about 13 db more. This will make the reverberant speech levels about equal to the spectrum level of the ambient noise so that the articulation index will drop to approximately 0.65. Although the estimated word intelligibility at 60 ft will still be passable, the articulation index for seats more than 60 ft away will be lower because the number N decreases 6 db for each doubling of distance. For example, at a distance of 180 ft (the rear of the auditorium) with no audience, N will be about 23 db less than in column 7, yielding an articulation index of about 0.44. This corresponds to an estimated word intelligibility of 86 per cent for trained listeners or 45 per cent for naïve listeners, which is hardly satisfactory.

PART **XXXII** *Psychoacoustic Criteria*

Most acoustical design has as its purpose the satisfaction of some human need. The loudspeaker and microphone are designed to make music pleasant or to make speech intelligible to human beings. Noise control is for the comfort of individuals. We design acoustically our auditoriums, classrooms, and public and private buildings to satisfy the listening desires of people.

In this part those acoustical design criteria are presented which are at present most widely used.[12] In employing these criterion curves you

[12] L. L. Beranek, "Noise Control in Office and Factory Spaces," *Trans. Bull.* 18, 15*th Ann. Meeting Chem. Engs. Conf.*, pp. 26–33 (1950).

should bear in mind that prior to 1945 the only publicized criteria were those giving the optimum reverberation times for auditoriums and the desired response curves for music-reproduction systems. The criterion curves given here dealing with damage to hearing, speech interference, and residential noise levels have been developed recently. Hence, it must be expected that this branch of knowledge is in a fluid state, and the reader should consult the current literature to keep abreast of progress.

13.11. Levels Producing Damage to Hearing.[2,13,14] An important consideration in most industries is that the noise levels in factory spaces be low enough so that no damage should occur to the hearing of employees who are exposed to the noise over a long period of time. Tentative criteria have been established for no damage to hearing based on an

TABLE 13.7.   Damage Risk Criteria†

Frequency, cps	Pure tone levels or critical band levels of continuous noise	Octave-band frequencies, cps	Octave-band levels of continuous noise	Half-octave-band frequencies, cps‡	Half-octave-band levels of continuous noise	Third-octave-band frequencies, cps‡	Third-octave-band levels of continuous noise
50	110	37.5– 75	110	37.5– 53	108	45– 57	107
100	95	75– 150	102	75– 106	100	90– 114	99
200	88	150– 300	97	150– 212	94	180– 228	93
400	85	300– 600	95	300– 425	91	360– 456	90
800	84	600–1200	95	600– 850	91	720– 912	90
1600	83	1200–2400	95	1200–1700	91	1440–1824	90
3200	82	2400–4800	95	2400–3400	91	2880–3648	90
6400	81	4800–9600	95	4800–6800	91	5760–7296	90

† Tentative criteria for negligible risk of damage to hearing. Levels in decibels not to be exceeded for reasonable assurance of no permanent damage. Unit: Decibels *re* 0.0002 microbar.

‡ Only part of the bands are listed here in order to save space.

This table is valid for long periods (years) of exposure. The numbers are not to be taken too literally. Numbers 10 db lower would involve negligible risks indeed, while numbers 10 db higher would result in significant hearing loss. The levels apply to exposure noise that has a reasonably continuous time character with no substantial sharp energy peaks.

analysis of all reliable information on the subject in the literature. Briefly, this criterion says that, in the frequency region above 300 cps, the sound levels in any one critical band shall not exceed approximately

[13] K. D. Kryter, The Effects of Noise on Man, *J. Speech Hearing Disorders, Mon. Suppl.* 1 (September, 1950).

[14] *Proc. 2d and 3d Ann. Natl. Noise Abatement Symposia,* **2** (1951), **3** (1952), National Noise Abatement Council, New York.

85 db *re* 0.0002 microbar.[13] Below 300 cps, only sketchy data are available so that only tentative estimates can be made.[12] These estimated levels of pure tones or of critical bandwidths of continuous noise for no damage to hearing are shown as a function of frequency in column 2 of Table 13.7.

A very common way of measuring noise is to use a set of wave filters that divide the frequency scale up into 8, 16, or 24 regions, each region being about one octave, or one-half octave, or one-third octave in width. In measuring a noise having a continuous frequency spectrum with these filters, the levels in each octave band will be higher than the values in column 2 of Table 13.7 by about the number of decibels given in the formula

Added number of decibels in using a wide-band filter as compared
$$\text{with a filter with critical bandwidth} = 10 \log_{10} N \quad (13.2)$$

where N equals the number of critical bands lying within the wider band.

Application of Eq. (13.2) for the wider frequency bands of Table 13.7 yields approximately the levels shown. For example, in the 1200 to 2400-cps band, two-ear listening, there are approximately 16 critical bands. From Eq. (13.2) we get $10 \log_{10} 16 \doteq 12$ db. This number added to 83 db of column 2 yields 95 db.

The numbers of Table 13.7 are intended to mean that if the noise levels lie below those tabulated in *all* the frequency bands by, say, as much as 5 db, the chance is very small that people in such an environment will suffer a hearing loss *due to the noise* even if they work in the noise 8 hr a day, 50 weeks a year, and for periods up to 5 years. What data were available for establishing these numbers did not include known noise situations that people had been in for more than 5 years. If, on the other hand, the noise level in any *one* band lies above the criterion number for that band by, say, as much as 5 db, there is a fair chance that *some* people will suffer a hearing loss due to the noise if they work in that environment for periods of more than a year.

It is known that people often recover a substantial part of their hearing loss after an extended rest. The extent of hearing damage should probably be judged after the person exposed has had a rest of about a week and has arrived at a stable threshold of audibility.

Furthermore, there is great variability among people, and some people become deaf even doing normal outdoor work (like gardening) owing to age, disease, and other causes. Furthermore, deafness, like death, is a statistical matter, and any particular set of criterion numbers will fail to predict hearing damage for a few people. The numbers of Table 13.7 must, therefore, be viewed as reasonable ones to use in acoustical design.

Employers will do well to have a positive program for checking the hearing of employees at the time of employment and for rechecking their hearing at frequent intervals if there is any substantial noise in the plant —even though this noise may lie below the numbers indicated in Table 13.7 in all frequency bands.

13.12. Speech-interference Levels. In the preceding part we discussed some of the factors governing speech intelligibility. We said that the speech energy below 200 cps and above 7000 cps contributes almost nothing to speech intelligibility. We also showed that if the frequency scale is properly divided into bands of equal contribution to articulation index, one can determine the effect of noise on speech intelligibility by finding the average difference between the peak levels (in decibels) of speech and the rms noise levels in the bands and dividing this average by 30. To illustrate this average difference, assume that the peak levels of speech are 80, 75, 70, and 65 in four bands and the rms noise levels are 60, 55, 50, and 45, respectively, in these same four bands. Then the average difference is 20 db, and the articulation index for these four bands is $2\%_{30} = 0.67$. However, a difference in any one band that is greater than 30 db is called 30 db.

Reference to Fig. 13.14 shows that if we desire to divide the frequency scale into *three* bands of equal contribution to speech intelligibility, using available analyzing equipment, we should divide it into the frequency ranges 300 to 1200 cps, 1200 to 2400 cps, and 2400 to 4800 cps. However, because the articulation-index frequency scale is more nearly linear below 1000 cps than logarithmic (see Fig. 13.14), an intensity average in the 300- to 1200-cps band is not correct. Moreover, usual available analyzing equipment includes the bands 300 to 600, 600 to 1200, 1200 to 2400 and 2400 to 4800 cps.

To a sufficiently close approximation, we can, if the level in the 300- to 600-cps band is not more than 10 db above that in the 600- to 1200-cps band, use the 600- to 1200-cps band as the first band and then define the *speech-interference level as the arithmetic average of the sound pressure levels in the three bands* 600 *to* 1200, 1200 *to* 2400, *and* 2400 *to* 4800 *cps.*[15,16] However, if the levels in the 300- to 600-cps band are more than 10 db above those in the 600- to 1200-cps band, the average of the levels in the *four* bands between 300 and 4800 cps should be used instead.

It must be emphasized that the speech-interference level for a noise at a particular location does not say anything about how intelligible speech will be to a listener at that location because it does not describe the level of the speech itself. If the levels of the peaks of speech are also known at that location, an estimate can be made of the articulation index by the following procedure:

[15] L. L. Beranek, Airplane Quieting II—Specification of Acceptable Noise Levels, *Trans. ASME,* **67**: 97–100 (1947).

[16] L. L. Beranek and H. W. Rudmose, Sound Control in Airplanes, *J. Acoust. Soc. Amer.,* **19**: 357–364 (1947).

1. Using the "slow" scale (highly damped circuit) on the octave-band analyzer, determine the rms long-time average sound pressure levels of the noise in the bands 600 to 1200, 1200 to 2400, and 2400 to 4800 cps. Then take the arithmetic average of the noise levels in these three bands. This average is the speech-interference level.

2. Using the "fast" scale on the octave-band analyzer, determine the peaks of the sound pressure levels in the bands 600 to 1200, 1200 to 2400, and 2400 to 4800 cps. The noise levels must be sufficiently low so as not to contribute to the peak readings, or else some means must be used to separate the two—such as turning the air-conditioning system off, or waiting until an hour when the background noise has died down. Find the arithmetic average of these three band peak levels. The sound-level meter does not respond rapidly enough to indicate the true (1 per cent) peak levels, so that about 4 db should be added to the figure just determined. This gives the adjusted peak levels of speech.

3. The difference between the adjusted peak levels of speech and the speech-interference level yields a number that when divided by 30 approximates the articulation index. For a more accurate calculation, the more detailed procedures of the previous part should be used.

If two men are standing, facing each other in a noise field, the maximum speech-interference levels that just permit reliable communication at various voice levels and distances are as shown in Table 13.8. In making up this table, average male voices and good hearing are assumed, as well as unexpected word material. If the vocabulary is limited or if sentences only are spoken, the permissible speech-interference levels may be increased by about 5 db. If a woman is speaking, the permissible levels should be decreased by about 5 db. Absence of reflecting surfaces is assumed.

TABLE 13.8. Speech-interference Levels (in Decibels *re* 0.0002 Microbar) That Barely Permit Reliable Word Intelligibility at the Distances and Voice Levels Indicated. No Reflecting Surfaces to Aid the Direct Speech Are Assumed†

Distance, ft	Voice level (average male)			
	Normal	Raised	Very loud	Shouting
0.5	71	77	83	89
1	65	71	77	83
2	59	65	71	77
3	55	61	67	73
4	53	59	65	71
5	51	57	63	69
6	49	55	61	67
12	43	49	55	61

† After L. L. Beranek, Reference 15.

13.13. Criteria for Residential Areas. The management of factories located near residential areas often needs to consider the noises made by its operations. Although this problem has always existed, the advent of the jet-aircraft engine brought the question of neighborhood noise control into the foreground. Several acoustics groups have recently played important parts in the establishment of these criteria, but, as stated earlier, particular numbers are likely to be revised as more experience with this type of criterion accumulates.†

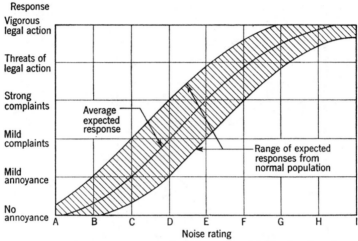

Fig. 13.22. Relation between response of residents in a neighborhood to the rating of the noise causing the response. (*After Rosenblith and Stevens, Handbook of Acoustic Noise Control, Vol. II, Noise and Man.*)

To arrive at a criterion curve for a given neighborhood, use is made of Figures 13.22 and 13.23, and Table 13.9.[2] First, decide what response from the neighborhood you wish to achieve, *e.g.*, no annoyance, mild annoyance, mild complaints, strong complaints, threats of legal action, or vigorous legal action. Then enter Fig. 13.22, and choose the letter (*A* to *I*) that results from this choice. For example, assume that no annoyance whatsoever is desired. Then noise rating *A* applies.

Next use Table 13.9 to adjust the rating upward or downward by a certain number of letters. For example, let us assume a case where we have a continuous-spectrum noise (0); an impulsive noise such as from a drop forge (+1); 10 to 60 exposures per hour (−1); an urban neighborhood near some industry (−2); daytime only (−1); considerable previous exposure (−1). This combination totals −4. The correct level rank

---

† Acoustics groups developing neighborhood criteria include Armour Research Foundation, Technology Center, Chicago, Ill.; Bolt Beranek and Newman, Consultants in Acoustics, Cambridge, Mass.; and Acoustics Research Group, General Engineering Labs., General Electric Co., Schenectady, N.Y.

for this neighborhood and this type of noise is, therefore, 4 letters less severe than A, that is to say, E.

The neighborhood noise-level criterion is found, with the aid of Fig. 13.23, from the level rank just determined. For example, for level rank E, the neighborhood noise level in each octave frequency band should not

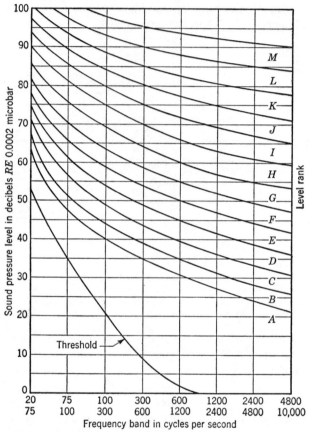

Fig. 13.23. Relation between the octave-band frequency spectrum and the level rank of the noise in a neighborhood. The level rank for the total noise is equal to the highest level rank achieved in any frequency band. (*After Rosenblith and Stevens, "Handbook of Acoustic Noise Control," Vol.* II, *Noise and Man. WADC Tech. Rept.,* 52–204.)

lie above the curve drawn above the letter E on Fig. 13.23. If an additional engineering factor of safety is desired, the levels should not lie above the curve drawn below the letter E in Fig. 13.23.

In cases where the noise spectrum is already known and the noise rating is desired, the above procedure is reversed. The measured octave-band noise spectrum is superimposed on Fig. 13.23. The level rank of the noise is given by the highest area into which the spectrum intrudes in any band. This procedure implies that the noise level in a single band can determine the level rank uniquely.

Then, Table 13.9 is used to obtain a correction number to be applied to the level rank to give the noise rating. Finally, using Fig. 13.22, the noise rating is translated into the probable neighborhood reaction to the noise.

TABLE 13.9. List of Correction Numbers to Be Applied to Level Rank to Give Noise Rating

Influencing factor	Possible conditions	Correction No.
Spectrum character...............	Continuous	0
	Pure-tone components	+1
Peak factor......................	Continuous	0
	Impulsive	+1
Repetitive character (20- to 30-sec exposures assumed)	One exposure per min (or continuous)	0
	10–60 exposures per hr	−1
	1–10 exposures per hr	−2
	4–20 exposures per day	−3
	1–4 exposures per day	−4
	1 exposure per day	−5
Background noise.................	Very quiet suburban	+1
	Suburban	0
	Residential urban	−1
	Urban near some industry	−2
	Area of heavy industry	−3
Time of day.....................	Daytime only	−1
	Nighttime	0
Adjustment to exposure............	No previous exposure	0
	Considerable previous exposure	−1
	Extreme conditions of exposure	−2

Noises created in sleeping quarters in houses or apartments due to sources in other parts of the same building should lie in all frequency bands about 5 db below the lower curve of the appropriate region of Fig. 13.23.

13.14. Criteria for Office Spaces. In office spaces, satisfactory speech communication is generally the principal reason for noise control. In a survey of noise conditions in a large metals-processing company, a radio-manufacturing company, and an educational institution, it was found that when office workers were asked to rate noise on a scale ranging from "very quiet" to "intolerably loud," they did so in approximate proportion to the speech-interference level. It was also revealed that the rating depended on the function of the office space, *e.g.*, private office, secretarial office, and so forth. The results are shown in Fig. 13.24.[17] The noise ratings are shown on the vertical scale, the speech-interference levels on the horizontal scale, and the satisfactoriness of telephone usage on the upper horizontal scale.

When questioned, employees said that point *A*, in Fig. 13.24, corresponded to normal voice at 9 ft and point *B* corresponded to a slightly raised voice at 3 ft. Comparison of these statements with the numbers

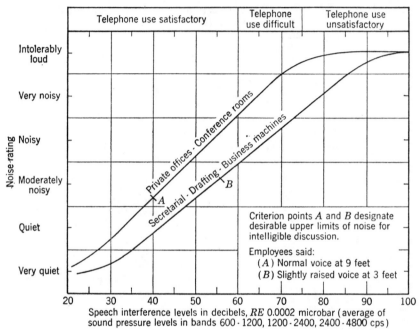

Fig. 13.24. Rating scale for noises in office spaces as a function of speech-interference levels. (*After Beranek and Newman, Rating of Noises in Rooms, presented at a meeting of the Acoustical Society of America, Pennsylvania State College, June, 1950.*)

given in Table 13.8 shows that the employees desire speech-interference levels about 6 db lower than the table would indicate. Table 13.8 applied to average men's voices. Women's voices are lower by several decibels, and also the spread of men's voices from the average is about ±8 db, which may account for this difference.

The rating scale for telephone conversation shown along the top of Fig. 13.24 is based on intercity calls using the F-1 telephone set (not the latest set with numbers outside the dial). For calls within a single modern exchange the permissible speech-interference levels are about

[17] L. L. Beranek and R. B. Newman, Rating of Noises in Rooms, presented at a meeting of the Acoustical Society of America, Pennsylvania State College, June, 1950.

5 db greater than those shown. For the latest telephone set, intra-exchange conditions generally exist for intercity calls.

13.15. Criteria for Auditoriums. *Reverberation Time T vs. Volume and Frequency.* Traditionally, the criterion for the design of rooms and auditoriums has been the reverberation time T defined on page 306. It is based on audience judgments of the acoustic quality of existing rooms and auditoriums.[18,19]

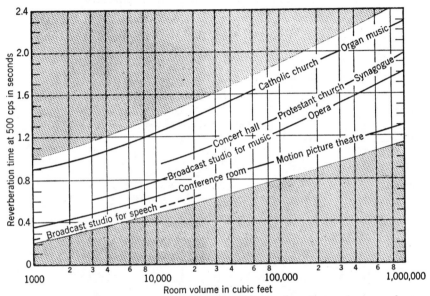

Fig. 13.25. Optimum reverberation times for rooms of various volumes and uses. (*Compiled from the literature and from the experience of Bolt Beranek and Newman, Consultants in Acoustics.*)

A reasonable summary of currently accepted optimum reverberation times is given in Fig. 13.25. These values are generally used at a frequency of 500 cps. The optimum reverberation time at other frequencies relative to that at 500 cps is shown in Fig. 13.26. Two curves are given, one for speech and the other for music.

Room Constant R. A more recent way of designating the optimum reverberation characteristics of a room is in terms of the room constant R defined on page 312. The optimum values of R for three categories of rooms are shown in Fig. 13.27 in terms of room volume. Some modern concert halls, such as the Royal Festival Hall in London, are designed for high definition largely because the music of the past hundred years has

[18] V. O. Knudsen, "Architectural Acoustics," pp. 408–414, John Wiley & Sons, Inc., New York, 1932.

[19] V. O. Knudsen and C. M. Harris, "Acoustical Designing in Architecture," pp. 192–195, John Wiley & Sons, Inc., New York, 1950.

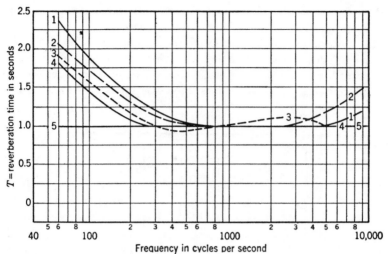

Fig. 13.26. Various recommended curves of reverberation time vs. frequency for studios and auditoriums relative to the reverberation time at 1000 cps. (1) MacNair, 1930; (2) Morris and Nixon, 1936; (3) Danish broadcasting studios for music; (4) Richmond and Heyda, 1940; (5) v. Békésy and others for speech rooms. Curve 3 below 1000 cps and curve 5 (flat) above 1000 cps are recommended by the author for music rooms. Curve 5 (flat) is recommended at all frequencies for speech rooms.

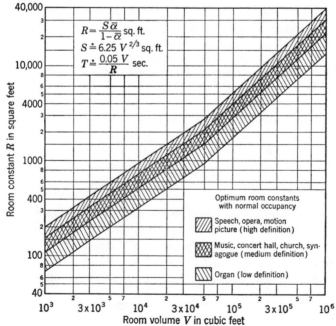

Fig. 13.27. Optimum values of room constant R at 1000 cps as a function of room volume for three catagories of rooms.

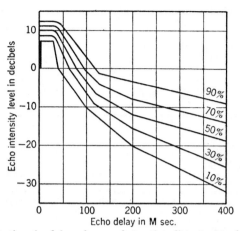

FIG. 13.28. A tentative (and largely unsubstantiated) criterion for annoyance due to echo. The contours are for constant percentage disturbance, and the coordinates are echo intensity level vs. echo delay. [*After Bolt and Doak, A Tentative Criterion for the Short Term Transient Response of Auditoriums, J. Acoust. Soc. Amer.*, **22**: 507–509 (1950).]

TABLE 13.10.   Criteria for Noise Control†

Type of room	SC criterion curve
Broadcast studios	15–20
Concert halls	20–25
Legitimate theaters (500 seats, no amplification)	25
Music rooms	25
Schoolrooms (no amplification)	25
Homes (sleeping areas)	25
Conference room for 50	25
Assembly halls (amplification)	25–30
Conference room for 20	30
Motion-picture theaters	30
Hospitals	30
Churches	30
Courtrooms	30
Libraries	30
Small private office	40
Restaurants	45
Coliseums for sports only (amplification)	50
Secretarial offices (typing)	55
Factories	40–65

†. L. L. Beranek, J. L. Reynolds, and K. E. Wilson, Apparatus and Procedures for Predicting Ventilation System Noise, *J. Acoust. Soc. Amer.*, **25**: 313–321 (1953).

become more intricate and detailed, requiring a reasonably dead hall for its faithful presentation.

Echoes.   Echoes are known to occur whenever an isolated reflected wave arrives at the ear of a listener more than $\frac{1}{15}$ sec (67 msec) after the time of arrival of the original sound.   In auditoriums, there are many reflected waves that arrive at the ear at time intervals greater than 70 msec because the reverberation time may be as high as several seconds.

Recent investigations show that long-time reflections are not troublesome if their intensity is sufficiently below that of the initial sound.[20] In Fig. 13.28 we show a set of curves that relate the approximate per cent of listeners noticing something wrong with the acoustics of the hall to

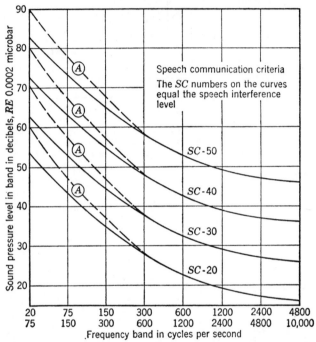

FIG. 13.29. Speech-communication and annoyance criteria for rooms. SC stands for speech communication, and the number following it is the speech-interference level for that contour. The dashed *A* curves give the permissible low-frequency levels if the levels in the 600- to 4800-cps bands are actually equal to the SC levels written on the curves. If not, the low-frequency levels must be lower. This graph must be used in connection with Table 13.10. [*After Beranek, Reynolds, and Wilson, Apparatus and Procedure for Predicting Ventilation System Noise, J. Acoust. Soc. Amer.*, **25**: 313–321 (1953).]

the sound pressure level of the echo as a function of the delay time, following the arrival of the direct sound. For example, if a reflection occurs 400 msec after the arrival of the direct sound at the ears of a group of listeners and if its level is 20 db below that of the direct sound, about half of the listeners will be aware that something undesirable has occurred.

13.16. Noise Backgrounds in Buildings. Freedom from noise is one of the most important considerations in the design of rooms and audi-

---

[20] R. H. Bolt and P. E. Doak, A Tentative Criterion for the Short-term Transient Response of Auditoriums, *J. Acoust. Soc. Amer.*, **22**: 507–509 (1950).

toriums.   With Table 13.10 and Fig. 13.29 we may arrive at criterion curves giving the maximum permissible noise levels in any frequency band for rooms of various types.[21]

For example, in a large assembly hall with a good speech-reinforcement system, we see from Table 13.10 that the 25-db SC criterion curve should be selected.   If the noise is concentrated in a portion of the frequency range, the appropriate solid curve of Fig. 13.29 is used.   If the noise levels principally lie near the appropriate curve in bands between 300 and 4800 cps, then, because of masking effects, the noise in the lowest three frequency bands may be higher.   For this case, the dashed curves A may be used in design.   In other words, more noise is permissible at low frequencies if there is high frequency masking noise than if there is not.

[21] L. L. Beranek, J. L. Reynolds, and K. E. Wilson, Apparatus and Procedures for Predicting Ventilation System Noise, *J. Acoust. Soc. Amer.*, **25**: 313–321 (1953).

PROBLEMS

Chapter 1

1.1. (*a*) Calculate the wavelength in meters and in feet for sound in air for 72°F for $f = 100, 500$ cps.

(*b*) Calculate the frequency of a sound whose wavelength in air at 72°F is 1 m, 1 ft, 1 in.

1.2. What is the characteristic impedance of air in mks units for a barometric pressure of 0.75 m Hg and $T = 0$°F, 32°F, 50°F, 90°F?

1.3. Show how to convert the following ratios to decibels without use of a slide rule or tables. Find first the number of decibels as if these were pressure ratios, and then as if these were intensity ratios. Remember that approximately $\log_{10} 2 \doteq 0.3$ and $\log_{10} 3 \doteq 0.5$.

(*a*) 2/1	(*d*) 0.002/1
(*b*) $\sqrt{10}/200$	(*e*) 8/0.05
(*c*) 30/1.33	(*f*) 17.32/1.6

What are the exact values of $\log_{10} 2$ and $\log_{10} 3$?

1.4. A microphone is connected to an amplifier with an input resistance of 500 ohms. The output of the amplifier is connected to a nondirectional loudspeaker with an electrical resistance of 32 ohms. If the amplifier produces 11 volts (rms) across the terminals of the loudspeaker when the microphone produces 0.013 volts (rms) across the input of the amplifier, what is the power gain in decibels of the amplifier? The voltage gain?

1.5. The sound intensity level measured at a distance of 100 ft from a nondirectional loudspeaker is 70 db *re* 10^{-16} watt/cm^2. What is the total power radiated by the loudspeaker? Find the sound intensity and sound intensity level at a distance of 200 ft.

1.6. Given n different sounds of sound intensity level IL_n, find the total sound intensity level IL_t (assume that the various sound intensities add arithmetically).

1.7. The sound intensity level (*re* 10^{-16} watt/cm^2) of each of various sounds at a certain location outdoors is given below.

Sound	db
A	68
B	73
C	60
D	70

What is the total sound intensity level due to all of these sounds?

1.8. Given the noise levels measured in the frequency bands indicated below for an airplane with a random noise spectrum, calculate the spectrum level (sound pressure level in decibels for bands one cycle wide) of the noise in each band, assuming that it is uniform throughout each band. Plot the datum point for each band on two-cycle semilog paper at the geometric mean frequency f_g of that band, *i.e.*, $f_g = \sqrt{f_1 f_2} \doteq 1.4 f_1$, and connect the points with a curve.

Frequency bands, cutoff frequencies, cps		Sound pressure level, db, re 0.0002 microbar
f_1	f_2	
75	150	95
150	300	101
300	600	97
600	1200	91
1200	2400	86
2400	4800	79
4800	9600	70

1.9. Spectrum levels of a continuous spectrum noise are tabulated below for the frequency range 100 to 2500 cps. An analyzer having half-octave filter bands is used to measure this noise, the datum point for each band being at the geometric mean frequency. Determine the noise level readings obtained in successive half-octave bands, starting with the bands 100–141 cps, 141–200 cps, 200–282 cps, etc. Note that each band has a width extending from f to $\sqrt{2}\,f$ and that the geometric frequency equals $f \sqrt[4]{2}$

Frequency, cps	Spectrum level, db, re 0.0002 dyne/cm²
100	84
200	85
300	82
500	80
700	77
1000	71
1500	66
2000	62
2500	57

1.10. A noise spectrum is measured with a General Radio sound analyzer. This analyzer has a filter whose mean frequency is continuously variable and whose bandwidth is equal to $0.02f$. Convert the measured sound pressure levels (re 0.0002 microbar) to spectrum levels. Plot a curve of spectrum level vs. frequency on two-cycle semilog paper.

Frequency, cps	Measured sound pressure levels, db
75	72
150	75
300	78
700	74
1100	69
2200	63
4800	59
7500	53

1.11. Plot the *spectrum level* (plot db for a 1-cps bandwidth vs. frequency on two-cycle semilog paper) of the electrical noise that will produce an output of 0.1 volt independent of frequency, from the following perfect filters. As a reference voltage level use 1 volt.

(a) A sound analyzer of the constant percentage bandwidth type. This instrument has a bandwidth equal to 0.02f, where f is the frequency to which the analyzer is tuned.

(b) A wave analyzer of a constant bandwidth type. This instrument has a constant bandwidth of 3 cps.

(c) An octave filter set with the following contiguous bands:

Limiting frequencies of the bands,
cps
75–150
150–300
300–600
600–1200
1200–2400
2400–4800
4800–9600

For case (c) plot the data at the geometric mean frequency of each band, *i.e.*,

$$f_g = \sqrt{f_1 f_2}.$$

1.12. You are asked to measure a continuous noise spectrum extending over a frequency range of 20 to 15,000 cps. You find that it will take two analyzers, one for the range from 20 to 5000 cps and the other from 5000 to 15,000 cps. The bandwidth of the low-frequency analyzer is a constant equal to 100 cps. The bandwidth of the other analyzer varies as a function of frequency and is equal to 0.01f, where f is the mid-band frequency. If the following data are obtained, what values would you plot on a spectrum-level vs. frequency graph?

Low-frequency Analyzer

f	SPL
100	60
200	63
2000	64
4000	68

High-frequency Analyzer

f	SPL
5000	65
6000	66
7000	67
10,000	69

1.13. A sound wave is propagated in a room. At a particular point the sound pressure is $1.4\underline{/30°}$ and the particle velocity is $0.02\underline{/-30°}$. Find the specific acoustic impedance.

1.14. A source of sound is located in free space. A series of 12 measurements of intensity level is made at 12 points on a hypothetical spherical surface of radius of 10 m with center at the source. The points of measurement are so located that they lie at the centers of equal areas that divide the surface of the sphere into 12 parts. The 12 intensity levels are 90, 89, 88, 87, 86, 85, 84, 83, 82, 81, 80, and 70 db *re* 10^{-16} watt/cm². Assuming that the intensity of the sound is uniform over each part, calculate (a) the total power radiated by the source; (b) the power level; (c) the average intensity; (d) the average sound pressure; and (e) the average sound pressure level.*

---

* The area of a sphere is $4\pi r^2$.

Chapter 2

2.1. On the conventional pressure vs. volume (PV) diagram, plot the behavior of a volume of perfect gas contained in a cylinder when the piston closing one end is driven so that its displacement as a function of time is a square wave. What power, in watts, is dissipated by the moving piston for any repetition frequency f? (A graphical solution for a typical case is acceptable.) If power is dissipated where does it go?

2.2. A cylindrical tube having a radius of 3 cm is closed at one end. The diameter of this tube is small compared to the wavelength of the sound in it, which means that plane waves only will be found. If the tube is 38 cm long, calculate the acoustic, specific acoustic, and mechanical impedances at the open end for frequencies of 60, 110, and 490 cps. What is the percent error if calculations are made at 60 cps, using the approximate formula (single-term expansion of the cotangent) for the impedance of a short tube?

2.3. Given a cylindrical tube of length L with one end open and the other end closed with a piston with velocity $U_0 e^{j\omega t}$. The diameter of the tube is small compared to a wavelength of the sound in it, which means that only plane waves are found. It is assumed that negligible sound power is radiated from the open end, and consequently the sound pressure at this end is approximately zero. Obtain a solution of the one-dimensional wave equation under these conditions and derive an expression for the specific acoustic impedance looking into the tube at the piston end.

2.4. For the tube of Fig. 2.3, the specific acoustic reactance at a plane x_0 along its length is -500 mks rayls at 1000 cps. What is the specific acoustic reactance at values of X that are greater than x_0 by 0.05, 0.10, and 0.20 m?

2.5. How does the sound pressure level in a plane free-traveling wave vary as the temperature varies from 0 to 1000°C if the intensity level is held constant at 100 db?

2.6. A point source operating at 1000 cps radiates 1 acoustic watt. Determine, at a distance of 10 m, (a) intensity; (b) sound pressure; (c) intensity level; (d) sound pressure level; (e) rms particle velocity; (f) rms particle displacement; (g) rms incremental density; and (h) rms incremental temperature.

2.7. A sound intensity level of 80 db is measured 10 m from a spherical source of 100 cps radiating into free space. What is the total power radiated in all directions? What is the power level? For this case plot the intensity level in decibels as a function of r on semilog paper with r on the logarithmic axis. How many decibels does the intensity level decrease for each doubling of distance? Does the sound pressure level decrease at the same rate?

2.8. Discuss the variation of sound intensity and of energy density with distance r in a spherical free-progressive wave. If these quantities vary in different ways, explain the physical meaning of the difference in the variations.

2.9. The one-dimensional wave equation in cylindrical coordinates is

$$\frac{\partial^2 p}{\partial r^2} + \frac{1}{r}\frac{\partial p}{\partial r} = \frac{1}{c^2}\frac{\partial^2 p}{\partial t^2}$$

Here we have assumed that pressure is independent of the coordinates θ, z and depends on r and t only. Show that

$$p = \frac{P_0 e^{jk(ct \pm r)}}{\sqrt{r}}$$

satisfies this equation for $4r^2 k^2 \gg 1$. What is the significance of the $\pm$ sign in the exponent?

2.10. Find the specific acoustic impedance at a distance r from the axis of a long cylindrical source. Assume that the source produces a sound field that can be approximated by

$$p = \frac{P_0 e^{jk(ct-r)}}{\sqrt{r}}$$

To what value does this impedance reduce for $4k^2r^2 \gg 1$? See Table 2.1 for obtaining u in the steady state. Substitute r for x in the equation of motion.

Chapter 3

3.1. Convert the mechanical system shown in Fig. P3.1 into a mobility-type circuit and into an impedance-type circuit. Motion is possible only along the x axis. The supporting springs have a total compliance of $2C_{M2}$ at each of the four positions where they occur.

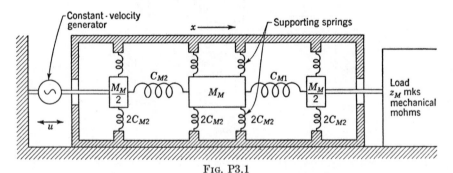

Fig. P3.1

3.2. (a) For the mechanical system shown in Fig. P3.2, draw the mobility-type and the impedance-type circuits. Motion is possible only along the y axis.

(b) If the constant velocity source u is represented by $u = \sqrt{2} |u_{rms}| \cos \omega t$, find the rms magnitude and the phase (relative to u) of the velocity of M_{M1}.

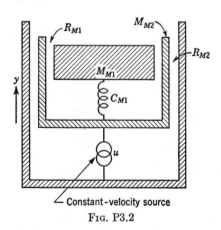

Fig. P3.2

3.3. In the sketch of a mechanical system shown in Fig. P3.3, the mass M_{M1} is supported by four springs each with a compliance C_{M1}. The mechanical resistance R_{M2}

represents the total friction between M_{M1} and M_{M2}. Motion is possible only along the y axis.

(a) Draw the mechanical schematic.

(b) Draw the mobility-type circuit.

(c) Draw the impedance-type circuit.

(d) If the constant velocity generator is represented by $u = \sqrt{2}\, A \cos \omega t$, determine the frequency at which the velocity of M_{M1} will be a maximum, assuming R_{M1} to be very small.

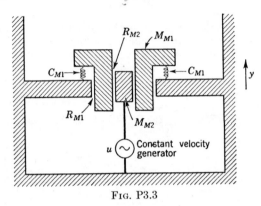

Fig. P3.3

3.4. An idealized drawing of an automobile is shown in Fig. P3.4. Suppose one wheel goes over a series of bumps in the road, giving that tire a velocity, vertically,

$$u = \sqrt{2}\, u_0 \sin \omega t$$

Assume that only vertical motion is possible. Draw the mobility-type and impedance-type circuits.

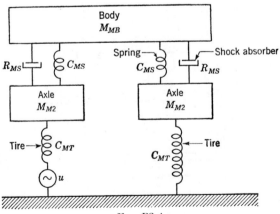

Fig. P3.4

3.5. Draw the mobility-type circuit for the agitator shown in Fig. P3.5. Indicate on the circuit diagram the velocity u_p of the small particle M_{M4} that is *falling* through the viscous fluid held inside a test tube T. The mass of the wall of the test tube is M_{M3}. The mechanical resistance between the particle and the walls of the test tube is R_{M3}. Neglect the small "d-c" downward velocity of the small particle. Construct the mobility- and impedance-type circuits, assuming $C_{M1} = C_{M2}$.

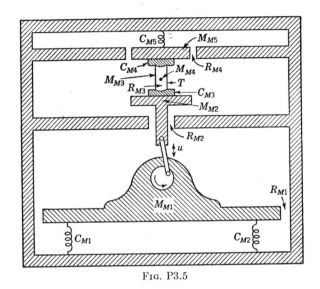

FIG. P3.5

3.6. (a) Draw a mobility-type circuit and sketch a mechanical system which would have the analogous electrical circuit shown in Fig. P3.6, where i is analogous to force f.

(b) Where must the ground point of this circuit be located if the mechanical system is to be physically realizable?

(c) Give the magnitude of each mechanical element in terms of the constants of the above circuit and show *where* and in *what direction* the force f is acting.

(d) Obtain the dual of the electrical circuit. Label the elements of the dual in terms of the given circuit.

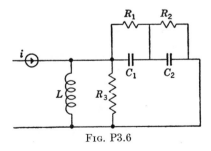

FIG. P3.6

3.7. (a) Using the impedance-type analogy draw a circuit for the acoustical system shown in Fig. P3.7, where M_{M1} is in kg; M_{A1}, M_{A2}, M_{A3} are in kg/m⁴; C_{A2} is in m⁵/newton; and R_{A1}, R_{A2} are in mks acoustic ohms.

(b) Convert this circuit to a mobility-type circuit.

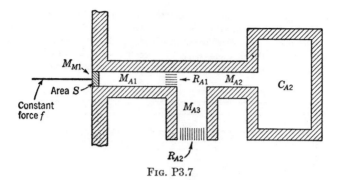

FIG. P3.7

3.8. Sketch a possible acoustic realization of the electrical filter shown in Fig. P3.8. Show the acoustic equivalent of placing a resistive load across the last LC branch.

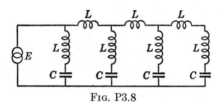

FIG. P3.8

3.9. In the device of Prob. 3.2 the element sizes are as follows:

$$M_{M1} = 0.01 \text{ kg}$$
$$M_{M2} = 0.02 \text{ kg}$$
$$C_{M1} = 10^{-3} \text{ m/newton}$$
$$R_{M1} = 1 \text{ mks mechanical ohm}$$
$$R_{M2} = 2 \text{ mks mechanical ohms}$$

If the velocity generator has a frequency of 100 radians/sec and a velocity 1 m/sec, find the active and reactive power for each element and the total active and reactive power.

3.10. The device of Prob. 3.7 has the following constants:

$$M_{M1} = 0.001 \text{ kg}$$
$$S = 3 \text{ cm}^2$$
$$M_{A1} = 0.001 \text{ kg/m}^4$$
$$R_{A1} = 10 \text{ mks acoustic ohms}$$
$$M_{A3} = 0.002 \text{ kg/m}^4$$
$$M_{A2} = 0.003 \text{ kg/m}^4$$
$$R_{A2} = 20 \text{ mks acoustic ohms}$$
$$C_{A2} = 10^{-7} \text{ m}^5/\text{newton}$$

If the generator has a frequency of 1000 cps and an rms force of 1 newton, determine the active and reactive power in each element and the total active and reactive power.

Chapter 4

4.1. A very small loudspeaker in a small enclosed baffle behaves like a simple spherical source of the same area at low frequencies. If its rms amplitude of motion (*i.e.*, rms displacement) is 0.1 cm at 100 cps, and its diameter is 10 cm, determine (*a*) sound pressure at 1 m; (b) intensity at 1 m; (c) total power radiated; (d) sound pressure level; (e) sound power level.

4.2. The sound power level of a nondirectional source of sound is 130 db. Find, at a distance of 70 ft, (a) intensity; (b) sound pressure; (c) sound pressure level. Also find the peak-to-peak displacement at 100 cps of the spherical source if its radius is 0.5 ft.

4.3. Verify the directivity indexes given in Fig. 4.5 for the cases of $\lambda/2$, λ, and $3\lambda/2$. Assume that the $0°$ axis is an axis of symmetry in three-dimensional space.

4.4. For the source of Fig. 4.11 determine the sound pressure level at 3 m on the $\theta = 0$ axis for $ka = 0.5$, 1.0, 2.0, and 3, with a PWL = 130 db *re* 10^{-13} watt.

4.5. Compute the directivity factor and directivity index for a microphone from the following data taken at 3000 cps:

θ, deg	Output voltage level minus output voltage level at $\theta = 0°$, db
0– 30	0
30– 60	−10
60– 90	−20
90–120	−30
120–150	−40
150–180	−30

4.6. Given a loudspeaker with the following directivity characteristic at 2000 cps:

θ, deg	Acoustic response in db relative to acoustic response at $\theta = 0°$
0– 10	0
10– 20	− 1.5
20– 30	− 6.0
30– 40	−10.0
40– 50	−17.0
50– 60	−19
60– 70	−21
70– 80	−22
80– 90	−23
90–100	−25
100–110	−27
110–120	−29
120–130	−32
130–140	−47
140–150	−35
150–160	−30
160–170	−29
170–180	−27

Calculate the directivity factor Q and the directivity index DI. Assume that at 2000 cps the sound pressure level at a distance of 15 ft from the loudspeaker measured in an anechoic chamber is 111 db *re* 0.0002 dyne/cm² for an available electrical power input of 50 watts. What is the power available (PAE) efficiency level in db at 15 ft? Would this answer be different if a distance of 40 ft were assumed as a basis for calculation, assuming, as before, that 111 was the level measured at 15 ft?

Chapter 5

5.1. A symmetrical loudspeaker is mounted in a wall between two anechoic rooms. Its radius is 0.1 m and the rms velocity of the diaphragm is 1 cm/sec. Determine the total acoustic power radiated at 100 and 1000 cps. Determine the intensity on the axis of the loudspeaker at 3 m distance at 1000 cps, taking into account directivity.

5.2. Assume the source of Fig. 4.11 with a radius of 30 cm for which directivity indexes are given in Figs. 4.12 and 4.20. Calculate for cases (a) and (f) shown in Fig. 4.12 the rms velocity of the diaphragm necessary to produce a constant sound pressure level of 100 db at a distance of 10 ft on the $\theta = 0$ axis.

5.3. A piston of radius a in an infinite baffle is vibrating with an rms velocity u. At $\omega_i = \dfrac{1}{2}\dfrac{c}{a}$ the power radiated is (0.12) $\pi a^2 \rho_0 c u^2$. Approximately what is the power radiated at $\omega_2 = \frac{1}{10}\omega_1$ and at $\omega_3 = 10\omega_1$, assuming that u remains unchanged?

5.4. The acoustic device shown in Fig. P5.4 is used as the load on one side of the diaphragm of a transducer.

(a) Draw the acoustic impedance-type circuit.

(b) Alter the circuit diagram so that it may be coupled through the area S to a mechanical mobility-type circuit.

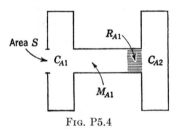

Fig. P5.4

(c) Attach an electrostatic transducer to the area S. Assume that it has an electrical capacitance C_E, a coupling constant τ, and a mechanical compliance C_M (see Fig. 3.37).

(d) Attach an electromagnetic transducer to the area S. Assume that it has an electrical winding impedance Z_E, a coupling constant Bl, and a diaphragm with a mass M_M (see Fig. 3.35).

5.5. Sketch an equivalent circuit for the mechano-acoustical system shown in Fig. P5.5. The volumes V_1, V_2, V_3; the acoustic resistance and acoustic mass R_A, M_A; and the effective mechanical mass M_M, mechanical compliance C_M, and area S of the diaphragm are in mks units. Give the relation between each of the element sizes on your circuit diagram and the mechanical or acoustical quantities given here.

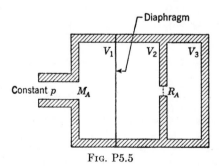

Fig. P5.5

5.6. A loudspeaker diaphragm has an air cavity behind it and a resonator that is connected into one side of the cavity. Find the mechanical impedance load at 100 cps on the rear side of the diaphragm if the diaphragm area is 50 cm², the volume of the air cavity behind the diaphragm is 1000 cm³, and the resonator comprises a tube 5 cm long and 2 cm in diameter connecting to a second volume of 100 cm³.

5.7. A loudspeaker and a microphone are placed on opposite sides of a circular hole in a perfectly rigid wall as shown in the diagram in Fig. P5.7. The wall may be con-

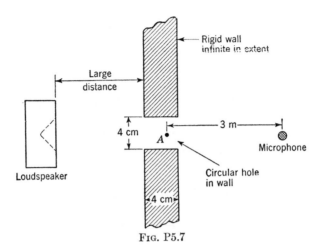

Fɪɢ. P5.7

sidered infinite in extent in all directions. If the hole is *plugged* with a rigid material, the loudspeaker produces a pressure *p* of 94 db *re* 0.0002 dyne/cm² on the left-hand surface of the wall at a frequency of 500 cps.

(*a*) Compute the power that emerges from the hole on the microphone side of the wall when the hole is opened.

(*b*) Assume that a hemispherical wave (centered at point *A*) is propagated to the right of the wall. Compute the open-circuit voltage of the microphone, assuming that it has a sensitivity of −50 db *re* 1 volt for a sound pressure of one microbar.

(*c*) How does the open-circuit microphone voltage vary as the frequency is decreased? Assume that *p* (with the hole plugged) remains at 94 db *re* 0.0002 dyne/cm² and that the microphone voltage is the same at all frequencies if it is in a sound field of constant pressure.

5.8. Draw the equivalent circuit of the device shown in Fig. P5.8 in both the mobility-type and the impedance-type analogies, using acoustic impedances. The areas of the tubes at the right and the left of the cavity are both *S*. Assume the

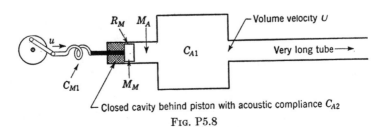

Fɪɢ. P5.8

device can be represented by lumped elements and assume that end corrections may be neglected. Show the volume velocity of the diaphragm U on your circuits.

5.9. Figure P5.9 is an analogous circuit for a mechano-acoustical system. The elements within the dotted boxes are mechanical elements. Sketch a possible arrangement of acoustical and mechanical elements that has an analogous circuit like that shown in the figure.

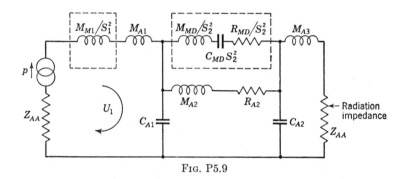

FIG. P5.9

5.10. An acoustical element made from a perforated sheet with a thickness of 0.0025 m and an area of 0.01 m² is needed that has a "Q_A" of 2.0 and an acoustic mass of 135 kg/m⁴ at an angular frequency of 100 radians/sec. Find the number of holes required in such an element. *Hints:* Assume that the large fraction at the right of Eq. (5.58) is approximately unity. Assume $(1 - a/b) \doteq 1$ in Eq. (5.57). At the end, test the percentage error in "Q_A" and M_A made by these assumptions. Note also that the number of holes n equals the area divided by b^2.

Chapter 6

6.1. A moving-coil microphone has a diaphragm with an area of 5 cm². When this diaphragm is replaced by a rigid block of steel of the same shape the sound pressure produced at 500 cps at its surface in a free field by a particular source is 1 dyne/cm². The acoustic impedance of the diaphragm (when in place on the microphone) as measured at its external surface is $(20 + j0)$ cgs acoustic ohms. What sound pressure is produced at the surface of the diaphragm?

6.2. A source of sound with a very high mechanical impedance is imbedded in the inner surface of a closed cavity. The cavity has a volume of 20 cm³. A microphone with a diaphragm impedance of $200 - j(10^6/\omega)$ cgs acoustic ohms is imbedded in another part of the inner surface. If the source diaphragm has an rms displacement of 0.05 cm³, what is the complex rms velocity of the microphone diaphragm at 500 cps if it has an area of 4 cm²?

6.3. The response of a capacitor microphone is -52 db re 1 volt for a sound pressure of 1 microbar at the diaphragm. The capacitance when the microphone diaphragm is exposed to the air $(p_D \doteq 0)$ is 50.0 micromicrofarad $(\mu\mu f)$. The capacitance when the microphone diaphragm is blocked $(u_D = 0)$ is 48.0 $\mu\mu f$. The air volume behind the diaphragm is 0.2 cm³ and the diaphragm area is 3 cm². Draw an equivalent circuit for the microphone at low frequencies and give the values of the elements in mks units.

6.4. A capacitor microphone whose diaphragm is flush with one side of a small cavity has the following characteristics:

$$\text{Diaphragm compliance} = 5 \times 10^{-4} \text{ m/newton}$$
$$\text{Compliance of air behind diaphragm} = 2.2 \times 10^{-5} \text{ m/newton}$$
$$\text{Polarizing voltage} = 200 \text{ volts}$$
$$\text{Spacing of diaphragm and backplate} = 10^{-4} \text{ m}$$
$$\text{Area of diaphragm} = 3 \text{ cm}^2$$
$$\text{Electrical capacitance of the microphone} = 35 \ \mu\mu f$$
$$\text{Load resistance} = 20 \text{ megohms}$$

Neglect the mass and resistance of the diaphragm. A small constant-velocity piston imbedded in one side of the cavity generates a pressure of 0.1 newton/m² throughout the cavity with the microphone diaphragm fixed so that it cannot move. The mechanical compliance of the cavity is 9×10^{-3} m/newton. Compute the voltage across the load resistance at an angular frequency of 3000 radians/sec.

6.5. A crystal microphone of the type shown in Fig. 6.40 is used in the circuit shown in Fig. P6.5. The X-cut Rochelle-salt crystal element inside the microphone is like that shown in Fig. 6.35a, with dimensions $1 \times 1 \times \frac{1}{16}$ in. The plates are connected

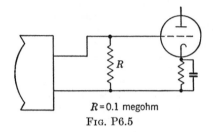

$R = 0.1$ megohm
FIG. P6.5

in parallel electrically. Assume that for a given sound pressure the microphone produces an *open-circuit* voltage level of -60 db *re* 1 volt at frequencies of 100 and 1000 cps and $T = 40°C$. What will the voltage levels be at the same frequency if the sound pressure is held constant and the temperature is set at each of the following values: 10, 25, and 30°C?

6.6. A barium titanate unit is to be used in a microphone. Assume that the unit will be the bender Bimorph type, series-connected. If the length is 10 cm, the width 2.5 cm, and the thickness 0.2 cm, determine the open-circuit voltage produced by the microphone when a rms force of 0.01 newton at $\omega = 1000$ radians/sec is applied to the proper place on the bender Bimorph unit.

6.7. A tentative design for a Rochelle-salt-pressure microphone calls for an aluminum diaphragm connected to one corner of a square-torque Bimorph crystal that has been made from an X-cut shear plate (see Fig. 6.40). Physical constants are:

$$\text{Diaphragm diameter} = 4 \text{ cm}$$
$$\text{Diaphragm mass} = 5 \times 10^{-4} \text{ kg}$$
$$\text{Diaphragm compliance} = 4.5 \times 10^{-5} \text{ m/newton}$$
$$\text{Crystal area} = 1 \text{ cm}^2$$
$$\text{Crystal thickness} = 2 \text{ mm}$$
$$\text{Volume of cavity behind diaphragm} = 10^{-5} \text{ m}^3$$

The plates are connected electrically in parallel. Disregard the cavity between the diaphragm and sintered plate.

(*a*) Compute the low-frequency open-circuit pressure sensitivity in db *re* 1 volt per microbar, disregarding the effect of the sintered metal plate.

(*b*) Disregarding the air load, calculate the resonance frequency with the electrical terminals open-circuited.

(c) What would be the effect on sensitivity and frequency response of reducing the volume of the cavity behind the diaphragm?

Chapter 7

7.1. A small direct-radiator loudspeaker is to be used both as a microphone and as a loudspeaker in an intercommunication set. It is calibrated as a microphone in the middle of the audio-frequency region where, to a rough approximation, all mass and stiffness reactances may be neglected and the specific radiation impedance on each side of the diaphragms is $\rho_0 c$.

If the open-circuit free-field sensitivity as a microphone is -80 db re 1 volt for a blocked-diaphragm sound pressure of 0.1 newton/m^2, what will be the maximum power available efficiency of the device operating as a loudspeaker, assuming a generator resistance of 0.2 ohm, and radiation into half-space?

Let voice-coil resistance $R_{ED} = 0.2$ ohm; flux density in air gap $B = 2.2$ webers/m^2; length of wire in coil $l = 2$ m; effective radius of diaphragm $= 0.04$ m.

7.2. A public address system with an over-all "gain" at 1000 cps of 15 db is desired. The "gain" of the system, here, is 20 times the logarithm to the base 10 of the ratio of the sound pressure produced at a distance of 50 ft by the loudspeaker in an anechoic room to that produced at the microphone by the person talking. You have available a microphone with a free-field open-circuit sensitivity of -90 db below 1 volt per microbar and an internal resistance of 30 ohms. You also have a loudspeaker that has a maximum available power response of 96 db re 0.0002 microbar and 1 watt measured at 1000 cps and 20 ft. The loudspeaker impedance is $8 + j0$ ohms and the amplifier has an input impedance of $30 + j0$ ohms and an output impedance of $26 + j0$ ohms. What matched-impedance power gain in decibels must your amplifier have at 1000 cps?

7.3. The power available efficiency of a loudspeaker operating in an infinite baffle is 5 per cent, assuming radiation into half-space. At a distance of 20 ft from the speaker there is a dynamic microphone. The open-circuit voltage calibration of the microphone is -90 db re 1 volt for a sound pressure level of 1 microbar. Assume that the speaker radiates a spherical wave. Also assume that the output impedance of the microphone is essentially resistive and equal to 30 ohms. If the power available to the loudspeaker is 10 watts, what is the power in a 30-ohm resistor connected to the microphone?

7.4. In a certain loudspeaker there is only a limited space for the voice coil. Some of the constants and dimensions of this loudspeaker are as follows:

Cone diameter $= 0.25$ m
Mass of the cone $= 0.03$ kg
Voice-coil diameter $= 0.05$ m
Maximum radial thickness of the voice coil winding $= 0.0025$ m
Winding length of voice coil $= 0.0125$ m
Number of turns on the voice coil $= 10$

(a) Compute the ratio of the loudspeaker efficiency for an aluminum voice coil to the loudspeaker efficiency for a copper voice coil. Consider only the range where $ka < 0.3$.

(b) What is the d-c resistance of each voice coil?

7.5. For frequencies below 500 cps, determine the resonance frequency and the total "Q_T" of a direct-radiator loudspeaker with the physical constants given below, assuming an infinite baffle mounting. Let $R_g = 10$ ohms; $R_E = 8$ ohms; $B = 2.1$ weber/m^2; $l = 10$ m; $M_{MD} = 0.02$ kg; $C_{MS} = 2 \times 10^{-4}$ m/newton; and effective diameter $= 0.2$ m.

7.6. An idealized direct-radiator loudspeaker has the following characteristics:

$$\text{Mass of the cone and voice coil } = 0.025 \text{ kg}$$
$$\text{Compliance of suspension } = 10^{-5} \text{ m/newton}$$
$$\text{Air-gap flux } = 1 \text{ weber/m}^2$$
$$\text{Length of the voice-coil winding } = 10 \text{ m}$$

Assume that the impedance of the voice coil is negligible at all frequencies of interest and that the specific acoustic impedance of the air load on the cone is 407 newton-sec/m^3

(a) At what frequency does the loudspeaker resonate?

(b) If the mass of the air load had been taken into consideration, how would the resonance frequency of the system have been affected?

(c) Sketch the response of this speaker as a function of frequency. Discuss any differences between this curve and the usual direct-radiator loudspeaker response curves.

7.7. Given a single-cone single-coil direct-radiator loudspeaker with the following characteristics: effective diameter = 0.3 m; mass of the cone = 0.025 kg; mass of the voice coil = 0.005 kg; mechanical compliance of suspension = 6×10^{-5} m/newton; mechanical resistance of suspension = 1.5 mks mechanical ohms; air-gap flux = 1.0 weber/m^2; voice-coil resistance = 8 ohms; voice-coil inductance = negligible; diameter of voice coil = 0.05 m; number of turns on voice coil = 50; and generator resistance = 8 ohms. Find the power available efficiency at 500 cps, assuming radiation to one side of an infinite baffle.

7.8. Given a single-cone single-coil direct-radiator loudspeaker with the following characteristics: diameter = 8 in.; mass of cone = 25 g; mass of voice coil = 2 g; mechanical compliance of suspension = 5×10^{-7} cm/dyne; mechanical resistance of suspension = 2000 mechanical ohms; air-gap flux gauss = 15,000; voice-coil resistance = 8 ohms; voice-coil inductance = negligibly small; diameter of voice coil = 2.0 in.; and number of turns on voice coil = 50.

Assume that a constant voltage of 1 volt rms is held across the voice-coil terminals. Compute the acoustic power output at 1000 cps, and determine the efficiency at this frequency, assuming radiation to one side of an infinite baffle.

Chapter 8

8.1. Design and give brief instructions for building a closed-box baffle for a loudspeaker with an advertised diameter of 12 in. The resonance frequency of the speaker in the closed box should be only 10 per cent higher than the resonance frequency without baffle.

Assume that the compliance of suspension, C_{MS}, equals 2.1×10^{-4} m/newton and the mass of diaphragm and voice coil, M_{MD}, equals 0.012 kg. What would be the resonance frequency if the speaker radiated into the air on both sides of an infinite baffle?

8.2. The loudspeaker of Example 8.3 in the text, when mounted in a closed-box baffle with a volume of 0.3 m^3, has a Q_T of 0.27 for $R_g = 3$ ohms, and a resonance frequency of 50 cps.

(a) Draw the response curve at frequencies below 400 cps.

(b) What is the effect on the response of the loudspeaker in this frequency range resulting from rigidly attaching three "domes" (cones with a height of 1 cm and a diameter of 10 cm) to the cone so as to increase the mass of the diaphragm? Calculate two cases where each of the three domes has a mass, respectively, of 0.02 and 0.1 kg.

Note: In the calculations assume that $Q_T = M_A/R_A$ is the same at all frequencies as its value at 100 cps.

8.3. The loudspeaker of Example 8.3 in the text, when mounted in a closed-box baffle with a volume of 0.3 m³, has a Q_T of 0.27 and a resonance frequency of 50 cps for $R_g = 3$ ohms.

(a) Draw the response curve at frequencies below 500 cps.

(b) What is the effect on the response of the loudspeaker of mounting a short section of an exponential horn in front of the loudspeaker that loads the front side of the diaphragm with the following specific acoustic impedances (divided by $\rho_0 c$) in place of the front-side radiation impedance?

Frequency, cps	$R_S/\rho_0 c$	$X_S/\rho_0 c$
20	0.0005	0.03
40	0.002	0.05
75	0.008	0.1
150	0.03	0.2
250	0.1	0.4
300	0.15	0.5
400	0.25	0.6
500	0.4	0.7

8.4. A particular 15-in. low-frequency direct-radiator loudspeaker is placed in a closed box and the sound pressure on axis (in the far-field) relative to the mid-band pressure is as shown in Fig. 8.15, under the conditions that

$$R_g = 14 \text{ ohms}$$
$$R_E = 5.5 \text{ ohms}$$
$$Bl = 25 \text{ webers/m}$$
$$R_{MS} = 2.3 \text{ mks mechanical ohms}$$
$$M_{MD} = 0.045 \text{ kg}$$
$$C_{MS} = 2.82 \text{ m/newton}$$
$$\text{Volume of box} = 0.3 \text{ m}^3$$
$$S_D = 8.03 \times 10^{-2} \text{ m}^2$$
$$S_D/L^2 = 0.2 \text{ (see Fig. 8.6)}$$

(a) Assume that we wish to convert the closed box into a vented enclosure. What is the resonance frequency of the driver and its air loads (the series part of the circuit of Fig. 8.18)?

(b) If we construct the port by cutting a 20-cm-diameter hole in the box and inserting in it a tube, what must the length of the tube be in order to have the enclosure resonate at the frequency found in (a)?

(c) What are the new frequencies of cone resonance (*i.e.*, ω_L and ω_H for maximum cone velocity)?

(d) Refer to Figs. 8.22 to 8.24. Plot three points and sketch the on-axis pressure responses for $Q_2 = 0.5$, 3.0, and 10.0. In plotting the curve, take into account the change in efficiency in region C, if any occurs, relative to that for the closed box at 500 cps with $R_g = 0$.

8.5. At frequencies below 500 cps, analyze the behavior of the small bass-reflex assembly sketch shown in Fig. P8.5, assuming the following information:

$$
\begin{aligned}
\text{Inside volume of box} \\
\text{not occupied by loudspeaker} &= 0.5 \text{ ft}^3 \\
\text{Number of holes} &= 15 \\
\text{Diameter of each hole} &= 0.5 \text{ in.} \\
\text{Thickness of box side walls} &= 0.5 \text{ in.} \\
R_g &= 0 \text{ ohm} \\
R_E &= 2 \text{ ohms} \\
M_{MD} &= 0.0025 \text{ kg} \\
C_{MS} &= 3.6 \times 10^{-4} \text{ m/newton} \\
R_{MS} &= 1.0 \text{ mks mechanical ohm} \\
\text{Advertised diameter} &= 6 \text{ in.} \\
Bl &= 5 \text{ webers/m}
\end{aligned}
$$

Fig. P8.5

8.6. Determine the length of air gap necessary in the permanent-magnet structure of a direct-radiator loudspeaker for producing an undistorted sound pressure level of 100 db at 20 ft in free space at 40 cps. Assume an advertised diameter of 12 in., voice-coil length of $\frac{1}{4}$ in., and a closed-box baffle. Would it be reasonable to construct a loudspeaker with this size of air gap?

Chapter 9

9.1. An exponential horn is to be constructed with a throat diameter of 1 in. and mouth dimensions of 3×5 ft.

(a) Determine the lowest cutoff frequency that can be used if the permissible length of the horn is 6 ft.

(b) Compute the flare constant of the horn length for this frequency.

9.2. An exponential horn is to be constructed with a cutoff frequency f_c of 70 cps. A driving unit with a diameter of 10 cm is available. Design the horn so that it will be as small as possible consistent with producing a response within ± 3 db between 100 and 400 cps. Assume that the specific acoustic impedance looking back into the driving unit is $\rho_0 c$ at all frequencies.

9.3. A loudspeaker system is to be designed for the reproduction of music. It is to have a low-frequency section and a high-frequency section with separate driver units. The harmonic distortion of the radiated sound wave is to be not greater than 5 per cent at the maximum acoustic power output desired of 10 watts. The lowest frequency to be passed is 90 cps. The crossover frequency is to be 700 cps and you may assume that the power divides evenly between the two units.

Determine for each of the horns (a) diameters of the mouths (bells); (b) cutoff frequencies; (c) throat areas; (d) flare constants; (e) lengths; and (f) the upper frequency limit on the high-frequency horn.

9.4. Analyze the horn of Fig. 9.18 in an effort to learn (a) its cutoff frequency; (b) its flare constant; and (c) the frequencies at which the bends in the horn should produce irregularities in the response curve. Assume that the height is 48 in., the over-all width is 30 in., and the over-all depth is 40 in.

9.5. Design the shortest possible length of exponential horn to be placed in front of a direct-radiator loudspeaker that will provide a mass load of 0.1 kg at 40 cps and a specific acoustic impedance load of nearly $\rho_0 c$ at 400 cps to the diaphragm whose area is 0.12 m^2. *Hint:* At 40 cps, the horn can probably be treated as a cylindrical tube containing the same mass of air.

9.6. Estimate the power available efficiency at several frequencies of a typical 15-ft.-diameter loudspeaker if it is terminated on the front side by the horn of the previous example and on the rear side by a closed box with a compliance equal to that of the suspension.

Chapter 10

10.1. In a certain rectangular room the (1,0,0) mode occurs at 10 cps, the (0,2,0) mode at 30 cps, and the (0,0,3) mode at 50 cps, when the air temperature is +20°C.

(a) What is the frequency of the (2,3,1) mode at 20°C?

(b) What would the frequencies of the (3,0,0), (0,2,0), and (0,0,1) modes be at −20°C?

(c) Sketch the contours of equal pressure for the (1,0,3) mode.

10.2. Find the angles of incidence of the traveling waves on the three walls of a rectangular room with dimensions 3 × 2 × 1 m for the (10,17,3) and (3,17,10) modes of vibration.

10.3. Plot the resonance curve and determine the "Q" for a normal mode of vibration with a reverberation time of 3 sec and a resonance frequency of 1000 cps.

10.4. Determine the room constant for the living room of your home or for your dormitory room, assuming all doors and windows closed and two persons occupying the room.

10.5. A source of sound with a frequency of 1000 cps and a power level of 100 db *re* 10^{-13} watt is located in the center of a large irregular room with a room constant of 1000 ft^2 and approximate dimensions of 50 × 40 × 15 ft. If the source has a directivity index that is symmetrical about the $\theta = 0°$ axis and is equal to +10 db at $\theta = 0°$; +3 db at $\theta = 45°$; −8 db at $\theta = 90°$; −30 db at $\theta = 135°$; and −20 db at $\theta = 180°$, plot the rms sound-pressure-level contour curves on a horizontal plane parallel to the floor and intersecting the source.

10.6. You are asked to treat acoustically a small lecture room with dimensions 40 × 23 × 11 ft. The walls are gypsum plaster on metal lath, the ceiling is painted concrete, and the floor is covered with linoleum tile. The window areas are 100 ft^2. Prescribe suitable acoustical treatment for the room, assuming 10 people seated in wooden chairs and a desired reverberation time of 1.1 sec at all frequencies. Is air absorption important at 4000 cps?

10.7. A lecture room having dimensions 40 × 30 × 18 ft seats 50 students in wooden chairs. The walls and ceiling are gypsum plaster on hollow tile, and the floor is painted concrete. Two-thirds of the area of one of the 18 × 40 ft walls is occupied by windows; the blackboard area is 240 ft^2. The desired reverberation time is 1.2 sec at 500 cps.

(a) Determine the reverberation time of the room at 500 cps when empty, half full, and full.

(b) Prescribe acoustical correction, assuming half-capacity seating, and compute the new reverberation time with full capacity.

10.8. The pressure response characteristic of a loudspeaker measured on the axis at 10 ft in an anechoic chamber is as follows:

Frequency, cps	*Response in db re 0.0002 microbar* *and a power available of 1 watt*
100	98
200	100
400	102
1000	97
2000	100
4000	104

The directivity factor Q is found from Fig. 4.24.

This loudspeaker is to be used in a room with a volume of 10^4 ft^3, an area of 5000 ft^2 and a reverberation time of 1.2 sec at all frequencies. Calculate the sound pressure level, at the six frequencies above, that the loudspeaker will produce in the room at 30 ft for a constant power available of 1 mw.

10.9. A lecture hall in which a public-address system is to be used is to be treated with acoustical material for optimum acoustical properties. The dimensions and other specifications of the room are as follows: length = 90 ft, width = 100 ft, height = 30 ft at front and 10 ft at back (see Fig. P10.9).

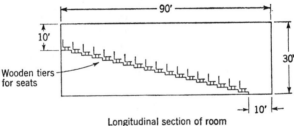

Longitudinal section of room

Fig. P10.9

The floor is wood; the walls and ceiling are lime plaster on metal lath; the 350 seats are wooden; and one-half of the ceiling, one-third of the side walls, and all of the rear wall are to be covered with the same acoustical material. Assuming a reverberation time of 1.4 sec at all frequencies for an audience of 250 persons, determine the absorption coefficients of the ceiling material, as a function of frequency.

10.10. A source of sound with a power level of 110 db re 10^{-13} watt is located in a large irregular room with a room constant of 1000. One wall of this room has an area of 600 ft^2 and a transmission loss of 30 db. Find the sound pressure level near the wall and across the adjoining room, assuming it has the same room constant.

Chapter 11

11.1. A centrifugal blower for a furnace produces, at the top of a smoke stack, a pure tone at 1000 cps with a power level of 140 db. People living at a distance of 1000 ft complain about the noise, and a court order requires the company to reduce the level to 25 db re 0.0002 microbar under all weather conditions. Prescribe a suitable sound treatment. Assume that the stack is 6 ft in diameter and that this open area must be preserved in the treatment.

11.2. A quiet conference room with dimensions 30 × 30 × 15 ft is to be built next to a trading center with dimensions 200 × 75 × 24 ft. The sound pressure levels in the conference room must not rise above an average of 40 db in each of the 600–1200, 1200–2400, and 2400–4800 cps bands. If 40 people shout at one time in the trading center, what amount and kind of acoustical treatment must be provided in each of the two rooms and what type of wall should be installed between them to produce the desired levels in the conference room?

11.3. Four tabulating machines are to be operated in a floor space of 15 × 15 ft in one corner of a large office measuring 50 × 90 × 14 ft. It is desired that the noise levels not exceed an average of 55 db in the 600–1200, 1200–2400, and 2400–4800 cps bands in the part of the office not occupied by the tabulating machines. Prescribe an acoustical treatment.

11.4. A hotel ventilating system uses a propeller that produces the sound power levels shown in column 2 below:

Frequency band, cps	Power levels re 10^{-13} watt, db	Criterion levels, db
37– 75	90	50
75– 150	90	40
150– 300	89	30
300– 600	85	25
600–1200	80	20

If the levels must be reduced to the values shown in column 3, prescribe an acoustical treatment that will provide the necessary noise reduction in a 6 × 6 ft duct leaving about 50 per cent or more of the duct area open for the air to move through.

11.5. A room in a building faces a street. The noise level at the plane of the windows would be 90 db if the building were not there. The sound waves arrive at the wall at many angles of incidence. The average transmission loss of the window glass is 29 db, the area of windows is 60 ft², and the room constant is 500 ft² without treatment and 3000 ft² with treatment. Find the level of the reverberant sound in the room arising from the traffic noise with and without acoustical correction in the room, assuming all noise enters through the glass windows.

11.6. In a small factory room having dimensions of 30 × 50 × 20 ft, the reverberant noise level with all machines running measures 94 db re 0.0002 microbar. The reverberation time of the room with all machines quiet is measured to be 2.8 sec.

(a) Compute the total absorption in sabins present in the room, the average absorption coefficient, and the room constant R.

(b) Compute the power output of a fictitious noise source that will produce the same reverberant noise level as the factory machines.

(c) If the walls and ceiling are covered with a material that increases the absorption coefficient of these surfaces to 0.7, find the resulting reverberation time and the reduction in the level of the reverberant sound field.

Chapter 12

12.1. A microphone is calibrated using the circuit shown in Fig. P12.1:

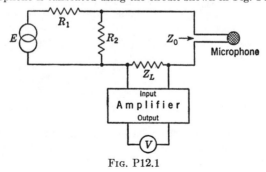

Fig. P12.1

The circuit values are $R_1 = 999$ ohms, $R_2 = 1.0$ ohm, $Z_L = 50$ ohms resistive, and $Z_0 = 50$ ohms resistive. The generator E is first short-circuited and the microphone is immersed in a sound field. With a free-field sound pressure of 74 db re 0.0002 microbar, the voltmeter V reads 10 volts. The sound field is then turned off. The short circuit is now removed from the generator. When E is adjusted to read 2 volts, the voltmeter V again reads 10 volts.

(a) Compute the open-circuit voltage of the microphone for the above sound pressure.

(b) The microphone is connected to an amplifier with a power gain of 90 db and a loudspeaker with a power available response of 75 db re 0.0002 microbar and 1 watt measured in an anechoic chamber 30 ft from the source. Assume matched impedances at the input of the amplifier. Compute the sound pressure level produced at this same distance with a free-field sound pressure level of 70 db re 0.0002 microbar at the microphone.

12.2. Indicate whether the following statements are true or false.

(a) Rochelle-salt-crystal microphones are unaffected by temperatures below 150°F.

(b) Moving-coil microphones must be operated into very high resistances rather than their nominal resistances to avoid reduction in the response at very high frequencies.

(c) The output of crystal microphones is lower at high frequencies than at low frequencies when 50 ft of coaxial cable is used.

(d) Graphic level recorders give one a knowledge of the wave form of a sound with a 1000-cps fundamental.

(e) The calibration of sound-level meters should be checked frequently.

12.3. A source of sound resting on the ground produces sound pressure levels of 90, 85, 80 and 75 db re 0.0002 microbar at the centers of four equal sectors of a hemispherical surface with a radius of 1 m centered at the source. Determine the total power level assuming standard temperature and barometric pressure. Also, find the four directivity indexes associated with the four areas.

12.4. The following data were taken with a constant percentage bandwidth analyzer on the noise from a continuous spectrum source. If the analyzer has a bandwidth that is 2 per cent of the mean frequency, determine the levels in the usual 8-octave bands.

Frequency, cps	Band level in db re 0.0002 microbar
30	33
50	36
100	44
200	55
500	52
1000	50
2000	50
5000	50
10,000	50

12.5. You are asked to measure the power-level spectrum of a large outdoor 60-cps power transformer equipped with large cooling fans. The purpose of the tests is to determine the detailed spectrum of the noise and the octave band levels. Assume that 1000 ft of cable will be necessary; that temperatures up to 100°F will be encountered and that noise levels between 40 and 80 db are to be measured.

Discuss how you would select your apparatus, how and at what positions you would take the data, and how you would record the data and plot the results.

12.6. Two identical capacitor microphones are to be calibrated by the reciprocity technique. If the open-circuit low-frequency response of each is -50 db re 1 volt for a sound-pressure field of 0.1 newton/m²; if the electrical capacitance of each is 50 $\mu\mu$f when the microphone is in free space; if the acoustical compliance of each diaphragm when the electrical terminals are open-circuited is 10^{-12} m^5/newton; and if the volume of the coupling airspace is 10^{-5} m^3, find the ratio of the open-circuit voltage produced by the No. 2 microphone to the voltage driving the No. 1 microphone when the diaphragms of both terminate in the coupling airspace.

Chapter 13

13.1. Determine the loudness level in phons of each of the following three sounds, and total loudness level of the combination: (a) a pure tone of 70 db SPL at 2000 cps; (b) a band of white noise of spectrum level equal to 50 db located between 800 and 1000 cps; (c) a pure tone of 80 db SPL at 40 cps.

13.2. Indicate whether the following statements are true or false.

(a) You are listening to a 1000-cps Morse-code signal in the presence of noise. A filter with a 100-cps bandwidth centered at 1000 cps will improve your ability to hear the signals.

(b) A pure tone of 60 db sound pressure level at 100 cps is less than half as loud as a pure tone of 60 db sound pressure level at 3000 cps. Assume that the comparison is made in a free field and that the transfer function of Fig. 13.11 is a measure of loudness.

(c) Pitch and the position coordinate of auditory nerve patches on the basilar membrane are related to frequency in approximately the same manner.

(d) In general an intermediate-frequency tone will be masked more by a low-frequency tone than by a high-frequency tone.

13.3. The spectrum level of a noise as a function of frequency passes through the following points:

Frequency, cps	Spectrum level, db, re 0.0002 microbar
50	66
350	53
700	49
1000	48
1600	51
2200	57
3000	55
4000	51
5000	49
8000	45
12,000	43

Plot a smooth curve on semilog graph paper through these points to obtain the plot of spectrum level vs. frequency.

Calculate the loudness level in 300-mel bands up to a pitch of 3300 mels. Find the total loudness level.

13.4. Given seven pure tones with the following frequency and sound pressure levels:

Frequency, cps	Sound pressure level, db, re 0.0002 microbar
40	85
100	80
300	78
700	82
1500	68
3500	75
9000	70

(a) On semilog graph paper make a plot of loudness level vs. frequency by drawing a vertical line at the frequency of each of the seven tones.

(b) Suppose the sound pressure of each of these tones is reduced by 30 db. Again plot loudness level vs. frequency and determine the loudness level of the combination.

(c) In the light of the above exercise, does adjustment of the volume control on a radio affect the faithfulness of what the listener hears?

13.5. (a) The noise in an automobile traveling at 40 mph is measured with a sound analyzer that has a filter which is continuously variable as a function of frequency. The bandwidth of the filter also varies and is equal to $0.02f$. Convert the measured sound pressure levels, given below, to spectrum levels and plot a curve of them on two-cycle semilog paper.

Frequency, cps	Sound pressure level db re 0.0002 microbar
100	64
200	67
400	69
800	65
1200	61
2500	52
5000	49
10,000	52

(b) A cooling fan, a vibrator, and a motor-generator set are needed in the automobile for a special purpose. The fan gives off a tone with a frequency of 150 cps, the vibrator gives off a 3000 cps tone, and the motor-generator set gives off a tone of 670 cps. How should the noise specifications on these devices read if all three are to be inaudible when the car is moving faster than 40 mph? Assume that the three devices are procured with just tolerable noise levels for each. What will be the loudness level of the three tones together when the car is standing still?

APPENDIX I

DECIBEL CONVERSION TABLES†

It is convenient in measurements and calculations in electroacoustics to express the ratio between any two amounts of electric or acoustic power in units on a logarithmic scale. The *decibel* (⅒th of the *bel*) on the briggsian or base-10 scale is in almost universal use for this purpose.

Since voltage and sound pressure are related to power by impedance, the *decibel* can be used to express voltage and sound pressure ratios, if care is taken to account for the impedances associated with them.

Table I and Table II on the following pages have been prepared to facilitate making conversions in either direction between the number of *decibels* and the corresponding power, voltage, and sound pressure ratios.

To Find Values Outside the Range of Conversion Tables. Values outside the range of either Table I or Table II on the following pages can be readily found with the help of the following simple rules:

Table I: Decibels to Voltage, Sound Pressure and Power Ratios

NUMBER OF DECIBELS POSITIVE (+). Subtract +20 decibels successively from the given number of decibels until the remainder falls within range of Table I. *To find the voltage (sound pressure) ratio*, multiply the corresponding value from the right-hand voltage-ratio column by 10 for each time you subtracted 20 db. *To find the power ratio*, multiply the corresponding value from the right-hand power-ratio column by 100 for each time you subtracted 20 db.

Example—*Given:* 49.2 db

$$49.2 \text{ db} - 20 \text{ db} - 20 \text{ db} = 9.2 \text{ db}$$

Voltage (sound pressure) ratio: 9.2 db →

$$2.884 \times 10 \times 10 = 288.4$$

Power ratio: 9.2 db →

$$8.318 \times 100 \times 100 = 83{,}180$$

NUMBER OF DECIBELS NEGATIVE (−). Add +20 decibels successively to the given number of decibels until the sum falls within the range of Table I. *For the voltage (sound pressure) ratio*, divide the value from the left-hand voltage-ratio column by 10 for each time you added 20 db. *For the power ratio*, divide the value from the left-hand power-ratio column by 100 for each time you added 20 db.

Example—*Given:* −49.2 db

$$-49.2 \text{ db} + 20 \text{ db} + 20 \text{ db} = -9.2 \text{ db}$$

Voltage (sound pressure) ratio: −9.2 db →

$$0.3467 \times \tfrac{1}{10} \times \tfrac{1}{10} = 0.003467$$

† Courtesy of the General Radio Company, Cambridge, Mass.

455

Power ratio: −9.2 db

$$0.1202 \times \tfrac{1}{100} \times \tfrac{1}{100} = 0.00001202$$

Table II: Voltage and Sound Pressure Ratios to Decibels

FOR RATIOS SMALLER THAN THOSE IN TABLE. Multiply the given ratio by 10 successively until the product can be found in the table. From the number of decibels thus found, subtract +20 decibels for each time you multiplied by 10.

Example—*Given:* Voltage (sound pressure) ratio = 0.0131

$$0.0131 \times 10 \times 10 = 1.31$$

From Table II, 1.31 →

$$2.345 \text{ db} - 20 \text{ db} - 20 \text{ db} = -37.655 \text{ db}$$

FOR RATIOS GREATER THAN THOSE IN TABLE. Divide the given ratio by 10 successively until the remainder can be found in the table. To the number of decibels thus found, add +20 db for each time you divided by 10.

Example—*Given:* Voltage (sound pressure) ratio = 712

$$712 \times \tfrac{1}{10} \times \tfrac{1}{10} = 7.12$$

From Table II, 7.12 →

$$17.050 \text{ db} + 20 \text{ db} + 20 \text{ db} = 57.050 \text{ db}$$

TABLE I. Decibels to Voltage, Sound Pressure and Power Ratios†

To account for the sign of the decibel. For positive (+) values of the decibel: Both voltage and power ratios are greater than unity. Use the two right-hand columns. For negative (−) values of the decibel: Both voltage and power ratios are less than unity. Use the two left-hand columns. Use the voltage columns for sound pressures.

	Power ratio	Voltage ratio
Example: Given: ±9.1 db. *Find:*		
+9.1 db	8.128	2.851
−9.1 db	0.1230	0.3508

−		db	+		−		db	+	
Voltage ratio	Power ratio		Voltage ratio	Power ratio	Voltage ratio	Power ratio		Voltage ratio	Power ratio
1.0000	1.0000	0	1.000	1.000	0.7079	0.5012	3.0	1.413	1.995
0.9886	0.9772	0.1	1.012	1.023	0.6998	0.4898	3.1	1.429	2.042
0.9772	0.9550	0.2	1.023	1.047	0.6918	0.4786	3.2	1.445	2.089
0.9661	0.9333	0.3	1.035	1.072	0.6839	0.4677	3.3	1.462	2.138
0.9550	0.9120	0.4	1.047	1.096	0.6761	0.4571	3.4	1.479	2.188
0.9441	0.8913	0.5	1.059	1.122	0.6683	0.4467	3.5	1.496	2.239
0.9333	0.8710	0.6	1.072	1.148	0.6607	0.4365	3.6	1.514	2.291
0.9226	0.8511	0.7	1.084	1.175	0.6531	0.4266	3.7	1.531	2.344
0.9120	0.8318	0.8	1.096	1.202	0.6457	0.4169	3.8	1.549	2.399
0.9016	0.8128	0.9	1.109	1.230	0.6383	0.4074	3.9	1.567	2.455
0.8913	0.7943	1.0	1.122	1.259	0.6310	0.3981	4.0	1.585	2.512
0.8810	0.7762	1.1	1.135	1.288	0.6237	0.3890	4.1	1.603	2.570
0.8710	0.7586	1.2	1.148	1.318	0.6166	0.3802	4.2	1.622	2.630
0.8610	0.7413	1.3	1.161	1.349	0.6095	0.3715	4.3	1.641	2.692
0.8511	0.7244	1.4	1.175	1.380	0.6026	0.3631	4.4	1.660	2.754
0.8414	0.7079	1.5	1.189	1.413	0.5957	0.3548	4.5	1.679	2.818
0.8318	0.6918	1.6	1.202	1.445	0.5888	0.3467	4.6	1.698	2.884
0.8222	0.6761	1.7	1.216	1.479	0.5821	0.3388	4.7	1.718	2.951
0.8128	0.6607	1.8	1.230	1.514	0.5754	0.3311	4.8	1.738	3.020
0.8035	0.6457	1.9	1.245	1.549	0.5689	0.3236	4.9	1.758	3.090
0.7943	0.6310	2.0	1.259	1.585	0.5623	0.3162	5.0	1.778	3.162
0.7852	0.6166	2.1	1.274	1.622	0.5559	0.3090	5.1	1.799	3.236
0.7762	0.6026	2.2	1.288	1.660	0.5495	0.3020	5.2	1.820	3.311
0.7674	0.5888	2.3	1.303	1.698	0.5433	0.2951	5.3	1.841	3.388
0.7586	0.5754	2.4	1.318	1.738	0.5370	0.2884	5.4	1.862	3.467
0.7499	0.5623	2.5	1.334	1.778	0.5309	0.2818	5.5	1.884	3.548
0.7413	0.5495	2.6	1.349	1.820	0.5248	0.2754	5.6	1.905	3.631
0.7328	0.5370	2.7	1.365	1.862	0.5188	0.2692	5.7	1.928	3.715
0.7244	0.5248	2.8	1.380	1.905	0.5129	0.2630	5.8	1.950	3.802
0.7161	0.5129	2.9	1.396	1.950	0.5070	0.2570	5.9	1.972	3.890

TABLE I. Decibels to Voltage, Sound Pressure and Power Ratios.†— (*Continued*)

−		db	+		−		db	+	
Voltage ratio	Power ratio		Voltage ratio	Power ratio	Voltage ratio	Power ratio		Voltage ratio	Power ratio
0.5012	0.2512	6.0	1.995	3.981	0.3350	0.1122	9.5	2.985	8.913
0.4955	0.2455	6.1	2.018	4.074	0.3311	0.1096	9.6	3.020	9.120
0.4898	0.2399	6.2	2.042	4.169	0.3273	0.1072	9.7	3.055	9.333
0.4842	0.2344	6.3	2.065	4.266	0.3236	0.1047	9.8	3.090	9.550
0.4786	0.2291	6.4	2.089	4.365	0.3199	0.1023	9.9	3.126	9.772
0.4732	0.2239	6.5	2.113	4.467	0.3162	0.1000	10.0	3.162	10.000
0.4677	0.2188	6.6	2.138	4.571	0.3126	0.09772	10.1	3.199	10.23
0.4624	0.2138	6.7	2.163	4.677	0.3090	0.09550	10.2	3.236	10.47
0.4571	0.2089	6.8	2.188	4.786	0.3055	0.09333	10.3	3.273	10.72
0.4519	0.2042	6.9	2.213	4.898	0.3020	0.09120	10.4	3.311	10.96
0.4467	0.1995	7.0	2.239	5.012	0.2985	0.08913	10.5	3.350	11.22
0.4416	0.1950	7.1	2.265	5.129	0.2951	0.08710	10.6	3.388	11.48
0.4365	0.1905	7.2	2.291	5.248	0.2917	0.08511	10.7	3.428	11.75
0.4315	0.1862	7.3	2.317	5.370	0.2884	0.08318	10.8	3.467	12.02
0.4266	0.1820	7.4	2.344	5.495	0.2851	0.08128	10.9	3.508	12.30
0.4217	0.1778	7.5	2.371	5.623	0.2818	0.07943	11.0	3.548	12.59
0.4169	0.1738	7.6	2.399	5.754	0.2786	0.07762	11.1	3.589	12.88
0.4121	0.1698	7.7	2.427	5.888	0.2754	0.07586	11.2	3.631	13.18
0.4074	0.1660	7.8	2.455	6.026	0.2723	0.07413	11.3	3.673	13.49
0.4027	0.1622	7.9	2.483	6.166	0.2692	0.07244	11.4	3.715	13.80
0.3981	0.1585	8.0	2.512	6.310	0.2661	0.07079	11.5	3.758	14.13
0.3936	0.1549	8.1	2.541	6.457	0.2630	0.06918	11.6	3.802	14.45
0.3890	0.1514	8.2	2.570	6.607	0.2600	0.06761	11.7	3.846	14.79
0.3846	0.1479	8.3	2.600	6.761	0.2570	0.06607	11.8	3.890	15.14
0.3802	0.1445	8.4	2.630	6.918	0.2541	0.06457	11.9	3.936	15.49
0.3758	0.1413	8.5	2.661	7.079	0.2512	0.06310	12.0	3.981	15.85
0.3715	0.1380	8.6	2.692	7.244	0.2483	0.06166	12.1	4.027	16.22
0.3673	0.1349	8.7	2.723	7.413	0.2455	0.06026	12.2	4.074	16.60
0.3631	0.1318	8.8	2.754	7.586	0.2427	0.05888	12.3	4.121	16.98
0.3589	0.1288	8.9	2.786	7.762	0.2399	0.05754	12.4	4.169	17.38
0.3548	0.1259	9.0	2.818	7.943	0.2371	0.05623	12.5	4.217	17.78
0.3508	0.1230	9.1	2.851	8.128	0.2344	0.05495	12.6	4.266	18.20
0.3467	0.1202	9.2	2.884	8.318	0.2317	0.05370	12.7	4.315	18.62
0.3428	0.1175	9.3	2.917	8.511	0.2291	0.05248	12.8	4.365	19.05
0.3388	0.1148	9.4	2.951	8.710	0.2265	0.05129	12.9	4.416	19.50

TABLE I.   Decibels to Voltage, Sound Pressure and Power Ratios.†—
(*Continued*)

Voltage ratio	Power ratio	db	Voltage ratio	Power ratio	Voltage ratio	Power ratio	db	Voltage ratio	Power ratio
	−		+		−			+	
0.2239	0.05012	13.0	4.467	19.95	0.1496	0.02239	16.5	6.683	44.67
0.2213	0.04898	13.1	4.519	20.42	0.1479	0.02188	16.6	6.761	45.71
0.2188	0.04786	13.2	4.571	20.89	0.1462	0.02138	16.7	6.839	46.77
0.2163	0.04677	13.3	4.624	21.38	0.1445	0.02089	16.8	6.918	47.86
0.2138	0.04571	13.4	4.677	21.88	0.1429	0.02042	16.9	6.998	48.98
0.2113	0.04467	13.5	4.732	22.39	0.1413	0.01995	17.0	7.079	50.12
0.2089	0.04365	13.6	4.786	22.91	0.1396	0.01950	17.1	7.161	51.29
0.2065	0.04266	13.7	4.842	23.44	0.1380	0.01905	17.2	7.244	52.48
0.0242	0.04169	13.8	4.898	23.99	0.1365	0.01862	17.3	7.328	53.70
0.2018	0.04074	13.9	4.955	24.55	0.1349	0.01820	17.4	7.413	54.95
0.1995	0.03981	14.0	5.012	25.12	0.1334	0.01778	17.5	7.499	56.23
0.1972	0.03890	14.1	5.070	25.70	0.1318	0.01738	17.6	7.586	57.54
0.1950	0.03802	14.2	5.129	26.30	0.1303	0.01698	17.7	7.674	58.88
0.1928	0.03715	14.3	5.188	26.92	0.1288	0.01660	17.8	7.762	60.26
0.1905	0.03631	14.4	5.248	27.54	0.1274	0.01622	17.9	7.852	61.66
0.1884	0.03548	14.5	5.309	28.18	0.1259	0.01585	18.0	7.943	63.10
0.1862	0.03467	14.6	5.370	28.84	0.1245	0.01549	18.1	8.035	64.57
0.1841	0.03388	14.7	5.433	29.51	0.1230	0.01514	18.2	8.128	66.07
0.1820	0.03311	14.8	5.495	30.20	0.1216	0.01479	18.3	8.222	67.61
0.1799	0.03236	14.9	5.559	30.90	0.1202	0.01445	18.4	8.318	69.18
0.1778	0.03162	15.0	5.623	31.62	0.1189	0.01413	18.5	8.414	70.79
0.1758	0.03090	15.1	5.689	32.36	0.1175	0.01380	18.6	8.511	72.44
0.1738	0.03020	15.2	5.754	33.11	0.1161	0.01349	18.7	8.610	74.13
0.1718	0.02951	15.3	5.821	33.88	0.1148	0.01318	18.8	8.710	75.86
0.1698	0.02884	15.4	5.888	34.67	0.1135	0.01288	18.9	8.811	77.62
0.1679	0.02818	15.5	5.957	35.48	0.1122	0.01259	19.0	8.913	79.43
0.1660	0.02754	15.6	6.026	36.31	0.1109	0.01230	19.1	9.016	81.28
0.1641	0.02692	15.7	6.095	37.15	0.1096	0.01202	19.2	9.120	83.18
0.1622	0.02630	15.8	6.166	38.02	0.1084	0.01175	19.3	9.226	85.11
0.1603	0.02570	15.9	6.237	38.90	0.1072	0.01148	19.4	9.333	87.10
0.1585	0.02512	16.0	6.310	39.81	0.1059	0.01122	19.5	9.441	89.13
0.1567	0.02455	16.1	6.383	40.74	0.1047	0.01096	19.6	9.550	91.20
0.1549	0.02399	16.2	6.457	41.69	0.1035	0.01072	19.7	9.661	93.33
0.1531	0.02344	16.3	6.531	42.66	0.1023	0.01047	19.8	9.772	95.50
0.1514	0.02291	16.4	6.607	43.65	0.1012	0.01023	19.9	9.886	97.72
					0.1000	0.01000	20.0	10.000	100.00

TABLE I.   Decibels to Voltage, Sound Pressure and Power Ratios.†—
(*Continued*)

		db	+	
Voltage ratio	Power ratio		Voltage ratio	Power ratio
3.162×10^{-1}	10^{-1}	10	3.162	10
10^{-1}	10^{-2}	20	10	10^2
3.162×10^{-2}	10^{-3}	30	3.162×10	10^3
10^{-2}	10^{-4}	40	10^2	10^4
3.162×10^{-3}	10^{-5}	50	3.162×10^2	10^5
10^{-3}	10^{-6}	60	10^3	10^6
3.162×10^{-4}	10^{-7}	70	3.162×10^3	10^7
10^{-4}	10^{-8}	80	10^4	10^8
3.162×10^{-5}	10^{-9}	90	3.162×10^4	10^9
10^{-5}	10^{-10}	100	10^5	10^{10}

† To find decibel values outside the range of this table, see page 455.   Use the voltage ratio columns for sound pressure ratios.

TABLE II. Voltage and Sound Pressure Ratios to Decibels†

Voltage and sound pressure ratios. Use the table directly.

Power ratios. To find the number of decibels corresponding to a given power ratio: Assume the given power ratio to be a voltage ratio and find the corresponding number of decibels from the table. The desired result is exactly one-half of the number of decibels thus found.

Example: Given: a power ratio of 3.41. *Find:* 3.41 in the table:

$$3.41 \rightarrow 10.655 \text{ db} \times \tfrac{1}{2} = 5.328 \text{ db}$$

Voltage ratio	0.00	0.01	0.02	0.03	0.04	0.05	0.06	0.07	0.08	0.09
1.0	0.000	0.086	0.172	0.257	0.341	0.424	0.506	0.588	0.668	0.749
1.1	0.828	0.906	0.984	1.062	1.138	1.214	1.289	1.364	1.438	1.511
1.2	1.584	1.656	1.727	1.798	1.868	1.938	2.007	2.076	2.144	2.212
1.3	2.279	2.345	2.411	2.477	2.542	2.607	2.671	2.734	2.798	2.860
1.4	2.923	2.984	3.046	3.107	3.167	3.227	3.287	3.346	3.405	3.464
1.5	3.522	3.580	3.637	3.694	3.750	3.807	3.862	3.918	3.973	4.028
1.6	4.082	4.137	4.190	4.244	4.297	4.350	4.402	4.454	4.506	4.558
1.7	4.609	4.660	4.711	4.761	4.811	4.861	4.910	4.959	5.008	5.057
1.8	5.105	5.154	5.201	5.249	5.296	5.343	5.390	5.437	5.483	5.529
1.9	5.575	5.621	5.666	5.711	5.756	5.801	5.845	5.889	5.933	5.977
2.0	6.021	6.064	6.107	6.150	6.193	6.235	6.277	6.319	6.361	6.403
2.1	6.444	6.486	6.527	6.568	6.608	6.649	6.689	6.729	6.769	6.809
2.2	6.848	6.888	6.927	6.966	7.008	7.044	7.082	7.121	7.159	7.197
2.3	7.235	7.272	7.310	7.347	7.384	7.421	7.458	7.495	7.532	7.568
2.4	7.604	7.640	7.676	7.712	7.748	7.783	7.819	7.854	7.889	7.924
2.5	7.959	7.993	8.028	8.062	8.097	8.131	8.165	8.199	8.232	8.266
2.6	8.299	8.333	8.366	8.399	8.432	8.465	8.498	8.530	8.563	8.595
2.7	8.627	8.659	8.691	8.723	8.755	8.787	8.818	8.850	8.881	8.912
2.8	8.943	8.974	9.005	9.036	9.066	9.097	9.127	9.158	9.188	9.218
2.9	9.248	9.278	9.308	9.337	9.367	9.396	9.426	9.455	9.484	9.513
3.0	9.542	9.571	9.600	9.629	9.657	9.686	9.714	9.743	9.771	9.799
3.1	9.827	9.855	9.883	9.911	9.939	9.966	9.994	10.021	10.049	10.076
3.2	10.103	10.130	10.157	10.184	10.211	10.238	10.264	10.291	10.317	10.344
3.3	10.370	10.397	10.423	10.449	10.475	10.501	10.527	10.553	10.578	10.604
3.4	10.630	10.655	10.681	10.706	10.731	10.756	10.782	10.807	10.832	10.857
3.5	10.881	10.906	10.931	10.955	10.980	11.005	11.029	11.053	11.078	11.102
3.6	11.126	11.150	11.174	11.198	11.222	11.246	11.270	11.293	11.317	11.341
3.7	11.364	11.387	11.411	11.434	11.457	11.481	11.504	11.527	11.550	11.573
3.8	11.596	11.618	11.641	11.664	11.687	11.709	11.732	11.754	11.777	11.799
3.9	11.821	11.844	11.866	11.888	11.910	11.932	11.954	11.976	11.998	12.019
4.0	12.041	12.063	12.085	12.106	12.128	12.149	12.171	12.192	12.213	12.234
4.1	12.256	12.277	12.298	12.319	12.340	12.361	12.382	12.403	12.424	12.444
4.2	12.465	12.486	12.506	12.527	12.547	12.568	12.588	12.609	12.629	12.649
4.3	12.669	12.690	12.710	12.730	12.750	12.770	12.790	12.810	12.829	12.849
4.4	12.869	12.889	12.908	12.928	12.948	12.967	12.987	13.006	13.026	13.045

TABLE II. Voltage and Sound Pressure Ratios to Decibels.†—(*Continued*)

Voltage ratio	0.00	0.01	0.02	0.03	0.04	0.05	0.06	0.07	0.08	0.09
4.5	13.064	13.084	13.103	13.122	13.141	13.160	13.179	13.198	13.217	13.236
4.6	13.255	13.274	13.293	13.312	13.330	13.349	13.368	13.386	13.405	13.423
4.7	13.442	13.460	13.479	13.497	13.516	13.534	13.552	13.570	13.589	13.607
4.8	13.625	13.643	13.661	13.679	13.697	13.715	13.733	13.751	13.768	13.786
4.9	13.804	13.822	13.839	13.857	13.875	13.892	13.910	13.927	13.945	13.962
5.0	13.979	13.997	14.014	14.031	14.049	14.066	14.083	14.100	14.117	14.134
5.1	14.151	14.168	14.185	14.202	14.219	14.236	14.253	14.270	14.287	14.303
5.2	14.320	14.337	14.353	14.370	14.387	14.403	14.420	14.436	14.453	14.469
5.3	14.486	14.502	14.518	14.535	14.551	14.567	14.538	14.599	14.616	14.632
5.4	14.648	14.664	14.680	14.696	14.712	14.728	14.744	14.760	14.776	14.791
5.5	14.807	14.823	14.839	14.855	14.870	14.886	14.902	14.917	14.933	14.948
5.6	14.964	14.979	14.995	15.010	15.026	15.041	15.056	15.072	15.087	15.102
5.7	15.117	15.133	15.148	15.163	15.178	15.193	15.208	15.224	15.239	15.254
5.8	15.269	15.284	15.298	15.313	15.328	15.343	15.358	15.373	15.388	15.402
5.9	15.417	15.432	15.446	15.461	15.476	15.490	15.505	15.519	15.534	15.549
6.0	15.563	15.577	15.592	15.606	15.621	15.635	15.649	15.664	15.678	15.692
6.1	15.707	15.721	15.735	15.749	15.763	15.778	15.792	15.806	15.820	15.834
6.2	15.848	15.862	15.876	15.890	15.904	15.918	15.931	15.945	15.959	15.973
6.3	15.987	16.001	16.014	16.028	16.042	16.055	16.069	16.083	16.096	16.110
6.4	16.124	16.137	16.151	16.164	16.178	16.191	16.205	16.218	16.232	16.245
6.5	16.258	16.272	16.285	16.298	16.312	16.325	16.338	16.351	16.365	16.378
6.6	16.391	16.404	16.417	16.430	16.443	16.456	16.469	16.483	16.496	16.509
6.7	16.521	16.534	16.547	16.560	16.573	16.586	16.599	16.612	16.625	16.637
6.8	16.650	16.663	16.676	16.688	16.701	16.714	16.726	16.739	16.752	16.764
6.9	16.777	16.790	16.802	16.815	16.827	16.840	16.852	16.865	16.877	16.890
7.0	16.902	16.914	16.927	16.939	16.951	16.964	16.976	16.988	17.001	17.013
7.1	17.025	17.037	17.050	17.062	17.074	17.086	17.098	17.110	17.122	17.135
7.2	17.147	17.159	17.171	17.183	17.195	17.207	17.219	17.231	17.243	17.255
7.3	17.266	17.278	17.290	17.302	17.314	17.326	17.338	17.349	17.361	17.373
7.4	17.385	17.396	17.408	17.420	17.431	17.443	17.455	17.466	17.748	17.490
7.5	17.501	17.513	17.524	17.536	17.547	17.559	17.570	17.582	17.593	17.605
7.6	17.616	17.628	17.639	17.650	17.662	17.673	17.685	17.696	17.707	17.719
7.7	17.730	17.741	17.752	17.764	17.775	17.786	17.797	17.808	17.820	17.831
7.8	17.842	17.853	17.864	17.875	17.886	17.897	17.908	17.919	17.931	17.942
7.9	17.953	17.964	17.975	17.985	17.996	18.007	18.018	18.020	18.040	18.051
8.0	18.062	18.073	18.083	18.094	18.105	18.116	18.127	18.137	18.148	18.159
8.1	18.170	18.180	18.191	18.202	18.212	18.223	18.234	18.244	18.255	18.266
8.2	18.276	18.287	18.297	18.308	18.319	18.329	18.340	18.350	18.361	18.371
8.3	18.382	18.392	18.402	18.413	18.423	18.434	18.444	18.455	18.465	18.475
8.4	18.486	18.486	18.506	18.517	18.527	18.537	18.547	18.558	18.568	18.578

TABLE II.   Voltage and Sound Pressure Ratios to Decibels.†—(*Continued*)

Voltage ratio	0.00	0.01	0.02	0.03	0.04	0.05	0.06	0.07	0.08	0.09
8.5	18.588	18.599	18.609	18.619	18.629	18.639	18.649	18.660	18.670	18.680
8.6	18.690	18.700	18.710	18.720	18.730	18.740	18.750	18.760	18.770	18.780
8.7	18.790	18.800	18.810	18.820	18.830	18.840	18.850	18.860	18.870	18.880
8.8	18.890	18.900	18.909	18.919	18.929	18.939	18.949	18.958	18.968	18.978
8.9	18.988	18.998	19.007	19.017	19.027	19.036	19.046	19.056	19.066	19.075
9.0	19.085	19.094	19.104	19.114	19.123	19.133	19.143	19.152	19.162	19.171
9.1	19.181	19.190	19.200	19.209	19.219	19.228	19.238	19.247	19.257	19.266
9.2	19.276	19.285	19.295	19.304	19.313	19.323	19.332	19.342	19.351	19.360
9.3	19.370	19.379	19.388	19.398	19.407	19.416	19.426	19.435	19.444	19.453
9.4	19.463	19.472	19.481	19.490	19.499	19.509	19.518	19.527	19.536	19.545
9.5	19.554	19.564	19.573	19.582	19.591	19.600	19.609	19.618	19.627	19.636
9.6	19.645	19.654	19.664	19.673	19.682	19.691	19.700	19.709	19.718	19.726
9.7	19.735	19.744	19.753	19.762	19.771	19.780	19.789	19.798	19.807	19.816
9.8	19.825	19.833	19.842	19.851	19.860	19.869	19.878	19.886	19.895	19.904
9.9	19.913	19.921	19.930	19.939	19.948	19.956	19.965	19.974	19.983	19.991

	0	1	2	3	4	5	6	7	8	9
10	20.000	20.828	21.584	22.279	22.923	23.522	24.082	24.609	25.105	25.575
20	26.021	26.444	26.848	27.235	27.604	27.959	28.299	28.627	28.943	29.248
30	29.542	29.827	30.103	30.370	30.630	30.881	31.126	31.364	31.596	31.821
40	32.041	32.256	32.465	32.669	32.869	33.064	33.255	33.442	33.625	33.804
50	33.979	34.151	34.320	34.486	34.648	34.807	34.964	35.117	35.269	35.417
60	35.563	35.707	35.848	35.987	36.124	36.258	36.391	36.521	36.650	36.777
70	36.902	37.025	37.147	37.266	37.385	37.501	37.616	37.730	37.842	37.953
80	38.062	38.170	38.276	28.382	38.486	38.588	38.690	38.790	38.890	38.988
90	39.085	39.181	39.276	39.370	39.463	39.554	39.645	39.735	39.825	39.913
100	40.000									

† To find ratios outside the range of this table, see page 455.

Conversion Factors

The following values for the fundamental constants were used in the preparation of the factors:

$$1 \text{ m } = 39.37 \text{ in.} = 3.281 \text{ ft}$$
$$1 \text{ lb (weight)} = 0.4536 \text{ kg} = 0.03108 \text{ slug}$$
$$1 \text{ slug} = 14.594 \text{ kg}$$
$$1 \text{ lb (force)} = 4.448 \text{ newtons}$$
$$\text{Acceleration due to gravity} = 9.807 \text{ m/sec}^2$$
$$= 32.174 \text{ ft/sec}^2$$
$$\text{Density of H}_2\text{O at } 4^\circ\text{C} = 10^3 \text{ kg/m}^3$$
$$\text{Density of Hg at } 0^\circ\text{C} = 1.3595 \times 10^4 \text{ kg/m}^3$$
$$1 \text{ U.S. lb} = 1 \text{ British lb}$$
$$1 \text{ U.S. gallon} = 0.83267 \text{ British gallon}$$

TABLE C.1 Conversion Factors

To convert	Into	Multiply by	Conversely, multiply by
acres	ft^2	4.356×10^4	2.296×10^{-5}
	miles^2 (statute)	1.562×10^{-3}	640
	m^2	4,047	2.471×10^{-4}
	hectare (10^4 m^2)	0.4047	2.471
atm	in. H$_2$O at 4°C	406.80	2.458×10^{-3}
	in. Hg at 0°C	29.92	3.342×10^{-2}
	ft H$_2$O at 4°C	33.90	2.950×10^{-2}
	mm Hg at 0°C	760	1.316×10^{-3}
	lb/in.^{2}	14.70	6.805×10^{-2}
	newtons/m^2	1.0132×10^5	9.872×10^{-6}
	kg/m^2	1.033×10^4	9.681×10^{-5}
°C	°F	(°C × 9/5) + 32	(°F − 32) × 5/9
cm	in.	0.3937	2.540
	ft	3.281×10^{-2}	30.48
	m	10^{-2}	10^2
circular mils	in.^{2}	7.85×10^{-7}	1.274×10^6
	cm^2	5.067×10^{-6}	1.974×10^5
cm^2	in.^{2}	0.1550	6.452
	ft^2	1.0764×10^{-3}	929
	m^2	10^{-4}	10^4
cm^3	in.^{3}	0.06102	16.387
	ft^3	3.531×10^{-5}	2.832×10^4
	m^3	10^{-6}	10^6
deg (angle)	radians	1.745×10^{-2}	57.30
dynes	lb (force)	2.248×10^{-6}	4.448×10^5
	newtons	10^{-5}	10^5
dynes/cm^2	lb/ft^2 (force)	2.090×10^{-3}	478.5
	newtons/m^2	10^{-1}	10
ergs	ft-lb (force)	7.376×10^{-8}	1.356×10^7
	joules	10^{-7}	10^7
ergs/cm^3	ft-lb/ft^3	2.089×10^{-3}	478.7
ergs/sec	watts	10^{-7}	10^7
	ft-lb/sec	7.376×10^{-8}	1.356×10^7
ergs/sec-cm^2	ft-lb/sec-ft^2	6.847×10^{-5}	1.4605×10^4
fathoms	ft	6	0.16667
ft	in.	12	0.08333
	cm	30.48	3.281×10^{-2}
	m	0.3048	3.281
ft^2	in.^{2}	144	6.945×10^{-3}
	cm^2	9.290×10^2	0.010764
	m^2	9.290×10^{-2}	10.764
ft^3	in.^{3}	1728	5.787×10^{-4}
	cm^3	2.832×10^4	3.531×10^{-5}
	m^3	2.832×10^{-2}	35.31
	liters	28.32	3.531×10^{-2}

TABLE C.1   Conversion Factors—Continued

To convert	Into	Multiply by	Conversely, multiply by
ft H_2O at $4°C$	in. Hg at $0°C$	0.8826	1.133
	lb/in.^{2}	0.4335	2.307
	lb/ft^2	62.43	1.602×10^{-2}
	newtons/m^2	2989	3.345×10^{-4}
gal (liquid U.S.)	gal (liquid Brit. Imp.)	0.8327	1.2010
	liters	3.785	0.2642
	m^3	3.785×10^{-3}	264.2
gm	oz (weight)	3.527×10^{-2}	28.35
	lb (weight)	2.205×10^{-3}	453.6
hp (550 ft-lb/sec)	ft-lb/min	3.3×10^4	3.030×10^{-5}
	watts	745.7	1.341×10^{-3}
	kw	0.7457	1.341
in.	ft	0.0833	12
	cm	2.540	0.3937
	m	0.0254	39.37
in.^{2}	ft^2	0.006945	144
	cm^2	6.452	0.1550
	m^2	6.452×10^{-4}	1550
in.^{3}	ft^3	5.787×10^{-4}	1.728×10^3
	cm^3	16.387	6.102×10^{-2}
	m^3	1.639×10^{-5}	6.102×10^4
kg	lb (weight)	2.2046	0.4536
	slug	0.06852	14.594
	gm	10^3	10^{-3}
kg/m^2	lb/in.^{2} (weight)	0.001422	703.0
	lb/ft^2 (weight)	0.2048	4.882
	gm/cm^2	10^{-1}	10
kg/m^3	lb/in.^{3} (weight)	3.613×10^{-5}	2.768×10^4
	lb/ft^3 (weight)	6.243×10^{-2}	16.02
liters	in.^{3}	61.03	1.639×10^{-2}
	ft^3	0.03532	28.32
	pints (liquid U.S.)	2.1134	0.47318
	quarts (liquid U.S.)	1.0567	0.94636
	gal (liquid U.S.)	0.2642	3.785
	cm^3	1000	0.001
	m^3	0.001	1000
$\log_e n$, or $\ln n$	$\log_{10} n$	0.4343	2.303
m	in.	39.371	0.02540
	ft	3.2808	0.30481
	yd	1.0936	0.9144
	cm	10^2	10^{-2}
m^2	in.^{2}	1550	6.452×10^{-4}
	ft^2	10.764	9.290×10^{-2}
	yd^2	1.196	0.8362
	cm^2	10^4	10^{-4}
m^3	in.^{3}	6.102×10^4	1.639×10^{-5}
	ft^3	35.31	2.832×10^{-2}

TABLE C.1 Conversion Factors—Continued

To convert	Into	Multiply by	Conversely, multiply by
m^3 *(Cont.)*	yd^3	1.3080	0.7646
	cm^3	10^6	10^{-6}
microbars (dynes/cm^2)	lb/$in.^2$	1.4513×10^{-5}	6.890×10^4
	lb/ft^2	2.090×10^{-3}	478.5
	newtons/m^2	10^{-1}	10
miles (nautical)	ft	6080	1.645×10^{-4}
	km	1.852	0.5400
miles (statute)	ft	5280	1.894×01^{-4}
	km	1.6093	0.6214
$miles^2$ (statute)	ft^2	2.788×10^7	3.587×10^{-8}
	km^2	2.590	0.3861
	acres	640	1.5625×10^{-3}
mph	ft/min	88	1.136×10^{-2}
	km/min	2.682×10^{-2}	37.28
	km/hr	1.6093	0.6214
nepers	db	8.686	0.1151
newtons	lb (force)	0.2248	4.448
	dynes	10^5	10^{-5}
newtons/m^2	lb/$in.^2$ (force)	1.4513×10^{-2}	6.890×10^3
	lb/ft^2 (force)	2.090×10^{-2}	47.85
	dynes/cm^2	10	10^{-1}
lb (force)	newtons	4.448	0.2248
lb (weight)	slugs	0.03108	32.17
	kg	0.4536	2.2046
lb H_2O (distilled)	ft^3	1.602×10^{-2}	62.43
	gal (liquid U.S.)	0.1198	8.346
lb/$in.^2$ (weight)	lb/ft^2 (weight)	144	6.945×10^{-3}
	kg/m^2	703	1.422×10^{-3}
lb/$in.^2$ (force)	lb/ft^2 (force)	144	6.945×10^{-3}
	N/m^2	6894	1.4506×10^{-4}
lb/ft^2 (weight)	lb/$in.^2$ (weight)	6.945×10^{-3}	144
	gm/cm^2	0.4882	2.0482
	kg/m^2	4.882	0.2048
lb/ft^2 (force)	lb/$in.^2$ (force)	6.945×10^{-3}	144
	N/m^2	47.85	2.090×10^{-2}
lb/ft^3 (weight)	lb/$in.^3$ (weight)	5.787×10^{-4}	1728
	kg/m^3	16.02	6.243×10^{-2}
poundals	lb (force)	3.108×10^{-2}	32.17
	dynes	1.383×10^4	7.233×10^{-5}
	newtons	0.1382	7.232
slugs	lb (weight)	32.17	3.108×10^{-2}
	kg	14.594	0.06852
slugs/ft^2	kg/m^2	157.2	6.361×10^{-3}
tons, short (2,000 lb)	tonnes (1,000 kg)	0.9075	1.102
watts	ergs/sec	10^7	10^{-7}
	hp (550 ft-lb/sec)	1.341×10^{-3}	745.7

APPENDIX III
SUPPLEMENTARY BIBLIOGRAPHY
PREPARED FOR 1986 REPRINT

CHAPTER 2

BOOKS:

D. S. Jones, *Acoustic and Electromagnetic Waves* (University of Dundee, U.K., 1985).

E. Zwicker and M. Zollner, *Elektroakustik, Hochschultext* (Springer-Verlag, Berlin, New York, 1984).

S. Temkin, *Elements of Acoustics* (Wiley, New York, 1982).

L. E. Kinsler, A. R. Frey, A. B. Coppens, and J. B. Sanders, *Fundamentals of Acoustics*, 3rd Ed. (Wiley, New York, 1982).

M. Heckl and H. A. Mueller, *Taschenbuch der Technischen Akustik* (Springer-Verlag, Berlin, 1975).

J. R. Pierce, *Almost All About Waves* (MIT Press, Cambridge, MA, 1974).

E. Meyer and E. Neumann, *Physical and Applied Acoustics* (Academic Press, New York, 1972).

E. Skudrzyk, *Foundations of Acoustics* (Springer-Verlag, Vienna, 1971).

W. Reichert, *Grundlagen der Technischen Akustik* (Akademische Verlags Gesellschaft Geest und Portig K.-G., Leipzig, 1968).

R. W. B. Stephens and A. E. Bate, *Acoustics and Vibrational Physics* (Edward Arnold, London, 1966).

ARTICLE:

C. M. Harris, "Absorption of Sound in Air versus Humidity and Temperature," J. Acoust. Soc. Am. **40**, 148–159 (1966).

CHAPTER 3

BOOKS:

J. Merhaut, *Theory of Electroacoustics* (McGraw-Hill International, New York, 1981).

H. F. Olson, *Dynamical Analogies*, 2nd Ed. (van Nostrand, New York, 1958).

F. V. Hunt, *Electroacoustics* (Wiley, New York, 1954).

ARTICLES:

L. L. Beranek, "Some Remarks on Electro-Mechano-Acoustical Circuits," J. Acoust. Soc. Am. **77**, 1309–1313 (1985).

Z. Jagodzinski, "Adequacy of Equivalent Circuits for Piezomagnetic Transducers," Acustica **21**, 283–287 (1969); see also, "Comments" by R. S. Woollett and Y. Yumamoto, **23**, 351–355 (1970) and "Replies," **23**, 355–361 (1970) and **24**, 175 (1971).

CHAPTER 4

BOOKS:

L. Cremer and M. Heckl, *Koeperschall* (Springer-Verlag, Berlin, 1967).

F. A. Fischer, *Fundamentals of Electroacoustics* (Interscience, London, 1955).

ARTICLES:

G. R. Harris, "Transient field of a Baffled Planar Piston Having an Arbitrary Vibration Amplitude Distribution," J. Acoust. Soc. Am. **70**, 186–204 (1981).

P. R. Stepanishen, "Radiation Impedance of a Rectangular Piston," J. Sound Vib. **55**, 275–288 (1977).

G. F. Lin, "Acoustic Radiation from Point Excited Rib-Reinforced Plate," J. Acoust. Soc. Am. **62**, 72–83 (1977).

M. C. Gomperts, "Sound Radiation from Baffled, Thin, Rectangular Plates," Acustica **37**, 93–102 (1977).

H. Heckl, "Abstrahlung von ebenen Schallquellen," Acustica **37**, 155–166 (1977).

G. C. Lauchle, "Radiation of Sound From a Small Loudspeaker Located in a Circular Baffle," J. Acoust. Soc. Am. **57**, 543–549 (1975).

C. E. Wallace, "Radiation Resistance of a Rectangular Panel," J. Acoust. Soc. Am. **51**, 946–952 (1970).

D. C. Greene, "Vibration and Sound Radiation of Damped and Undamped Flat Plates," J. Acoust. Soc. Am. **33**, 1315–1320 (1961).

R. L. Pritchard, "Mutual Acoustic Impedance Between Radiators in an Infinite Rigid Plane," J. Acoust. Soc. Am. **32**, 730–737 (1960).

R. L. Lyon, "On the Low Frequency Radiation Load of a Bass-Reflex Speaker," J. Acoust. Soc. Am. **29**, 654 (1957).

R. V. Waterhouse, "Output of a Sound Source in a Reverberation Chamber and Other Reflecting Environments," J. Acoust. Soc. Am. **30**, 4–13 (1958).

U. Ingard and G.L. Lamb, "Effect of a Reflecting Plane on the Power Output of Sound Sources," J. Acoust. Soc. Am. **29**, 743–744 (1957).

CHAPTER 5

BOOKS:

J. Merhaut, *Theory of Electroacoustics* (McGraw-Hill, New York, 1980).
F. A. Fischer, *Fundamentals of Electroacoustics* (Interscience, London, 1955).

ARTICLES:

Olaf Jacobsen, "Some Aspects of the Self and Mutual Radiation Impedance Concept with Respect to Loudspeakers," J. Audio Eng. Soc. **24**, 82–92 (1976).
M. Mongy, "Acoustical Properties of Porous Materials," Acustica **28**, 243–247 (1973).
J. Merhaut, "Contribution to the Theory of Electroacoustic Transducers Based on Electrostatic Principle," Acustica **19**, 283–292 (1967/68).

CHAPTER 6

BOOKS:

M. L. Gayford, *Electroacoustics—Microphones, Earphones and Loudspeakers* (American Elsevier, New York, 1971).
F. A. Fischer, *Fundamentals of Electroacoustics* (Interscience, London, 1955).
H. F. Olson, *Acoustical Engineering* (van Nostrand, New York, 1957).
F. V. Hunt, *Electroacoustics* (Wiley, New York, 1954).

ARTICLES:

J. L. Flanagan and R. A. Kubli, "Conference Microphone with Adjustable Directivity," J. Acoust. Soc. Am. **77**, 1946–1949 (1985).
R. Lerch, "Electroacoustic Properties of Piezopolymer Microphones," J. Acoust. Soc. Am. **69**, 1809–1814 (1981).
R. Zahn, "Analysis of the Acoustic Response of Circular Electret Condenser Microphones," J. Acoust. Soc. Am. **69**, 1200–1203 (1981); also Acustica **57**, 191–300 (1985).
A. J. Zuckerwar, "Theoretical Response of a Condenser Microphone," J. Acoust. Soc. Am. **68**, 1278–1285 (1978).
J. Ohga, S. Shirai, O. Ochi, T. Takagi, A. Yoshikawa, H. Nagai, and Y. Mizushima, "Granule Microphone Using Selenium-Tellurium Alloy," J. Acoust. Soc. Am. **64**, 988–994 (1978).
G. R. Hruska, E. B. Magrab and W. B. Penzes, "Environmental Effects on Microphones of Various Constructions," J.Acoust. Soc. Am. **61**, 206–210 (1977); also **62**, 1315 (1977).

W. R. Kundert, "Comments on Hruska *et al.* Paper," J. Acoust. Soc. Am. **61**, 1647 (1977).

D. P. Egoff, "Mathematical Modeling of a Probe-Tube Microphone," J. Acoust. Soc. Am. **61**, 200–205 (1977).

G. G. Parfitt and B. N. Gyang, "New Form of Condenser Microphone," Acustica **37**, 125–126 (1977).

G. M. Sessler and J. E. West, "Electret Microphones," a series of papers, J. Acoust. Soc. Am. **35**, 1354–1357 (1963); **40**, 1433–1440 (1966); **46**, 1081–1086 (1969); and **58**, 273–278 (1975).

E. Villchur and M. C. Killion, "Probe-Tube Microphone Assembly," J. Acoust. Soc. Am. **57**, 238–240 (1975).

J. S. Wang and M. J. Crocker, "Tubular Windscreen Design for Microphones for In-Duct Fan Sound Power Measurements," J. Acoust. Soc. Am. **55**, 568–575 (1974).

H. S. Madsen, "Optimization of a Ridge Backplate for Electret Microphones," J. Acoust. Soc. Am. **53**, 1616–1619 (1973).

C. W. Reedyk, "Noise Cancelling Electret Microphone for Light-Weight Head Telephone Sets," J. Acoust. Soc. Am. **53**, 1609–1615 (1973).

F. W. Fraim, P. V. Murphy, and R. J. Ferran, "Electrets in Miniature Microphones," J. Acoust. Soc. Am. 1601–1608 (1973).

J. L. Collins and G. E. Ellis, "Electrokinetic, Acoustic Pressure Transducer," J. Acoust. Soc. Am. **36**, 1808–1860 (1964).

CHAPTER 7

BOOKS:

Loudspeakers: An Anthology of Articles on Loudspeakers from the Pages of the Journal of the Audio Engineering Society Vols. 1–25, (1953–1977) (Audio Engineering Society, 60 E. 42nd Street, New York, NY 10017, 1978). A partial list of the articles contained pertinent to this chapter is:

Analysis of Decoupled-Cone Loudspeakers, J. M. Kates (Jan/Feb 1977)

Radiation from a Dome, J. M. Kates (Nov 1976)

Low-Frequency Loudspeaker Assessment by Nearfield Sound-Pressure Measurement, D. B. Keele, Jr. (Apr 1974)

Gradient Loudspeakers, H. F. Olson (Mar 1973)

Modulation Distortion in Loudspeakers, P. W. Klipsch, Part III, (Dec 1972); Part II (Feb 1970); Part I (Apr 1969)

Simplified Loudspeaker Measurements at Low Frequencies, R. H. Small, (Jan/Feb 1972)

Development of a Sandwich Construction Loud Speaker System, D.A. Barlow (June 1970)

Experimental Determination of Low-Frequency Loudspeaker Parameters, J. R. Ashley and M. D. Swan (Oct 1969)

Loudspeaker Phase Characteristics and Phase Distortion, R. C. Heyser, Parts 1 and 2 (Jan and Apr 1969)

High-power, Low-Frequency Loudspeakers, J. K. Hilliard (July 1965)

Low-Frequency Response and Efficiency Relationships in Direct-Radiator Loudspeaker Systems, R. F. Allison (Jan 1965)

Trends in Loudspeaker Magnet Structures, R. J. Parker (July 1964)

Interrelation of Speaker and Amplifier Design, V. Brociner and D.R. von Recklinghausen (Apr 1964)

A Method of Testing Loudspeakers with Random Noise Input, E. M. Villchur (Oct 1962)

Application of Negative Impedance Amplifiers to Loudspeaker Systems, R.E. Werner and R. M. Carrell (Oct 1958)

An Electrostatic Loudspeaker Development, A. A. Jenszen (Apr 1955)

Recent Developments in Direct-Radiator High Fidelity Loudspeakers, H. F. Olson, J. Preston and E. G. May (Oct 1954).

A. P. Dowling and J. E. Ffowcs-Williams, *Sound and Sources of Sound* (Wiley, New York, 1983).

T. F. Heuter and R. H. Bolt, *Sonics* (Wiley, New York, 1955).

ARTICLES:

R. A. Greiner and T.M. Sims, Jr., "Loudspeaker Distortion Reduction," J. Audio Eng. Soc. **32**, 956–963 (1984).

L. J. van der Pauw, "The Trapping of Acoustic Energy by a Conical Membrane and its Implications for Loudspeaker Cones," J. Acoust. Soc. Am. **68**, 1163–1168 (1980).

D. A. Kleinman and D. F. Nelson, "The Photophone—An Optical Telephone Receiver," J. Acoust. Soc. Am. **59**, 1482–1494 (1976); **60**, 240–250 (1976).

J. E. Benson, "Synthesis of High-Pass Filtered Loudspeaker Systems," AWA Tech. Rev. **15**, 115–128 (1974); also **15**, 143–148 (1974) and **16**, 1–11 (1975).

D. F. Nelson, K. W. Wecht, and D. A. Kleinman, "Photophone Performance," J. Acoust. Soc. Am. **60**, 251–255 (1976).

CHAPTER 8

BOOKS:

Loudspeakers: An Anthology, Chapter 7 Bibliography. A partial list of articles pertinent to this chapter is:

The Use of Fibrous Materials in Loudspeaker Enclosures, L.J.S. Bradbury (Apr 1976)

Vented-Box Loudspeaker Systems, R. H. Small, Parts I through IV, (Jun–Oct 1973)

Closed-Box Loudspeaker Systems, R. H. Small, Parts I and II (Dec and Jan/Feb 1973)

Loudspeakers in Vented Boxes, A. N. Thiele, Parts I and II (May and Jun 1971)

Phase and Delay Distortion in Multiple-Driver Loudspeaker Systems, R. H. Small (Jan 1971)

Loudspeaker Enclosure Walls, P. W. Tappan (Jul 1962)

Corner Loudspeaker Placement, P. W. Klipsch (Jul 1959)

Performance of Enclosures for Low-Resonance, High-Compliance Loudspeakers, J. F. Novak (Jan 1959)

Problems of Bass Reproduction in Loudspeakers, E. M. Villchur (Jul 1957).

ARTICLES:

W. M. Leach, Jr. "Active Equalization of Closed-Box Loudspeaker Systems," J. Audio Eng. Soc. **29**, 405–407 (1981).

A. D. Broadhurst, "Loudspeaker Enclosure to Simulate an Infinite Baffle," Acustica **39**, 316–322 (1978).

J. R. Macdonald, "Loudspeakers," J. Electrochem Soc. **124**, 1022–1030 (1977).

CHAPTER 9

BOOKS:

Loudspeakers: An Anthology, Chapter 7 Bibliography. The papers contained therein that are pertinent to this chapter include:

Horn-Loaded Electrostatic Loudspeaker, J. Merhaut (Nov 1971)

Design Factors in Horn-Type Speakers, D. J. Plach (Oct 1953).

ARTICLES:

S. Morita, N. Kyouno, T. Yamabuchi, Y. Kagawa, and S. Sakai, "Acoustic Radiation of a Horn Loudspeaker by the Finite Element Method," J. Audio Eng. Soc. **28**, 482–489 (1980); see also, W. M. Leach, Jr., "Comments" and S. Morita, *et al.*, "Reply," **28**, 900 (1980). Also see **30**, 896–905 (1982).

P. W. Klipsch, "A Low Frequency Horn of Small Dimensions," J. Audio Eng. Soc. **27**, 141–148 (1979).

W. M. Leach, Jr., "Specification of Moving-Coil Drivers for Low-Frequency Horn-Loaded Loudspeakers," J. Audio Eng. Soc. **27**, 950–959 (1979).

R. W. Pyle, Jr., "Effective Length of Horns," J. Acoust. Soc. Am. **57**, 1309–1317 (1975).

CHAPTER 10

BOOKS:

Y. Ando, *Concert Hall Acoustics* (Springer-Verlag, Berlin and New York, 1985).

L. Cremer and H. A. Mueller, *Principals and Applications of Room Acoustics. Vols.1 & 2*, translated by T. J. Schultz (Applied Science, London and New York, 1982).

H. Kuttruff, *Room Acoustics*, 2nd Ed. (Update, New York, 1981).

N. A. Grundy et al., *Practical Building Acoustics* (Halsted, Wiley, New York, 1976).

V. S. Mankovsky, *Acoustics of Studios and Auditoria*, edited by C. Gilford (Hastings House, New York, 1971).

M. D. Egan, *Concepts in Architectural Acoustics* (McGraw-Hill, New York, 1972).

L. Beranek, *Music, Acoustics, and Architecture* (Wiley, New York, 1962).

ARTICLES:

R. N. Miles, "Sound Field in Rectangular Enclosure With Diffusely Reflecting Boundaries," J. Sound Vib. **92**, 203–226 (1984).

M. Kleiner, "Audience Induced Background Noise Level in Auditoria," Acustica **46**, 82–88 (1980).

M. H. Jones, "Reverberation Enforcement—an Electro-Acoustical System," Acustica **27**, 357–363 (1972).

R. J. Hawkes and H. Douglas, "Subjective Acoustic Experience in Concert Auditoria," Acustica **24**, 235–250 (1971).

G. G. Sacerdote, "A Recent Story of the Ceiling of the Scala Theatre of Milan," Acustica **22**, 54–58 (1969/70).

H. Kuttruff and M. J. Jusofie, "Messungen des Nachhallverlaufes in mehreren Raumen, ausgefuehrt nach dem Verfahren der integrierten Impulsantwort," Acustica **21**, 1–9 (1969).

C. M. Harris, "Absorption of Sound in Air versus Humidity and Temperature," J. Acoust. Soc. Am. **40**, 148–159 (1966).

CHAPTER 11

BOOKS:

J. D. Erwin and E. R. Graf, *Industrial Noise and Vibration Control* (Prentice-Hall, Englewood Cliffs, NJ, 1979).

R. G. White and J. G. Walker, *Noise and Vibration* (Wiley, New York, 1982).

D. Reynolds, *Engineering Principles of Acoustics: Noise and Vibration Control* (Allyn and Bacon, Boston, MA, 1981).

H. Lord, W. S. Gatley and H. A. Evensen, *Noise Control for Engineers* (McGraw-Hill, New York, 1980).

Handbook of Noise Control, 2nd Ed., edited by C. M. Harris (McGraw-Hill, New York, 1979).

Shock and Vibration Handbook, 2nd Ed., edited by C. M. Harris and C. E. Crede (McGraw-Hill, New York, 1976).

E. B. Magrab, *Environmental Noise Control* (Wiley, New York, 1975).

Noise and Vibration Control for Industrialists, edited by S. A. Petrusewicz and D. K. Longmore (American Elsevier, New York, 1974).

Noise and Vibration Control Engineering, edited by M. J. Crocker (Purdue U., Lafayette, Indiana, 1972).

W. T. Thompson, *Theory of Vibration with Applications* (Prentice-Hall, Englewood Cliffs, NJ, 1973).

Noise and Vibration Control, edited by L. Beranek (McGraw-Hill, New York, 1971).

Noise Reduction, edited by L. Beranek (McGraw-Hill, New York, 1960).

P. H. Parkin and H. R. Humphreys, *Acoustics, Noise and Buildings* (Praeger, New York, 1959).

CHAPTER 12

BOOKS:

A. P. G. Peterson, *Handbook of Noise Measurement* (GenRad, Concord, MA, 1974).

ARTICLES:

M.A.N. de Araujo and S. N. Yousri Gerges, "Sound Power From Sources Near Reverberation Chamber Boundaries," J. Sound Vib. **91**, 471–477 (1983).

K. Bodlund, "Sound Power in Reverberation Chambers at Low Frequencies," J. Sound Vib. **55**, 563–590 (1977).

K. Bodlund, "Diffusion in Reverberation Chambers," J. Sound Vib. **50**, 253–283 (1977).

T. E. Vigran and S. Sorsdal, "Comparison of Methods for Measurement of Reverberation Time," J. Sound Vib. **48**, 1–13 (1976).

CHAPTER 13

BOOKS:

R. P. Hamerick, D. Henderson, and R. Salve, *New Perspective on Noise Induced Hearing Loss* (Raven, New York, 1982).

P. E. Cunniff, *Environmental Noise Pollution* (Wiley, New York, 1977).

D. Henderson, R. P. Hamerick, D. S. Dosanijh, and J. H. Mills, *Effects of Noise on Hearing* (Raven, New York, 1976).

S. Williams, Jr., E. Leyman, S. A. Karp, and P. T. Wilson, *Environmental Pollution and Mental Health* (Information Resources, Washington, D.C., 1973).

Effects of Noise on Hearing, edited by D. Henderson, R. P. Hamernik, D. S. Dosanjh, and J. H. Mills (Raven, New York, 1976).

ARTICLES:

N. Broner and H. G. Leventhall, "Modified PNdB for Assessment of Low Frequency Noise," J. Sound Vib. **73**, 271–277 (1980); also **84**, 443–448 (1982).

T. J. Schultz, "Synthesis of Social Surveys on Noise Annoyance," J. Acoust. Soc. Am. **64**, 377–405 (1978); also **65**, 849 (1979).

V. J. Krichagin, "Health Effects of Noise Exposure," J. Sound Vib. **59**, 65–71 (1978).

G. J. Thiessen, "Disturbance of Sleep by Noise," J. Acoust. Soc. Am. **64**, 216–222 (1978).

C. Hayashi, S. Kondo, and H. Kodama, "Psychological Assessment of Aircraft Noise Index," J. Acoust. Soc. Am. **63**, 815–822 (1978).

F. J. Langdon and I. B. Buller, "Road Traffic Noise and Disturbance to Sleep," J. Sound Vib. **50**, 13–28 (1977).

M. A. Jenkins and J. Pahl, "Measurement of Freeway Noise and Community Response," J. Acoust. Soc. Am. **58**, 1222–1231 (1975).

J. S. Lukas, "Noise and Sleep," J. Acoust. Soc. Am. **58**, 1232–1242 (1975).

B. Berglund, U. Berglund, and T. Lindvall, "Scaling Loudness, Noisiness and Annoyance of Aircraft and Community Noise," J. Acoust. Soc. Am. **57**, 930–934 (1975); also **60**, 1119–1125 (1976).

J. D. Miller, "Effects of Noise on People," J. Acoust. Soc. Am. **56**, 729–764 (1974).

A. Gabrielson, U. Rosenberg, and H. Sjogren, "Judgements of Perceived Sound Quality of Sound-Reproducing Systems," J. Acoust. Soc. Am. **55**, 854–861 (1974).

H. E. von Gierke and C. W. Nixon, "Human Response to Sonic Boom," J. Acoust. Soc. Am. **51**, 766–782 (1972).

L. L. Beranek, "Preferred Noise Criteria," J. Acoust. Soc. Am. **50**, 1223–1228 (1971); See also **28**, 833–852 (1956).

C. E. Williams, K. S. Pearsons, and M.H.L. Hecker, "Speech Intelligibility in the Presence of Time-Varying Aircraft Noise," J. Acoust. Soc. Am. **50**, 426–434 (1971).

J. C. Webster, "Speech Communications as Limited by Ambient Noise," J. Acoust. Soc. Am. **37**, 692–699 (1965).

R. W. Young, "Single-Number Criteria for Room Noise," J. Acoust. Soc. Am. **36**, 289–295 (1964).

INDEX